STATEMENT AND REFERENT

SYNTHESE LIBRARY

STUDIES IN EPISTEMOLOGY,

LOGIC, METHODOLOGY, AND PHILOSOPHY OF SCIENCE

VOLUME 224

D. S. SHWAYDER
University of Illinois at Urbana

STATEMENT AND REFERENT

An Inquiry into the Foundations of Our Conceptual Order

Part I: Statements are Products of Assertion

KLUWER ACADEMIC PUBLISHERS
DORDRECHT / BOSTON / LONDON

Library of Congress Cataloging-in-Publication Data

Shwayder, D. S., 1926-
Statement and referent : an inquiry into the foundations of our conceptual order : statements are products of assertion / D.S. Shwayder.
p. cm. -- (Synthese library ; v. 224)
Includes indexes.
ISBN 0-7923-1803-X (HB : acid free paper)
1. Metaphysics. 2. Act (Philosophy) 3. Language and languages--Philosophy. 4. Reference (Philosophy) I. Title. II. Series.
BD111.S548 1992
110--dc20 92-16652

ISBN 0-7923-1803-X

Published by Kluwer Academic Publishers,
P.O. Box 17, 3300 AA Dordrecht, The Netherlands.

Kluwer Academic Publishers incorporates
the publishing programmes of
D. Reidel, Martinus Nijhoff, Dr W. Junk and MTP Press.

Sold and distributed in the U.S.A. and Canada
by Kluwer Academic Publishers,
101 Philip Drive, Norwell, MA 02061, U.S.A.

In all other countries, sold and distributed
by Kluwer Academic Publishers Group,
P.O. Box 322, 3300 AH Dordrecht, The Netherlands.

Printed on acid-free paper

Printed in the Netherlands

For C. V.

CONTENTS

PREFACE

There are in this volume sentences written as long ago/ as 1957. What was then projected as the third part of a modest discussion of then current issues has, through some fifteen revisions, now expanded into its own three parts. Of the project as originally conceived, the first part, itself grown too large, was published (prematurely, I now believe) in 1965 (*Stratification of Behaviour*). The second part, which was to be on language proper, was abandoned around 1967; such materials on language as I need for the present work are now mostly compressed into Chapter 1, with some scatterings retained in Chapters 2 and 14.

My scheme discovered problems with which I have been much preoccupied. I have been less enjoyably delayed by missteps. Additions were put on and the renovations have been incessant. Even in the course of my ultimate revisions, I ran into slippery stretches and soft spots I could only gesture at repairing. But now time is running out and my energy is ebbing, and I must allow the work to come to its conclusion, with reservations certainly and not without a sense of despair. If the reception of this volume warrants, the two following parts will be wound up in what I hope may be fairly short order.

A swing in the direction of my thinking about my materials and in the development of the text occurred along about 1967. I had accepted a commission to write a short treatise of metaphysics. It seemed to me that I could do a book more serviceable to students by backing up my own speculations with discussions of the metaphysical doctrines of Plato and Aristotle. I then came to see my own enterprise as continuous with the tradition of First Philosophy, and judged that comparisons appropriately emplaced into this work would assist my presentation. My appreciation of "the tradition" deepened when, several years later, Arthur Melnick taught me something about the philosophy of Kant. These time-tested classics, in their contents, proved to be more instructive and more challenging to me for my endeavors and, I reckoned, for purposes of comparison and contrast, bound to be known to a wider and more enduring public than the contemporary literature I had been straining to keep abreast of. Some of my topics do indeed originate in the modern era, in the writings of Frege, Russell, Wittgenstein, Austin and their successors, and, for those topics, *those* writings are my classics. Other contemporary worthies are often noticed but little discussed, except where their writings have, for me, broken new ground, e.g. Dummett on *causation*, Grice on *meaning*, Kripke on *modality* and Urmson on *species*. I hope that these bits of autobiography which partially explain the postponements and the volume of this treatise may also work to spare me censure for inattention to still growing bodies of contemporary writing on the topics of this treatise.

My text, though full of commentary, is not a work of scholarship. I have not "researched the literature". My choice of authorities has been pretty much accidentally determined by what I already knew or through preparations for courses my departments have wanted me to teach. I use what I think I know of traditional doctrines both as sources of light and as reflecting surfaces, as points of reference and as parallels. Switching the figure once again, I hope to establish a line of credit from the texts themselves or from a bank of existing interpretations, but not to contribute to the fund of scholarship. I am of course liable for wrong readings.

This work is daunting in its size and complexity, and (I fear) heavy-handed and dull in its presentations; it's not "user friendly", as the publisher's referee found cause to remark. While I would like to believe that every part of the text will be of interest to someone other than myself, I do not think that there is anyone out there who could face the task of reading it straight through in order; the work has been composed with an apparatus attached and according to a plan calculated to dissuade anyone from so arduous an undertaking. I have also decided to present my treatise in three separate volumes, again with the hope that the reception of the first part will warrant the publication of the second and the third.

My three "parts" are explained by their titles. The first five chapters are groundwork for the rest. Subsequent chapters systematically depend upon their predecessors being brought to completion but draw little from the accomplishment and are pretty much self-contained. Summaries are set in the margins of the text, and I believe that any of the readership I envisage could get a pretty adequate idea of all the positions I hold by reading through these summaries, dipping into the main text only where they have a need for argument, illustration or amplification. Summaries of a like kind, extracted from an earlier version of Parts Two and Three, are assembled into a synopsis annexed to this volume, which may serve both to assist advance references and to give the interested reader an idea of where I'm headed.

In the text proper, resume's, comparisons with competing doctrines and traditional authorities, analyses of examples, responses to anticipated criticisms and off-track discussions of such large side issues as *perception* and *knowledge*--discussions I deem necessary to protect the integrity or to increase the plausibility of my systematic presentations--are either emplaced, in reduced font, as insets, or assembled into appendices. These passages are intended only for readers who may be particularly interested in the issues or comparisons brought under review. "Appendix D" is an inexact "formalization" of the materials of Chapter 3, and follow-ups will be included for Chapters 5 through 17. These "formalizations" have proven useful to me both for digesting and for checking my sundry proposals. Formula-haters are urged to skip them.

My presentation is jargon-ridden. The publisher's referee suggests a glossary, and I agree it could be helpful. I remain uncertain over how that glossary could most usefully and most economically be provided. My resolution is to include a few boldface glossary blurbs in the topical index under the appropriate headings, which happen to be mostly on *action* and its several varieties. I don't think this should be any more trouble to the reader than would be leafing forward in the main volume or sifting through a separate booklet.

Batches of the material now included in this volume and the two I hope will follow were, over many years, at several colleges and universities, presented to some two-dozen seminars and to at least as many classes. These captive audiences have invariably been most usefully forthcoming. Credits for some particular points made in discussions are recorded in footnotes. I simply cannot recollect all the contributors, but have found the following names in various seminar notes: Roger Ariew, Georgia and Paul Bassen, Tim Erdel, Tom Eudaly, Tim Griffin, Philip Hugly, David Israel, Dale Jordan, David Kolodny, Tom Norton-Smith, Gilbert Plumer, John Pollock, Shekhar Pradhan, Donald Riggs and Tom Sorrell. During the period of my final revisions, my department generously allowed me to present the materials one last time to a seminar, and the participants, Tim Griffin, Nancy Kendrick, Tim Ketcher and Jesus Illundain, all of them, made useful criticisms and suggestions. I thank all of these persons for their contributions and also those many others whom I cannot acknowledge particularly. I have also profited greatly in conversation and in correspondence with colleagues and friends. I give special thanks to William Alston, Charles Caton, Hugh Chandler, Tim McCarthy, Alfred MacKay, Robert Monk, Fred Schmitt, Michael Shapira, Manley Thompson, Robert Wengert and Fred Will. I came to my "formalization" of the theory of testing as a result of a brief but fruitful conversation with Dana Scott sometime around 1960, and latterly this part of my presentation has been greatly assisted by the criticisms and suggestions of Jose' Iovino. My thoughts about notions of space, brought together in Chapter 21 of the yet-to-be published third part of this work, largely owe to conversations with Ernest Adams on the materials for two seminars on space and time we jointly conducted at Berkeley; Adams has continued to be a generous correspondent, a valuable critic and an intimidating rival. My greatest debt is to Arthur Melnick for discussions that have invariably been challenging, brisk and freshening. Finally, I must tender thanks in abundance to the publisher's unnamed referee, who provided literally hundreds of criticisms and suggestions, no one of which went unregistered in my ultimate revisions. I thank the departments of philosophy at Berkeley, Urbana, Chapel Hill and Oberlin for the boon of classrooms, students and colleagues. I am grateful for grants of money and time to The Fulbright Commission, The Guggenheim Endowment, The National Endowment for the Humanities and to The Advanced Institute of the University of Illinois and, for clerical grants, to the Research Boards at Berkeley and Urbana. Finally, I gratefully and admiringly applaud Glenna Cilento for her patient decipherment and typing of several manuscript versions of this work and latterly for her enviable expertise as a word-processor.

Urbana, Illinois
Nov. 19, 1991

PART I

STATEMENTS ARE PRODUCTS OF ASSERTION

INTRODUCTION

STATEMENTS ARE PRODUCTS OF ASSERTION

My central concern in this work is with statements.

My central concern across the main body of this work will be with a range of *products* which, for want of a better name, I call "statements". Statements may be likened to and therefore also contrasted with such other items as promises and civil enactments. Promises are products of promising and civil enactments products of legislative activity, where promises may be brought under the general heading of undertakings and civil enactments under the general heading of laws. Statements are comprised among the products of acts of subjects' both saying and meaning what they think they know to be so, and may be described as flat formulations of putative fact. Promises, we know, may be kept or broken or sometimes neither, and civil enactments may be enforced or ignored or sometimes neither; statements, as I conceive them, may or may not be "true to the facts", and accordingly, in themselves, be true or false or perhaps sometimes neither. My interest in statements arises from their susceptibility to these "truth-value" determinations. While statements are not the most "primitive" bearers of "truth-value"--what philosophers curiously call "beliefs" may perhaps fill that role--they are, as I believe, for systematic purposes, the most "fundamental" vehicles of truth and falsity .

The interest is traditional and ancient.

Our interest is traditional and ancient, with connections to logic and metaphysics. Following leads that come from Plato and Aristotle, we shall move from a consideration of language into the territories of First Philosophy. The study of statements lies between the two.

First Philosophy is concerned with our conception of what is so. Statements are among the immediate products of successful attempts by subjects to say and mean what they think they know to be so. For convenience, call any act of meaning by saying, however conveyed--in script, sound, gesture, smoke or whatever--, an *utterance*. I believe that our conception of what is so is determined by our general conception of a statement in the qualified sense that anything we conceive could be so is also conceived to be resolvable without remainder into statement-formulable parts, where those statements are themselves producible in utterances.

Our conception of what is so is determined by our general conception of a statement.

Three objections answered: ineffability, generality and triviality.

I anticipate three immediate objections to the above thesis. First, our conception of what is so allows that there may be ineffable facts perhaps known to but altogether unformulable by us. I wish to say three things in response to this objection. First, although our apparatus will be designed for the representation of humanly producible statements, our general conception of a statement is meant to cover all flat formulations of fact, including ones not producible by us. We can conceive that there are such formulations though we cannot conceive what they are. Leibniz constructed his system of metaphysics according to principles he supposed governed the conceptions of deity; our inquiry is more modest; but still, our conception of what is so may coincide with his. Second, we do indeed right now have all sorts of ineffable *practical* knowledge, felt perhaps as dark forebodings, rather as our ancestors had nothing but ineffable knowledge of depressions of atmospheric pressure expressed as forebodings of storm, and I must concede that there are indeed hard questions about the relationships between our merely practical and our expressible "theoretical" knowledge of fact--questions we shall all too briefly touch upon in Appendix C. But, third, I don't yet see that this distinction in knowledge gives reason to think that there is anything in particular of what we conceive to be so that is not also conceived to be resolvable into parts formulable in statements sometime producible by some subject. The illustration of the objection is a case in point.

The second objection alleges as such a reason that there may be facts, peircian "general facts" perhaps, "too large" for statement. I allowed for as much in my cautious way of speaking of the "resolution" of what we conceive to be so. Now statements must be distinguished from generalizations, and we shall touch on the matter of this distinction in Appendix B. Still I do not see that there is anything in the distinction between statement and generalization to require that there be any element of a "general fact" not included within some statement-formulable part. I do concede that *that* condition, *viz* that the generalization should cover the whole general fact, is not itself formulable in a statement; the generalization does indeed express another *truth*, *viz* that nothing is excluded; but that condition of *nothing further*, I hold, is not a further fact.

I concede as a clarification prompted by both objections taken together that not all stateable facts are just anytime stateable by anyone, unless by God.

The third objection is that "flat formulations of putative fact" if true are trivial. Science and other worthwhile theoretical endeavors flourish on generalization, hypothesis, law, problem, proof and prediction with scarcely a side glance at statements. I agree, and my own investigation is no exception. I hold only that these other "interesting" and "progressive" productions couldn't stand without the continual support of statements and (as I shall argue) the analysis of concepts incorporated into all these various "illocutions" is best concentrated on statements. Statements are qualified formulations of fact and, within the mix, are (as I believe) also "most fundamental".

Statements may be true, false or neither. "Assertions" that, when successful, produce statements, produce those statements "as true". My usage of "statement" and "assertion" is technical and needs explanation.

Statements may be true, false or neither. Those utterances of saying and meaning what one thinks one knows to be so that, when successful, produce statements, produce those statements "as true". Such utterances serve to convey a speaker's sense of what is so.

It should now be evident that a statement in the usage I shall follow is not a kind of utterance or "use of language", but rather a product of utterance. My usage of "statement" is adapted from that vernacular idiom in which statements are "made", and it differs from that other vernacular usage in which statements are "makings". Statement-makings are called "assertions" in my lexicon.

Now my usage of "statement" is only adapted from the vernacular. Lacking talent for coinage, I use old metal in a contrived but not I hope illicit way to gain a measure of intellectual control over an engaging field of problems. I respect anyone's preconceptions about what statements might be, but am not beholden to them. Please do not plead your specimens as arguments against mine (see pp. 68f). My usage is technical and consequentially narrow. It must and will be explained. For now, it will perhaps be enough to observe that my usage notably does not cover statements "made to the press", which are submitted in proprietary capacity and may comprise announcements, declarations of policy, acceptances, registrations of opinion and much else not statements in my usage. Statements for me are only the flattest formulations of putative fact. I do not wish to put down other usages of "statement", and certainly not those which occur in everyday speech. Nor will I question the credentials of other products of language, such as generalizations, conjectures and hypotheses, which, though true or false, are not produced in assertions and are not statements in my narrow sense. Indeed, I hope to use my notion of statement as an instrument to advance our understanding of such other items as generalizations and conjectures.

Statements will be explained as products of assertion by defining assertion within a theory of language.

Statements, I repeat, are not assertions. They are, rather, products of assertion. However, statements, so taken, do depend upon assertion both in conception and in fact. I shall accomplish the task of saying what statements are by finding a place for them within an account of the products of language. I do that by finding a place for assertion within my theory of language.

What's to come.

Our understanding of the nature of language is heavily obstructed and I once thought it would be necessary to clear the way with a full theory of language before I could begin to elucidate *assertion.* I now believe that we shall need only a few select principles which I hope can be made plausible and comprehensible within the space of my Chapter 1. I shall then, in Chapter 2, use those principles in my explanation of *assertion.*

The main task of Chapter 3 will be to secure a general representation of statements in separation from the successful assertions that produce them. This concluding chapter of Part I is, both ideologically and systematically, "most central" to my enterprise. The second part of this treatise, comprising Chapters 4 through 14, will be devoted to the working out of a theory and of an apparatus for the representation and characterization of all humanly producible statements, and will involve consideration of what the Scholastics called *syncategoremata*[1]. That stretch of my exploration will make a very hard climb indeed. It will I hope finally bring us into position whence the main contours of First Philosophy and of Ontology can be surveyed and recorded in the several chapters of metaphysics that make up the third part of this work.

NOTE

[1]Approximately: "Greater Forms" (Plato), *"pros en* equivocals" (Aristotle), "Intellectual Ideas" (Leibniz), "Categories" (Kant).

CHAPTER 1

BEHAVIORAL AND LINGUISTIC PRELIMINARIES

1. STATEMENTS ARE PRODUCTS OF ASSERTION.

We now begin our task of establishing a technical usage for "statement". The endeavor would be worthless unless it covered a broad range of real cases; it will become evident in Chapter 2 that it does. To start, I make a "categorization" and adduce a trio of comparisons.

Statements are products not deeds.

Statements are products, not deeds. They stand in the same generic order of relationship to the assertional deeds that produce them as do promises, orders and conjectures to the acts of promising, ordering and conjecturing that produce them.

Analogy and disanalogy with "material products".

Products of language may, in first analogy, be likened to tables and chairs and contrasted with the acts of cutting and joining by which such "material products" are produced. Products, all of them, are among what I shall later (in Chapter 16) want to call "particularized forms". Disanalogies are obvious but not altogether easy to explain: promises, orders and statements, unlike tables, do not have a maker-independent "material basis" and are determined in every detail by the utterances that produce them. Statements are identified *in* utterance, whereas tables may be bodily identified *as* objects standing apart from activity. Now a material product is produced when a maker gives form to a body. The product continues to exist so long as the body that it is continues to have that form. The body may of course cease to have that form, e.g. by being transformed or reduced to pulp, or the body itself may be dismantled or even dispersed into its underlying stuff. In any such circumstance, the product, though not necessarily the body that it was, ceases to exist. The product could perhaps be remade by the original or another producer, but not unless the body which the product was had previously ceased to have the form of the product. Now statements are identified "in" the utterances that produce them, but are not to be identified "with" the utterances they are in. They may indeed be produced in different utterances without any one of those utterances or any other "material basis" first having to be transformed, dismantled, dispersed or anything of the kind. Statements and other "non-material products" may accordingly be remade without first being undone.

Analogy and disanalogy with "compositions".

Statements and other products of language, in a second, closer comparison, may be likened to passages of composed music or verse; such products may be subsequently identified in performance or recitation, just as products of language are identified in suitable utterances. The disanalogy is this. Listening to a passage of composed music normally takes place in separation from the composer's activity of writing the piece, and the music is usually identified as heard without the composer around to distract the performer. Statements, in contrast, are identified in the assertions that also produce them, and promises in the acts of promising that produce them. Passages of (composed) music might be but seldom are composed in performance, and performings are usually not composings. The sounding of music is normally separate from its composition.

The connection between a successful utterance and its product is so intimate, it may seem impossible to grasp the one without the other. One identifies the promise in the promising. The appearance of inseparability here will eventually bear hugely upon our attempt to say what utterance is, as a kind of action. Some may wish to call the product an "internal object" of successful utterance, and I have no objection to that. Just as we shoot shots as well as bullets and targets, we readily speak of promising promises and perhaps also of ordering orders; I shall likewise speak of asserting statements.

We must then distinguish the "internal" object from other affiliated objects. We may also be said to promise and to order actions and to predict events, and I allow myself to speak of asserting facts as well as

> statements. Promise-keepings, compliances and facts are indeed so-called in relation to formulations but are in themselves distinct from those formulations. Just as food is so-called in prospect of eating, so too, facts are prospective objects of assertion; but just as food may exist uneaten, so too facts may exist unasserted. Such objects of utterance as compliances and facts, unlike orders and statements, are not products of utterance.

We preferentially refer to products of language by aping the producing act.

Because they are identified in the acts which produce them, we normally and preferentially refer to such things as orders, promises and statements by aping the producing act: we use expressions that could be employed to produce the product for purposes of referring to the product. So, aping an updated version of Millikan's famous report, we speak of the statement that the charge on the electron is 1.602 x 10^{-19} coulomb. Non-imitative references to be sure are also available: we may refer to the just-mentioned statement as "Millikan's Updated Second Result". That last reference, however, depends on the possibility of the other. There could be no statement to refer to in the second way if no one could have produced the statement in terms then to be borrowed for making references of the first kind.

A distinction is required for reasons of replication and unsuccess.

Some will protest that this tightness of conception is oneness in fact: "internal objects" are aspects of action; there is no product, they may urge, because there is no distinction between the utterance and the product. A distinction is required, however, for reasons of replication and unsuccess. First, replication: the same promise or conjecture may be independently produced in different acts of promising or conjecturing. Second, unsuccess: an actual but faulted act of promising may produce no promise at all, as when the promiser unknowingly takes a statue for a qualified respondent.

Among products of language, statements lie somewhere "between" natural numbers and promises

My third and final comparison is between statements and such other "objects of thought alone" as the natural numbers. A natural number is something that can be counted-to, and I conceive natural numbers to be products of counting and therefore to be no less products of language than are promises or statements. But now the existence of a counting apparatus by which counts are made of itself assures us of the existence of a natural number corresponding to every counting numeral, for every such numeral can be taken as registering the count of its predecessors prefixed with "0". So we can know of the existence of natural numbers taken as products of counting even when they are not actually counted to. By way of contrast, promises could not be said to exist unless they were actually made. Statements, in this regard, may seem to be more like promises than like natural numbers: certain kinds of assertion are doomed to failure, in which event there is no statement to correspond, and we therefore cannot know from the "mere meaningfulness" of the assertion of the existence of a corresponding statement. We even know that there is no statement to correspond to the assertion *1/0 is a perfect square*, because "1/0" is demonstrably "undefined", as a natural number. In other instances, understood assertions may and do fail unbeknownst to anyone. There are, however, other examples that suggest that statements are more like natural numbers than promises on the point of comparison we are now pursuing: for example, all of us right now know of the existence of any number of false statements of the assertional form $n/p>n+1$, of which few have actually been asserted. Statements may, in a certain sense, be more "abstracted" from actual production than promises but less so than natural numbers.

2. ASSERTION IS PRODUCTIVE UTTERANCE.

A statement is a non-material product that may be remade without first being undone.

A statement, I say, is a product of successful assertion. Now products are produced by agents acting. There are of course many kinds of products. Chairs and transformers are products that cannot be remade (if at all) without first being undone. They are "material artifacts", in every instance *bodies* of some particular kind. We may call objects that satisfy that (formulatable) condition *material* products. Examples of the (undefinable) complement class of non-material products are passages of music and literature, editions of newspapers, titles to laws, corporations, promises, conjectures and statements. So a statement is a non-material product that may be remade without first being undone. Products of language are distinguished, as we noticed, even among non-material products, for their intimate dependence upon the acts that produce them, and it is on those activities that we shall concentrate in this chapter and the next.

Assertion is "productive action".

Not all action is "productive". Deeds may otherwise succeed, by bringing about some *effect* upon an independently existing object, as when one rings a bell, ties a shoe, smashes a cup or gives a gift. Again, a deed may succeed with the agent's *moving himself*, as when one wiggles one's ears, lifts an arm "directly", jumps a creek, gets into bed or performs a calculation.[1] Now my first stipulation about *assertion* is, simply, that it is a kind of productive action. Not all kinds of language are, e.g. *giving gifts* and *calculating* are not. However, since we are to be concerned almost entirely with *assertion* and other kinds of "productive" action, we shall not stop to examine interesting and difficult questions about effects and movement; it is enough to keep the distinctions in mind.

Assertion is utterance.

Assertion is also a "use of language" or a kind of utterance. Language for us is a classification of behavior, to be set in contrast with such others as *investigative behavior* (sniffing, looking, titrating), "replacement behavior" (thumb twiddling), *ingestive* and reproductive behavior.

Is language a generically distinctive kind of action?

That, too soon perhaps, brings us up against a major technical question: Is language or utterance a generically distinctive kind of action? I take genera in the aristotelian way, as preferred classifications that tell us what an object is.[2] So that thing is a horse and that other thing a mule and both are equines; again, $x^2+y^2 = 9$ is a circle and $3x+4y = 5$ is a line and both are conic sections; still again, an act may be a kick or a bite. Although it is no easy task to say what a generic classification is, I have no doubt but that things may be generically classified. Things may also be sorted and classified in ways that may tell us less or more than to what genus they belong. An act, for instance, may be said to be a crime, which does not yet tell us what kind of deed it was; or it may be said to be a murder which does tell us that it was a killing and more besides; a chase (in contrast with a stalk) is a fast pursuit. My ultimate conclusion will be that *language* or *utterance* is not a generically distinctive classification of action. To establish that, we must do a lot of stage-setting first.

Common classifications of every sort may be variously defined. Thus taxonomists and molecular biologists may differently provide functional or genetic definitions of common classifications by sex or species, and chemists provide quantum....

definitions of metals and other common stuffs. Those definitions, which may themselves stand as separate non-generic classifications, are always "responsible to" their common counterparts: The segmentation of genes along a chromosome would be merely arbitrary unless the theory were applied with an eye to the "phenotypical" features it explains.

Common classifications may be variously defined. Those definitions are "responsible to" their common counterparts.

We shall, in section #7, introduce a method for providing definitions of generic classifications of action. We shall go on to annex methods for defining other classifications of action and use all this to provide a definition of *utterance* as a (non-generic) classification of action.

We shall introduce a method for providing generic classifications of action.

Moods or (latterly) "illocutionary forces" are sub-classifications of language, commonly having familiar names. Thus *promising, conjecturing*, and *ordering* are moods of productive language; *cursing, giving gifts* and *calculating* moods of non-productive language. *Assertion* is also a mood, but a factitious one that must be defined before being put to use--the provision of that definition will be our chief task for Chapter 2. Assessment of that definition against the vernacular data is less direct than would be a definition of promising.

"Moods" are subclassifications of language. Assertion is a factitious mood that must be defined in advance.

Before proceeding another step toward the definitions of *language* and of *assertion* as sorts of action, a caveat: Not everything said about an object bears on its sorts. A cup, for example, may be said to be brown or broken. We may similarly say of a performance that it was deft, wrong, unsuccessful or mistaken; such remarks would tell us nothing about what sort of deed was done.

Objects have non-sortal features.

3. THREE GUIDING QUESTIONS ABOUT UTTERANCE.

We have asked whether the use of language or (as I prefer) utterance is a *generically* distinctive classification or kind of action. A more fundamental question--it is the central question of this chapter--asks whether utterance is in any way behaviorally distinctive, otherwise than in the means by which it is enacted. Russell, in his celebrated account of judgement, had the merit of being quite explicit on this point: he denied it. I believe that the

Whether utterance is behaviorally distinctive?

"tradition" from Plato onwards has leaned in that direction (we shall consider some apparent divergences at pp. 43f. below). Now certainly linguistic behavior is recognizable for what it is; its "distinctiveness" is a matter of theory. Those who have denied distinction to language have taken it to be simply an expression of the mind--of belief, attitude, emotion, want or will--conventionally and accidentally (even "arbitrarily") embodied in words, gestures, marks or other "expressions". I shall argue that that is wrong, and that utterance, together with some other behaviors for which language is readily substituted, does indeed have a distinctive "conventional quality" in itself and aside from the particularities of its embodiment. A vindication of *conventional behavior* will be essayed in #9 below.

Whether there are genera of action that are "distinctively conventional"?

A second, derivative question asks whether there are genera of action that are "distinctively conventional" in the sense I hope to establish. Giving a bone to a dog is rarely linguistically done, as giving an article to charity commonly is. *Giving*, generically taken, is obviously not "distinctively conventional". It seems otherwise, however, with *conjecturing*. We shall have to wonder about this in regard to *assertion* (at pp. 92f below).

Is there any kind of action that cannot be done by way of utterance?

I preface my third question with the observation that many kinds of action are sometimes done by utterance and sometimes not. So, for example, property rights may be created linguistically by bequest or non-linguistically by simple appropriation. Is there any kind of action that cannot be done linguistically or by utterance?

These three question will guide our efforts to fix *assertion* as a kind of behavior.

4. ON THE *DIFFERENTIA* OF UTTERANCE.

We aim to provide an honest definition of *assertion* as a sub-classification of utterance. We set out under the still unproven assumption that *utterance--"parole"* or "performance", in the jargon of some linguists--is distinctive. We shall provisionally assume that *assertion* is *"essentially conventional"*, at least if anything is.

Just a remark now about this weasel word "conventional". It has all manner of contrasting even conflicting applications (see pp. 42f below and pp. 281-285 of my *Stratification of Behavior*). Philosophical writers are apt to use it to cover almost any kind of co-operative "rule-considering" behavior such as rowing in a two-person boat (Hume's wonderful example). My usage is much narrower than that. I have misgivings about thus appropriating "conventional" for my own special purposes, and have considered introducing an harsh-sounding, overtly technical term for what I have in mind--"semasiological" was a candidate. I have finally decided that English is adaptable to my needs. I take guidance here from the naturalness of speaking of ritual, the use of money and language as conventional activities. It remains that we shall have to define this understanding of "conventional", if we are ever going to establish the interesting truth that utterance is "essentially conventional".

What I have just said raises another question about the *differentia* of utterance. Both ritual and ceremonial performances such as opening the ark of the covenant and the use of money for purposes of exchange are also naturally described as "conventional" in the same "right" sense. But these are non-linguistic. How then are we to distinguish utterance from them and their ilk?

Language is conventional action that is realizable in writing.

I find it natural to describe *linguistic performances* or utterances as *acts of meaning by saying*. "How d'ya do's and mathematical formulas, as it seems to me, "say something" and genuflections don't. The *differentia*, which I have worked out elsewhere (*ibid*, pp. 327ff.), is simply that language can be realized (not "recorded" or "transcribed") in writing, though certainly it is enormously more frequently "realized" in speech-cum-gesture. I shall want to check my characterization of conventional behavior against its several varieties (see p. 43 below). Otherwise I won't bother very much, for the distinction between language and other forms of assumedly conventional action is of really negligible consequence. Non-linguistic forms of conventional behavior can always be absorbed into wider groupings for which devices of linguistic performance can be made available. Deities may be supplicated verbally and goods conventionally exchanged by written acceptances in response to bids or offers: while using money is not language, acceptance may be; a display of the ark says nothing, but God's presence may be invoked in words, perhaps written out on paper and deposited on the altar.

5. TOWARD A DEFINITION OF CONVENTIONAL BEHAVIOR *PER GENUS ET DIFFERENTIAM.*

Conventional behavior is conformative, proficient action.

Our main task for this chapter is to provide the desiderated definition of *conventional behavior.* It is, I submit, a *kind* of action which can be done only by way of what I call *conformative behavior*, which is itself a kind of what I call *proficient behavior*, this in turn being a kind of action which cannot so much as be attempted unless the agent *knows how* to do something or other; *action*, finally, is a kind of *behavior.*

The just indicated "stratification of behavior"--there could no doubt be others--seizes upon distinctions you will have some sense of. Some such staging, if vindicated, will work to assist a step-by-step approach to an explanation of *conventional behavior.* Vindication is wanted; I have elsewhere done what I can to provide definitions that go some way toward making the case. The definitions of *behavior* and of *proficient* and *conformative behavior* are difficult and of interest in themselves; but, for brevity's sake, I eschew developing those definitions in this treatise, and shall pretty much rely upon your everyday understanding of "know how to" and upon your feel for the common though unexplained appeal to the theoretical notions of *behavior* and *rule.* I cannot, however, forego saying something about the idea of *action*, for without that my choice of a *differentia* for conventional behavior could not be understood at all.

Definitions of proficient and conformative behavior reached in *Stratification of Behavior* are briefly resumed on p. 41f. below. I worry about "false economy" in not redeveloping the definitions in this treatise, esp. that of proficient behavior. *Knowing-how-to* turns out to be of the profoundest importance for an endeavor such as ours. It is a "first step" from instinct, manifest surely in the behavior of all creatures that navigate. *Know-how* is of the essence of that transition which Locke called "abstraction" and of that kind of "generality" in which every word is general--not narrowly in the "predicative" sense of covering a possible generality of cases--but in the sense that we have a capacity also to make uniform repeated "singular" use of (e.g.) "this", capacity also to make uniform repeated "singular" use of (e.g.) "this", "that" and proper-names. A tenet that dominates this treatise is that the know-how evident in our use of words in assertoric context is grounded on a variety of investigative procedures, which are also species of know-how. Melnick has brought me to realize that Kant was a precedent.

Sensible intuition, in Kant's philosophy, is realized in investigative "tasks" and his Forms of Intuition--space and time--are procedural kinds. Kant took the Mathematical Antinomies to establish that the generality of procedure is not assimilable to the generality of *all* or totality, and that conclusion was central to the ultimate rendition of his "transcendental idealism" (see p. 260). I agree entirely that procedures are indispensable. There cannot be even anticipations of intellectual life without them, for there and only there lies the required "generality" of thought, and there and only there do we find a mechanism for bringing thought into touch with reality (see *ibid.*).

The conception of know-how is riddled with difficulties, some first taken written notice of by Aristotle in his celebrated examination of the virtues and latterly known from Wittgenstein's elaborated anxieties over "knowing-how" (e.g. to read, to count). Kripke's well-known "sceptical" redaction of Wittgenstein's argument is shattering. The difficulties can, I believe, be overcome by locating *know-how* within a general theory of behavior, such as ours. *Know-how* is a condition for doing deeds based on investigative mechanisms.

After all these uncertain and certainly uneconomical anticipations, I revert to a policy of paucity: *Know-how*, whether or not a fact, is an everyday notion on which we can provisionally rely in our effort to provide the required definition of *assertion*.

6. CONCEPTUAL EPIPHENOMENALISM: A FRAMEWORK FOR A THEORY OF ACTION.

Acts are movements in or of the bodies of live animals.

Acts are (physical) happenings of a certain kind; "action" names that kind of happening. Such happenings are movements in or of the bodies of live animals. Other kinds of happenings may also sometimes consist of movements in or of the bodies of animals. Just as a goldfish is one kind of body and a fishbowl another--the one is an animal, the other is not--so diving from a

Happenings may be variously classified. Our concern will be with characterizations of episodes as action.

Happenings, like other things, may be variously classified. The same one episode may be classified as a *hit*, a *shot* and a *score*. Some acts of diving are also cases of falling: "See that sky-diver falling toward his target!". Perhaps every movement in or of a live animal is *also* a neurological transaction.

So happenings or episodes resident in animals may be variously classified. Our interest is in *action*, and the immediate focus of our concern will be with the characterization of episodes as action. The immediate subject matter of our analysis will be, therefore, *not* those physical episodes that are acts and other things too, but rather our common ways of thinking about episodes, chiefly our everyday classifications of action.

Working from the presumption of broad agreement on specimens, we seek a general definition of action.

I assume that the reader commands considerable capacity for talking about animal behavior, and that we can agree that a speaker's employment of certain vocabulary items would ordinarily be taken to imply that he viewed the episode he was observing as a kind of action, e.g., such vocabulary items as "take a bite" (but not "eat"), "scratch" (but not "twitch"), "affirm", "look" (but not "see"), "move into position", "prepare one's racquet", "make a shot".

> *Eating* is an "activity" which may variously involve different actions; *twitching* unlike *scratching* may be "mere undirected movement"; *seeing*, is not movement but another kind of happening that may occur as the "measure of success" of *looking*.

We should stick by our presumed agreements. Uncertain specimens, of which perhaps *chewing* is an instance, may be reserved as test cases. Since the very same happenings may be severally classified as action and otherwise, we must find a way of identifying those characterizations which are distinctively action characterizations. What are those?

Acts are movements both meant and done, even when the acting subject did not do what he meant to do.

Acts are things creatures may be truly said *to do*. However, not everything a subject can be truly said to do characterizes his movements as action. "What did he do?", you ask; "He fell", say I. *Falling* is not a kind of action. I assume that an act is something which a subject animal can truly be said *to mean to do*. He meant to dive, but he didn't mean to fall. However, not everything an animal can be truly said to mean to do is a kind of action done.

The bank teller might really have meant to put that money back; but he didn't. The characterization of an episode as action implies that what happened was both meant and done. Yet an agent may not have done what he meant to do in doing an act, and will not have done so in case he fails: I meant to hit the target; what I did was to kill the cat. I shot but failed to score. An observer's identification on the episode as action *simpliciter* should not imply that the agent succeeded or that he failed; but only that he made the attempt.

Action-names provide answers to the question, "What was S meaning to do?" while implying answers to the question, "What did S do?"

With all this in mind, I find it useful to employ the following "verbal index" for selecting specimens of action characterization. The employment of a verbal expression *W* identifies a form of action if, in response to the question, "What was he meaning to do?", an observer could answer ""to"^W (e.g., "to hit the ball", "to move the king"), where that response would also imply an answer to the question, "What did he do?" (e.g., "swung at the ball", "tried to move the king"). A behavioral episode E is an act just in cases where E could be correctly characterized by the use of some *W*.

This theory is used to elucidate action-characterizations that may be given of subjects by observers.

I stress that this theory is to be used, not for characterizing acts themselves, but rather to elucidate action-characterizations that may be given of subjects by observers. Suppose someone (call him "We"), wishes to employ our theory to report data: *We* listens to another animal, *O*, who makes reports on the behavior of a subject animal *S*, where *S* may or may not be the same as either *We* or *O*. *We* wants to characterize what *O* says about *S*. *We* may, for example, infer from what *O* says that *O* is or is not imputing action to *S*. If *O* says that *S* is biting into an apple, *We* infers that *S* is doing an act; or, if *O* says that *S* blushes or stumbles, *We* infers that *S* did something other than an act; if *O* says that *S* is eating, *We* infers that *S* will soon have done a number of acts of various unspecified kinds such as cutting, sucking and spitting.

Herewith a picture to encapsulate the kind of investigation I shall be making:

S is an animal imagined to be moving in certain ways. *O* is an imagined observer of *S*, who reports what he sees *S* doing. *We* is a theoretical commentator of the reported observations of *O*. *We* is supposed to have available the apparatus of the ensuing theory. *We* applies that theory by drawing out the implications of *O*'s report. On the assumption that what *O* reports is true, *We*'s remarks contribute to a theoretical analysis of *S*'s movements, under *O*'s particular characterizations. *O* says that *S* is biting: *We* says that *S* is doing an act. *O* says that *S* is using chopsticks: *We* says that *S* must know how to use chopsticks. *O* says that *S* is sucking: *We* says that *S* thinks that there is liquid close by. *O* says that *S* thinks that there's liquid close by: *We* says that *S* may be meaning to suck it up. We should, within this schematism, be able to define all familiar ways in which episodes are characterized as action. We shall also find ourselves in the business of fabricating classifications of action, of which *assertion* is the chief case in point.

I call the "methodology" just described and illustrated *conceptual epiphenomenalism.* Conceptual epiphenomenalism seeks characterizations of reports. I shall take it that such reports report facts, notably facts that involve conscious subjects. Since such facts may be variously reported, our characterization of a report will have to say something about its particular "meaning" and distinguish it from other reports of the same fact. That meaning may implicate an explanation of the reported fact, and the explication of that meaning may, accordingly make mention of causes and other explanatory factors. I shall later argue (pp. 72ff.) that mentions of explanatory considerations are "opaque" in respect of the facts they explain. So, for main example, "causal attributions" of the *C caused E* form are "opaque" in the sense that the truth of what is said depends upon how the C and E happenings are referred to. It follows that we cannot use "cause" or other such explanatory predicates for purposes of defining or otherwise selecting C or E, for the selected conditions must remain the same however they are expressed. (My brother Herschel is "defined" by that name, but equally as my oldest brother and as my mother's first male child; again, 5 may be defined as the third prime or as the first natural number whose square is the sum of two squares.) It does not follow that we cannot use (e.g.) "cause" for purposes of defining the *meaning* of reports of the occurrence of C's and E's; that, as we noticed, may even be required. This frees us to make mention of the explanation of movements as part of our characterization of action. With that proviso, let us proceed:

Our "methodology": "Conceptual epiphenomenalism" seeks characterizations of reports. The explication of these "meanings" may make mention of causes and other explanatory factors.

The characterization of an episode as an action, e.g. as a dive, implies that the subject "meant to do" something. Other characterizations may carry no such implications; to say of subject that he fell, for example, does not. "Meaning to" suggests the presence of a "mental factor", whose observable presence prompted the observer to provide a characterization of subject's movements as an act, e.g. as a dive. That "mental factor", which is so far a mere blank in the formula, is present just in case the act-constituting movements satisfy a "certain condition" accessible to observation. The condition in question has something to do with subject's observable state of awareness and thought, hence a "mental" factor. More particularly, I suggest that an observer's action-report entails his judgment that the observed movement could have been explained by what the subject animal believed to..

An action report entails that an observed movement could have been explained by what the subject animal believed to be the case, where that belief was itself consequent upon the subject's awareness of something or other at the time of action.

be the case, where subject's beliefs are consequent upon his awareness of something or other at the time of action. Subject's awareness explains his having certain beliefs, and those beliefs, with others perhaps added in, explain his movements. The beliefs in question are identified as beliefs relative to the possibility of providing a certain kind of explanation. No one of the beliefs need actually be the desired explanation, and their presence is not belied if they aren't. Only this: a belief is so-called relative to the possibility of providing an explanation. The situation in regard to belief is on a par with the identification of bodies as *animals* relative to the possibility of providing functional explanations, and of the identification of an *eclipse* relative to the possibility of explaining darkness in terms of an occluding cause, and of the assignment of *weight* relative to the possibility of providing explanations in terms of gravitation.

Beliefs may also be other kinds of things, possibly even movements.

We are now familiar with the thought that this belief might have been otherwise characterized, and that even seems required by the condition of observability. Latter-day "physicalists" favor neurological characterizations. I see no reason why the belief might not sometimes actually be party to those movements which it explains. Take an analogy: the shot explains the score; but then the score *is* the shot. Of course shots and scores are different *kinds* of things. Furthermore, not all shots are scores, and scores need not be shots, e.g. they may be awarded by referees on account of infractions. Similarly, *belief* is a different *kind* of thing from *movement* even though beliefs may sometimes be party to the movements which they explain, as parts perhaps or inceptions or (as Pierce once suggested) as inhibitions.

> Aristotle long ago advanced some such a possibility in his stated opinion that actualized desire is the action it explains and the perception or belief, which actualizes the desire and thus contributes to the explanation of the movement, is nothing other than an early stage in the movement itself. (*De Anima*, 433b,17; 434a,20; also *De Mot. An.*, 701a,27-b1). I like this theory! I'm inclined to deny "action at a distance" in the tightest sense of maintaining that in all cases of *proximate* causal explanation the *explicans* and the *explicandum* are identical. I suspect that Aristotle would have agreed.

The relationship between belief and action underlies our understanding of a subject's *reason for* it's action. In saying what a subject meant to do, we invite a question The imputation of "purpose" to a subject invites the question "Why?", answered by giving "its reason for acting". To go on and assign a reason for subject's meaning to do as it did is to say something about what subject knew or believed to be the case: Her reason for selling was that the market was high (as she thought); his reason for hitting the ball over there was that his opponent was, as he thought, moving in the other direction; if she was meaning to type "A" on paper, then her reason for striking a particular combination of keys was, as she believed, that they would imprint "A". Here then is an alternative formulation of my proposal that an act consists of movements caused by subject's beliefs: An acting subject has various thoughts about what is so; saying what some of these thoughts are also identifies subject's reasons, real or mistaken, for doing as it did. A necessary (but not sufficient) condition for action is that the subject does what it does "for a (its) reason".

A necessary (but not sufficient) condition for action is that subject does what it does "for a (its) reason".

Just as belief contributes to the explanation of kinds of phenomena other than action, e.g. emotions and beliefs, so a subject has its reasons for its emotions and beliefs.

We have been speaking of subject's reasons for doing what it did; it may additionally "have" other reasons, some perhaps "believed" and others not..

Kinds ascribed to things only in relation to a prospective explanation are "epiphenomena" relative to what explains them or to what they explain.

I wish now to appropriate some old-style philosophical jargon. Those *kinds* that things are said to be only relative to the prospect of an explanation are called "epiphenomena" relative to what explains them or to what they explain.. A body's *weight* is an epiphenomenon relative to its "tendency to fall"; if a player scored because he made a good shot, the score is an epiphenomenon relative to the shot. Similarly, the belief one may observe in an act is an epiphenomenon relative to those act-constituting movements which the belief may also serve to explain.

Epiphenomena are "real". Explanation itself is "opaque".

The shot explains the score, not conversely. But the score is (=) the shot. The truth of our claim that one kind of thing explains the other is relative to how the same one episode is characterized, now as cause, as effect, or as neither: the shot caused the score, but the score didn't cause the shot. That does not carry over to the characterizations of the episode as being of the one kind or the other. The same one happening is both a score and a shot. The characterization of the episode as a score, which is given relative to the *possibility* of an explanation, is not on that account "opaque". So too with *action*: an episode either is or is not an act, regardless of how it is referred to. If it is an act, then the epiphenomenal factor of belief must be present. Epiphenomena are not unrealities or illusions. They are phenomena, called "epi" because brought to attention under a characterization that looks ahead to the possibility of an explanation.

Beliefs are observable only by observers capable of seeing why things happen.

Any observation of any phenomenon under any characterization depends of course on the intellectual attainments of the observer. The presence of belief is observable, but only by observers capable of seeing why happenings happen. Belief need be no more observable by the subject himself than are factors which explain his digestion, and will be observable by him only if he is capable of observing in others the causes which some of them may observe in him. The intellectual capacities required for this are no doubt considerable, but do not always require the use of language--a dog who sees another dog going to its dish will also see that the explanation for this activity is that its deluded companion thinks there's some food over there.

In summary: We are concerned with characterizations of episodes that imply the presence of cognitive factors identifiable by observers only relative to the possibility of providing a certain kind of explanation for those episodes. In what follows, I sometimes fall into the "material mode" and speak as if of *acts* as episodes actuated by belief, where accuracy would require me to speak of action characterizations as ones that imply the presence of such explanatory, cognitive factors.

7. A THEORY OF ACTION IN OUTLINE.

We suppose that an act is reported on in the terms we have stipulated

We suppose that an act is reported on in the terms we have stipulated, *viz*, the observer says "to"^W in answer to the question, "What was S meaning to do?" and that that answer implies an answer to the question, "What did S do?". The combination effectively separates S's *action* from other *kinds* of things S may have done at the same time.

Those beliefs in a subject that explain its action are, some of them, induced by that subject's awareness of something in its surroundings, e.g. the movements constituting a subject's act of meaning to open a door are in part explained by its belief that there's a door there, which belief is itself induced by subjects awareness of what may or may not be a door.

Action- explaining beliefs are induced by awareness.

Now, going back, an answer to our indexical question of "What was S meaning to do?" may or may not be the required answer to the question of "What did S do?"; an answer to the "meaning to" question of itself need imply nothing either way on the question of whether subject actually brought off what it meant to do. It does, however, imply that the question whether it succeeded could be asked and answered. From this we conclude that an acting subject may succeed or fail and, derivatively, that the act itself, when done, either succeeds or fails. (Nb: we are speaking now of failure *in* completed action, and not of failure *to* act or to complete an action.)

Acts have "success value". Animal movement explained by awareness need not be action.

We just noticed that the beliefs that may explain the movements constituting an act are themselves induced by awareness. Consider another kind of case, suggested by an example of Anscombe's: Observer reports that subject starts at the sight of a snake or that it wakens at a sound. Such a report implies that subject was aware of the snake or sound but not that what it did either succeeded or failed. Waking at a sound is another sort of animal movement explained by awareness; it is not action.

Action involves both awareness and success-value. A condition that must be met if an act is to succeed is a "condition of success" for that act.

Our conception of action brings together these two things--success or failure on the one hand, and awareness on the other. How so, one my wonder? The decisive consideration is whether awareness induces belief. Waking-up awareness doesn't, the awareness of action does. Awareness of itself is not true-or-false as are those beliefs that may be induced by the awareness. (Note well that a belief may be true without being a belief that something is true. Travelling abroad, I may believe that it's true that it's now dinnertime in this place; a pacing dog may simply believe that it's dinnertime.)

Now, failure in action is often traceable to erroneous belief: one thinks, for example, that the key can be more easily depressed than it can; again, one thinks that it's earlier than it is and therefore fails to find a person home when one phones. False belief may sometimes even necessarily result in action-failure. A typist attempting to print "A" on a piece of paper will think that his machine provides a combination of strikers for imprinting that letter e.g. a lambda and a bar, and he must fail if that is not so. Again, one could not successfully make a three o'clock visit if the hour had already passed. We call such a condition that must be met if an act is to succeed a *condition of success* for the act.

This "must" is innocuously "de dictu". That the condition is believed by subject and that the act fails if the condition is not met are implied by observer's imagined characterization of the episode.

An acting subject cannot disbelieve any of the conditions of success for its act.

One who believed that it was already four o'clock could not even so much as attempt a three o'clock visit; if one thought that there was no combination of strikers for the letter "A" in the only typewriter he had before him, he would not attempt to type "A" with that machine. That a subject should not thus disbelieve a condition of success for this or that kind of action is itself a condition for his attempting to do an act of that kind, and his disbelief would make it straight-out false to say that he was doing an act of that kind. The subject could not have either succeeded or failed in that kind of action.

Subject may know that he won't succeed, e.g. in his attempt to kick a 75 yard field-goal. He cannot, however, even make the attempt while disbelieving that a game is in progress, the goal's up and his side in possession of the ball. These are conditions of success, and that he should not disbelieve them is a condition for his kicking for a goal (= meaning to kick a goal).

A "truth condition" for the ascription of a kind of action to a subject is a "condition for doing" that kind of action.

More generally, we hold that a condition for S's attempting an act having a certain condition of success C, is that S should not disbelieve that C obtains. Still more broadly, we call any condition that must be met for an observer of the animal truly to say that that subject was (successfully or unsuccessfully) doing an act of sort A a condition for doing A.

I wish to stress the just noticed distinction between *conditions of success* for action and *conditions for doing* action, for it matters a lot in what follows. A subject doing an act of a certain kind cannot succeed if any of the conditions of success for his action are not satisfied; he will fail instead. In contrast, if any of the conditions for doing a certain kind of action were not satisfied in the person or the circumstances of a subject, then it would be simply false to say that that subject was meaning to do that kind of action; here the subject would not have attempted to do any such deed, and of course could not have succeeded or failed in any such effort.

A condition for doing an act of any kind is that subject should believe of any of that act's conditions of success that that condition is satisfied.

We have now cited as one prime condition for doing any kind of action that subject should not disbelieve any condition of success for the act. That condition does not suffice to distinguish between acts of trying to do A and acts of trying to find out whether A could be done--*meaning to try the door* is different from *meaning to open it.* This observation argues that mere absence of disbelief is not enough and for a stronger condition for doing any kind of action, *viz* that subject should, of any condition of success for its act, positively believe that that condition is satisfied. (There is no insurmountable difficulty in this "any". For a review of the situation, see my *Stratification of Behaviour*, pp. 126f.) We shall soon see that this conclusion--that a condition for truly saying that a subject did an act is that he should, of any condition of success of that act, have thought the condition satisfied--systematically provides for the definition of generic classifications of action.

There are other "conditions for doing"

There are other "condition for doing" acts--"truth-conditions" for reports of action--having to do with matters other than belief. So, for example, subject must *know how* to read if he is to read a passage, and he must be alive if he is to do any act at all.

The frequent occurrence of "can" and "could" in my presentation is "schematic". To say of subject that it "can" do A is not yet to formulate an *explicit* condition. It *may* formulate a variety of conditions--opportunity, training, strength, permission, good luck, etc. (This matter of the "amphiboly of 'can'" is discussed at pp. 151-155 of App.B and again at pp. 179f. of App. C.) Some of these are "conditions of success", others conditions for doing the act, and others neither one. It may even sometimes be possible for a subject to succeed in action when it would also be true to say that "He can't", e.g. hole a 90 foot putt in golf (Austin's example).

A true answer to action-indexical question of what subject was meaning to do, what we call a "specification of purpose", provides a generic classification of action.

I now revert to the vernacular classification of action. A true answer to the question, "What was subject meaning to do?" generically identifies what sort of act the subject was doing. I call that a "specification of purpose". Specified purposes are generic classifications of happenings as action. (Nb: a specified purpose is not a piece of the action additional to the constituting movements; it is a classification.)

We define genera of actions, or purposes by listing kinds of conditions of success.

Now, every sort of action--e.g., biting, kicking, making a move in a game of chess, promising--has characteristic kinds of conditions of success. For example, in making a promise--in contrast to a vow--a man must think that he is addressing himself to an intelligent respondent. I submit that we can always define genera of action (purposes) by listing characteristic kinds of conditions of success for acts of those sorts, *viz* for all acts having those purposes. Thus, that sort of action we call telephoning could be (partially) defined by saying that conditions of success for the purpose are that there is a telephone at hand, energized, etc. We define narrower more specific purposes by adding conditions regarding the existence and character of a respondent, the time of day, etc. A completed such list of conditions of success, if ever available, would define the action analogue of an aristotelian substantial species. In fact, such completed definitions are not obtainable.

Proof that action is "aristotelian" in respect of "purpose".

My theory of action classification will, I hope, be vindicated in application. Considerations now following may serve to rationalize the claims registered just above. Aristotle held that every substance is of a unique specific sort. Since *infimae* species are "problematic, we may revise the thesis to read that every substance falls within a unique nest of ever more specific genera. So revised, the doctrine seems plausible for organisms but less so for hunks and chunks in general. I now argue that *action*, in this matter of classification, is more than chunky and resembles animate life.

Every act is a behavioral episode consisting of movements explainable by reference to the fact that a subject believes that certain conditions obtain. These belief conditions include our conditions of success. Such conditions may be listed in their formulations. Any such list of formulated conditions are of kinds that defines a *purpose*. We can, therefore, assign a purpose to any act. But purposes are generic classifications of action. It follows that every act is of some sort or other. Different lists define different purposes. If an act fell within two such sorts thus defined, we could cover both with a single purpose defined by a consolidated listing of kinds of conditions of success. If, as we suppose, the aristotelian rule of unique speciation for substances comes down to the requirement that every specific determination is sub-generic, then we have established for *action* an analogous principle of unique speciation. I hope to have convinced you that my lists of conditions of success do indeed give the meaning of those "specifications of purpose" that provide answers to the action-indexical question of what the subject was meaning to do. Those answers would then indeed provide preferred generic classifications of action.

Genera of action are always distinguishable by differing kinds of conditions of success.

Purpose-defining lists of condition-kinds can be extended without limit. Thus purposes may be ever more specifically defined. Extending the lists also allows for the progressive generic differentiation of action. There is no nameable genus of action not distinguishable from any other by listings of kinds of conditions of success. We take on the hard challenge of defining the difference between *accepting* and *refusing* at the end of this section. Another testing case is to distinguish kicking for a goal from faking it (suppose gamblers have gotten to the kicker): acts of both sorts are doable in the same circumstances and are calculated to look exactly alike. How then to distinguish them? Not by "intention", which (in this context) is just another word for *purpose*. A difference is that a condition of success for "real" goal-kicking but not for faking is that a disqualifying penalty should not be called against the kicker's team. The reader is urged to try out my proposal on some examples of his own.

Summing up, I propose, as a definition, that an act is a behavioral episode the occurrence of which could be explained by mention of the fact that a subject believed that conditions were satisfied, where the conditions be instances of conditions of success for some purpose. Those beliefs at once account for and impart direction or purpose to the episode and, as viewed by the observer, give rise to the conjugate possibilities of attributing success and failure to the episode. The above analysis has now led us to a definition of *action* as a kind of behavior. A behavioral episode is qualified as action when some of its constituting movements could be explained by reference to the fact that a subject believed that certain conditions obtained, where those conditions are what we call "conditions of success". (The tortured sentence structure is calculated to obviate the implication that the subject should also think that the mentioned conditions are conditions of success.)

> This definition notably makes no appeal to *want, desire* or any other "major premiss" of an aristotelian "practical syllogism". The required presence of some such factor is screened by our early call for success-or-failure. What we shall presently dub the "measure of success" of an act satisfies the want from which the subject acted. There is no action in a subject's movements absent the presence of either want or belief in that subject, as causes of its movements. The belief brings the fact under the want as a reason for action. Either want or belief may be present and the other absent. A hungry sleeper or rueful wisher has a want with no belief to match; an insomniac may be unwantingly kept awake by an obsessive belief.

A killing objection? Since beliefs are identified in relation to formulations, it would seem to follow from the consideration that conditions admit of different formulations that we cannot define genera of action by listings of condition. Answer: Our characterization of action is in terms of KINDS of conditions of success, not actual conditions themselves.

Our doctrine as so far developed is open to an objection, which I must now try to answer. We hold that the identification of a subject's action is fixed by reference to that subject's beliefs. But now the believed conditions, whether called "of success" or by any other denomination, are simply facts that admit of various formulations. It would seem to follow from this that we do not characterize action simply by listings of conditions of success. An iron-smith may set about to forge an iron ring. A condition of success for any act he may attempt in pursuit of that enterprise is that he should have some iron to hand, as indeed he does believe. Now the actual condition of this stuff's being iron is no other than the condition of its being mostly Element #23; but surely no 18th century smith ever believed that his material was Element #23.

Answer: Our characterization of action is in terms of KINDS of conditions of success, not actual conditions themselves. A condition of success for my coming to town on Nov. 16,1984 is that today should be that date; a condition for my doing this deed is that I should believe that today is Nov. 16. Today is also Friday; and if today is Nov. 16, 1984, it couldn't but be Friday (a "logical truth"). But today's being Friday is not a success-condition-kind for my coming to town on Nov. 16. Similarly, this stuff's being iron is not the same kind of condition as it's being mostly Element #23[3].

Our doctrine is "spinozistic".

A second objection, which I cannot overcome, is that our doctrine of action is, at a second remove, haunted by dualism, on account of the need to select conditions and condition-kinds by reference to formulations. My own doctrinal inclinations are "materialist". I conceive *cognition, action* and other such human matters as facts continuous with the dynamic order. I am genuinely disturbed to think that my appeal to formulations--and I see no way around that--may suffuse this natural order with an ectoplasm of spirit. I see no way to dispel this spectra from our workaday conceptions of what we are and what we do. Let it be said, however, that the dualism by which my ideology is threatened is of the "double-aspect" "spinozistic" kind. I see no need to allow that there is anything in or out of mind that isn't *also* physical. Still, these happenings, most especially those that feed into our cognitional apparatus, in our dealings with them, are saliently identified in relation to mind-dependent formulations[4].

What must be observed to establish the success of an act is its "measure of success".

I now resume my outline of a theory of action with some observations about success-or-failure in action. Every completed act either succeeds or fails. It *must* fail if any of its conditions of success do not obtain. But it still *may* fail even when all those conditions are satisfied. All the conditions for one's kicking a field goal in a game of football may be met--there is an official ball, placed down, by a permitted player, in a permissible way, during a real game, with the posts up, no penalties assigned, etc--and the ball misses or for other cause the kick not be registered as a score. Generally, the "success value" of an act is established by separate observation. I call whatever must be observed to establish the success of an act the "measure of success" of that act. Depending upon the case, the measure of success for the kicker's deed would be the gameball flying through the posts or a referee's official declaration. An exposed negative is the measure of success of a photographic act; and a statement, I say, is the measure of success of an act of assertion. In every case, the measure of success is some combination of acted for effects, movements or products.

Objection: Success in action is "defeasible".

Reply: The exposed negative is a "fact". Even "institutional determinations" may be: games are won because scores are made. I allow that *observation*, which is "of" the measure of success, indeed often is "defeasible". Reason after reason may arise to make us wonder whether an observer did indeed see what he said.

Measures of success are specific but seldom peculiar or distinctive to purposes or action-kinds.

Identification of a measure of success implicates conditions for the purpose. So, for example, if the measure of success of an act is getting a certain party to know something, that party must have been conscious during or sometime after the act of communication if that act is to succeed; agent's believing that he has a sometime conscious respondent is a condition for his doing the act. Yet, since different kinds of action may have common measures of success--the implantation of an arrow in a target may result either from a successful shot or from simply punching the stick into the surface--action-kinds will commonly also have conditions of success not determined by their measures of success.

Again, success in action and the attainment of its "measure" may sometimes be without or even perhaps be in conflict with "intention": playing squash, I miff an intended shot into the front left corner, but luckily win the point by catching the nick; my ungainly and unmeant effort could hardly be reckoned a failure.

It is finally and furthermore to be noticed--a point that matters to our explanation of statements as products--that action kinds may have variant measures of success: knocking in a nail--a kind of action--may sometimes serve to secure a board and sometimes to produce a box.

Non-generic classifications of action may or may not involve the agent's beliefs. "By what means" characterization do involve the subject's beliefs.

All acts, we argued, are generically classified by purpose, which purposes are determined by the subject's beliefs. Action may also be otherwise classified. Thus, to call an act an "infraction" implies that it was rule-governed without yet telling us what kind of act it was. *Murder* is *wrongful* killing; a *thrust* is a movement *quickly* made, and a *blow* an attempt to displace matter *energetically and impulsively* enacted.

These various other classifications of action may or may not involve a subject's beliefs. A subject's decision to smash or push is determined in great part by his belief about how the object is fixed in its place. Quickness of execution, by contrast, may be due to high spirits.

Evidently, the beliefs from which one acts may affect not only what one does but other specific aspects of action as well--speed, "style" and the hopes with which he comes to the deed. One's beliefs may also affect *how* one does what he does, in that sense of "how" that coincides with one sense of "by what means": A burglar may simply walk into a house through an open door; or, thinking that all the doors are barred, he may "break in" through a window.

Problem: "By what means"-characterizations are also purpose-specifications, and it may therefore seem that an act could have several distinct purposes.

Problem: "By what means"-characterizations are also purpose-specifications, and it may therefore seem that an act could have several distinct purposes. An observer could truly report of this burglar that he *went in through the window.* But surely *going through a window* is a kind of action and a qualified answer to questions of what a subject was meaning to do. So our burglar was both entering the house and going through a window! I have argued that no act can be of two such separated generic sorts (pp. 26f), but now it seems it may. Of course, we say that burglar entered the house *by* going through the window. So there does seem to be a difference in how the two classifications are brought into place. What is that difference? Again: We do, as illustrated, systematically describe how acts are done by employing language available for the generic classification of action by purpose: Whence this borrowed way of speaking? We have two action-characterizations of the same one episode. One speaks of *what* was done; the other of *how.* But the characterization which tells us "how" this act was done might also have been used to tell us "what" some other deed was.

The apparatus we have contrived yields a fairly direct resolution of the problem. The conditions listed in the definition of a purpose cover a generality of cases differently realized in the persons and circumstances of action. The definition is unaffected by whether the characterization serves to tell *what* the action was or *how* it was done. The characterization would be correct either way, whenever a subject believed that the defining conditions were realized in his own person and in the circumstances of his action.

Solution: Some of the listed conditions that define the "how"-characterization do not cover the conditions of success of the act actually done; subject could then succeed even when those sorts of conditions do not obtain.

The difference in the "what" and "how" cases resides in whether those believed conditions are *also* conditions of success for the particular deed done. If they are, then the characterization tells us "what" the subject did; if not, then perhaps the characterization tells us "how" he did it. The defined purpose is a generic classification of the act just when the actual conditions of success for that act match the kinds of conditions listed in the definition. Subject might, however, believe that conditions of those kinds were satisfied without them having to be conditions of success for its actual act. So with the burglar who entered at a window instead of through an open door. He wanted to enter the house and, unknown to him, he could have attained that purpose otherwise than by going through a window, for the conditions of window-entry were not conditions of success for his gaining entry to the house. Nonetheless, his movements were explained by his belief that conditions of window-entry did obtain.

Different beliefs may serve to explain a subject's movements. Some of these may be of conditions of success for one sort of action, and others of another sort of action. Subject's beliefs that are of conditions of success for its actual act match condition kinds that would be listed to define that purpose whose specification tells what the subject was doing; other beliefs may match defining conditions for a purpose whose specification would, in that subject's particular case, tell us how or "by what means" he did the deed.

Purpose-specifying "How" characterizations of action are specifications of "collateral action".

In English we say that subject does one thing (entering the house) "by" doing another (going through the window). Two action-kinds or purposes are brought to attention and assigned to a single act. However, only one of these can be specified as *the* purpose of *that* act, to tell us what the subject meant to do. I shall say that the other "how"-characterization provides a "collateral description" of the act and that mention of that purpose is a specification of "collateral action" for that case. (I shall sometimes permit myself the inaccuracy of "collateral act", but that must be understood to mean the particular act in question thought of as given under a collateral characterization.)

In the upshot, we can separate what one does from the collateral means by which one does it solely in terms of conditions of success. If any of the conditions definitive of the "what"-kind of action subject performs do not obtain, subject must fail in its action; but if conditions definitive of the "how"-collateral action sort by which it does the act do not obtain, it may find other "means", e.g., it may enter through the window. Act characterizations are not collateral in and of themselves, but only in respect of particular acts with their particular conditions of success. The condition-kinds that define the collateral characterization of that act, unlike the condition-kinds for what the subject does, are not of conditions of success for the actual act done. An act is done by a sort of collateral means *C* just in cases where subject believes that *C*-defining kinds of conditions obtain, and these beliefs contribute to the explanation of the subject's movements, but (some of) those conditions are not conditions of success for the actual act.

A subject's reasons may comprise both conditions for what subject did and for the collateral means by which he proceeded.

Any condition the believing of which by a subject explains movements constituting an act done by that subject may be that subject's "reason for action", and those to include conditions for what subject did and for the collateral means by which he proceeded: The burglar's reason for going in is that this (as he thinks) is a house; his reason for his proceeding to the task as he does is that the window (as he thinks) is easily opened.

Certain sorts of action are more natural and more common than are others. Thus taking food into the mouth is a natural sort of action which goes on almost every day everywhere animals abide; hitting golf balls, though it goes on fairly frequently among "advanced peoples", is an unnatural action. Now people take food in different ways, by use of fork, straw or chopsticks. The means may be unnatural. Yet we learn to do these things. An unhungry chinese child may be drilled in the use of chopsticks even when he isn't eating. He learns to do that sort of thing. Lifting morsels with chopsticks is a sort of action or purpose normally assignable as the collateral means of taking food into the mouth. Now a Westerner, when first exposed to the Mysteries of the East may, when hungry, wish to eat his chinese food with chopsticks, and wouldn't be satisfied if offered a fork instead. His purpose, we may imagine, is to take food with chopsticks. The purpose in question is definable by a consolidation of the conditions for food-taking (e.g., the presence of food) and for using chopsticks (e.g., the presence of chopsticks). Food-taking is a most natural and common thing to do, and using chopsticks usually a collateral means thereto. But not in this case. What subject wants to do, his purpose, is to take-food-with-chopsticks. It is useful to be able to describe the one kind of action in terms of the other. I shall say that *taking food with chopsticks* is a "hybrid purpose"; this kind of action is hybridized from the "original purpose" of food-taking and the "original way" of using chopsticks. Generally put: A hybrid purpose is definable by consolidating two lists of kinds of conditions of success, the one defining what is commonly a collateral original way of attaining an original purpose defined by the other list.

A "hybrid purpose" is definable by consolidating two lists of kinds of conditions of success, the one defining what is commonly a collateral "original way" of attaining an "original" purpose defined by the other list.

The characterization of purposes as hybrids is of no great intrinsic interest because it is patently culture-bound and parochial. "Hybrid", in relation to purposes and action, resembles "offensive", for what is offensive in Urbana may be good manners in Rabat just as what is hybrid in Reno may be accustomed in Beijing. Hybrid action remains a useful notion, both for characterizing the purposes of our familiars and for anthropological commentary: We might wish to say of certain cultivated societies that stylized action sorts, always hybrid for earthier folk, are never so for them. We shall also have occasion to consider that certain action kinds that are not hybrid "might have been".

Accepting and Refusing. A challenging brace of cases for our account, and therefore one well suited to illustrate it, is that of *accepting* and *refusing*. These are fundamental and pervasive forms of action observable in the advances and withdrawals of brutes and in high level responses to verbal offers. Almost any sort of thing may be refused or accepted: bits of food, pills, chastisement, handshakes, jobs, and a lamb's request to suckle. We wish our analysis to separate accepting and refusing in terms of characteristic conditions of success; we wish, that is, to find conditions under which the one kind of action must fail and the other needn't. It is evident that many of the conditions are the same, most prominently that something has been offered or requested. What makes them look so different? It may but needn't be the direction of movement, which, anyway, by itself, would not serve our task of finding conditions. Nor can we non-circularly appeal to what subject wants to do for (to put the matter briefly) our method can be thought of as a scheme for explaining differences between wanting to do one thing rather than another. We wish to identify beliefs that differentiate wanting to receive and wanting not to receive something.

There is a significant lack of parallelism between *accepting* and *refusing* in their respective measures of success. A refusing subject must have failed in his act if what he believed to be offered goods were forcibly credited to his account. On the other hand, an act of accepting could succeed even in the opposed circumstances because the offerer failed to deliver the goods. I may have successfully accepted a piece of cake even though it crumbles on the way to my plate. But I have not succeeded in my refusal if the cake is forced on me, and left on my plate against my expressed wishes. That suggests that acceptance is not action, but inaction, a kind of merely "letting things happen", whereas refusal is "doing something about it". There is a lump of truth in that thought, as is confirmed by the consideration that, at higher levels of behavior, subject may positively express unwillingness to receive and still accept ("Oh, I can't let you pay."), whereas one cannot express willingness to receive and still refuse. An explanation of that would be that acceptance is not an opposed action where refusal is. There is a lump of untruth here too. Acceptance is something specific and always a positive kind of letting things happen, certainly a kind of action.

I first concentrate on offers, leaving requests for the end. Our observation suggests a distinction between accepting an object O and accepting an offer of O. Accepting O is a kind of taking possession of O; but not the only kind. The distinction is that accepting O is taking possession of O upon accepting the offer of O. Our earlier observations show that one might have accepted the offer of O and still not have taken possession. In the upshot, then, accepting O can be resolved as a "hybrid" of accepting an offer of O into taking possession of O. What is specifically acceptance about this is the "original way", the acceptance of an offer.

Similarly, successful refusal of an offered object is a case of failing to be in possession of O by cause of refusing the offer of O. However, there is that difference we have already observed: One cannot succeed in refusal of an offer and yet come to be in possession of the goods. Refusing objects is not a hybrid of refusing an offer into not taking possession. The explanation of the difference is, simply, that "not taking possession" cannot be the name of any kind of action. My "not taking possession" of your wallet is not something I am now doing, even though it is true that I am not now taking possession of your wallet. Non-possession, regarded as a result of action, can be due only to something like refusal. That suggests that what is distinctively refusal, unlike what is distinctively acceptance, must be governed by conditions bearing on the eventual possession of offered goods. Specifically, a distinguishing condition of success for refusal is that offered goods should not come into subject's possession by other concurrent cause. (Of course one might later change his mind, or buy the goods.) The counterpart condition, that offered goods should not be taken from subject's possession by other concurrent cause, is not a condition of success for acceptance.

I take it as confirmation of the foregoing paragraphs that the analysis seems to carry over easily to the matter of requests. The ewe can accept the lamb and the lamb frolic off; but the ewe will not have succeeded in its refusal if the lamb will not take no for an answer. (The lamb will likely not have succeeded either.) The acceptance unlike the refusal is not governed by conditions having to do with the subsequent location of the lamb. Similarly: I can't go on leave if the university successfully refuses my request; I needn't go on leave if they accept it.

8. REMARKS ON PROFICIENT ACTION AND ON CONFORMATIVE ACTION.

Conventional action is both conformative and proficient

The orders of proficient and conformative action, spoken of on pp. 14f., matter a lot for the upcoming account of utterance and conventional behavior. The use of language and other forms of conventional action cannot, I hold, be undertaken except by "rule considering", conformative means, and the existence of "rules" is a condition for doing such deeds. Conformative action is itself a sort of learned or proficient behavior. My actual definitions of these forms of action, which can be provided within the scheme of analysis set out in the previous section and which are attached at the end of this section, are rarely consulted in the developments that follow. It should be enough for our present purposes to post advance notices at points where misunderstandings and objections may arise. (We shall, in App. C at pp. 191f. and in Chap. 3, at pp. 228ff.. have occasion to speak at some further length about the importance of *knowing-how-to*.)

We assume that there are proficient genera of action.

First: Proficiency may be brought to bear on acts of any kind. Even scratching and sucking may be modified by learning and then done with greater or less skill. Still, not all acts of scratching or sucking manifest skill. There do, however, seem to be generic classifications of action that actually require some measure of skill, e.g. frisbee catching (by man or dog), reading Greek, tying one's shoestring in a bow knot, connecting wires in a switch box, playing the piano or squash, and integrating differential equations. Genera of action comprised within such activities would seem to have conditions of success that could not be believed except by skilled subjects. The data, I say, suggests that there are such genera, and I shall assume that it is so. My first remark is really a stipulation that only those action-kinds, e.g. frisbee-catching, reading Greek, playing the piano or integrating differential equations, are entitled as "proficient". Should the assumption fail, we shall still be able to say that acts are or are not done "proficiently", without or with more or less skill.

Proficiency is based on investigative proclivities.

Second: Proficiency in any sort of action, whether or not generically proficient, is based upon natural, non-proficient investigative proclivities to look around and sniff things out. These proclivities coincide with subject-seated investigative mechanisms, which are co-opted to the proficiencies they underlie. Skilled scratchers are good at finding the sites of their distress; frisbee-catching dogs are alert for the time of a toss and a pianist looks for middle-C.

Generically proficient kinds of action have kinds of conditions of success which could not be believed except by subjects possessed of investigative competencies.

Third: Generically proficient kinds of action or purposes (if there are such) are featured by conditions of success that no subject could come to believe unless competent to make an investigation, e.g. to determine where middle-C is or how a word is spelled (say by asking someone, "applying a rule" or consulting a dictionary).

The presence in a subject of some measure of "know-how" is a condition for its doing proficient action.

Fourth: the presence of "know-how" in some amount is a condition for doing and not a condition of success for genera of proficient action (if there are such kinds of action): "Can you play the piano?"; "I don't know; I never tried." A condition of success for reading Greek is the nearby presence of a passage of Greek; it seems reasonable to hold that no subject could take in that kind of information without at least knowing-how to recognize Greek for what it is; some such modicum of know-how accordingly seems to stand as a condition for a subject even so much as attempting to read a passage of Greek.

These remarks and also the definition that is attached below bear entirely on the notion of proficient behavior. None of them yet make any commitment on the troubled question of what proficiencies (skills, know-how's) *are*. I hazard and shall defend a speculation. Proficiencies, I submit, are sets or mechanisms in subjects verbally marked off as being apt for finding and utilizing information. The "utilizing" dockets these mechanisms as forms of knowledge (See App. C, pp. 182f., 192). Proficient "finding" mechanisms go beyond the underlying investigative mechanisms mentioned in the second observation above. A frisbee-catching dog who does not yet know that his master has the platter in hand is set to look to see whether that is so. A Greek reader is set to find out the meaning of unfamiliar or forgotten Greek words. Mathematicians are ready to derive, check-out and look up integrals. These mechanisms, like the valve mechanisms of a motor, are of course "seated" by doing things; they are "learned".

Know-how is knowledge, and action faulted by cause of deficient skill is always also faulted by cause of deficient information.

Know-how and knowledge-of things or facts come out by this analysis as different, related forms of knowledge. I later argue that any fault in action attributable to deficient know-how is also due to lack of knowledge of something (pp. 191f.). We observed that a proficient subject could not come to believe a condition of success for proficient action without some degree of skill. I now submit that this skill is a set or mechanism. The mechanism is judged to be impaired in a degree measured by the frequency with which subject fails in this kind of action for want of that kind of information, e.g. one's skill in squash is seen to be lacking in degree that he makes disadvantageous shots, not because of lack of strength or speed, but because of failure to take in and use information having to do with where his feet, his racquet, the ball and his opponent are.

My speculation that skills are sets or mechanisms will provoke an objection[5], which I halt to counter: "You describe know-how's as abilities; but abilities are 'dispositions' and dispositions are not conditions, as you claim." My answer to this is a series of contentions, as follows:

(i) The ascription of know-how to a subject is indeed a kind of "dispositional thing said".

(ii) Such things said may be true or false.

(iii) No fact is distinctively a disposition; otherwise, every disposition is also non-dispositionally classifiable.

(iv) Facts that make those dispositional things-said true are frequently what have been called "servo-mechanisms". *Ascriptions* of such mechanisms are not dispositional things-said, but the facts are the same.

To defend this, I must say something about the sense of "dispositional things said". First, as it seems to me, dispositional terms, e.g. "fragile", may usually be paraphrased as "can easily...", e.g. "can easily break". Second, "can easily" is schematic over a range of conditions (see pp. 153ff.). Third, a dispositional ascription is never proven or verified but at most confirmed by observing the subject to..., e.g. to break under conditions not explicitly specified. Fourth, confirmation is gained because there is a presumption that the mentioned result would occur under the still unspecified conditions by cause of the presence of an unspecified condition in the subject. Fifth, the paraphrastic "can easily" seems to be doubly schematic for the condition of the subject and for the conditions under which the mentioned result would occur. Sixth, the unspecified condition of the subject is the fact that "makes the dispositional ascription true". Seventh, the condition is always something in particular but differs from case to case, e.g. the fluid condition that is the fragility of a lightbulb is something other than is the crystalline condition that is the fragility of a pencil point. Eighth, the conditions under which the mentioned result is likely to occur are also particularities that vary from case to case, e.g. they may be droppings on the floor or immersions in cold water. An example I borrow from Melnick to illustrate these contentions is saying that an alarm clock is set to go off at 6. That is no doubt "dispositional": there is a mentioned response *viz* ringing, an indication of conditions, viz that it is 6 o'clock am or pm on some day or other; the occurrence of the response at some 6 am confirms the ascription but doesn't prove it; nor does the absence of the response disprove the ascription. The ascription asserts and is proved by the clock here and now being in a definite but so far unspecified state of being set. The presence of know-how, I believe, is something like being set. I can't prove it is so. However, pending objections, the thesis seems to me to be plausible on its face, once we have drawn the important distinction between the implications of *ascriptions* of know-how and the *facts* that make those ascriptions true.

I move now to conformative action. As a a *first* clarification: My use of "conformative action" covers both compliances and violations.

A condition for doing conformative action is that the subject believes itself "subject to" "rules" or a "system of expectations", which are various. Conformative action is proficient.

Second: A condition for a subject's undertaking conformative action is that it should act in the belief that there exists a "rule", but not in a literal sense of "rule"[6]. For our purposes, these "rules" may be vaguely thought of as systems of expectation within communities (more below).

Third: Conformative behavior is proficient. Subjects must become set to find out what is expected of them.

Fourth: An animal's behavior may be in compliance with or in violation of a rule even when it does not itself act to comply with or to violate the rule. Only acts of the latter kind-- where (as *we* can say) subject acts in the belief or with (not necessarily "for") the reason that a rule exists--fall within the order of conformative behavior. A condition for doing a conformative act is that subject believes that such and such is the rule, e.g. that one should stop at red lights, take off his hat in the presence of women, use "que" to mean *that*, etc.

Fifth: As I said, the existence of rules is a reason for conformative action, and subject's belief "That's the rule" contributes to the explanation of its movements. *What* is believed, *viz* that there is such and such "rule", may or may not be a condition of success for the act. The existence of orthographic rules are conditions of success for spelling words; rules of etiquette are not conditions of success for being polite. (I once thought that "enabling rules"--see just below--were always conditions of success for generic orders of conventional action. That, I now believe, is an "important falsehood", related to a point of concern in our examination of conventional behavior, at p. 54).

Finally, conformative behavior is multiform and we must at least notice and name a number of distinctions. *First distinction*: there would be no kind of action identifiable as making chess moves without the rules of chess; by contrast, rules of etiquette may affect only the manner in which subjects do otherwise non-conformative acts; e.g. food-taking. I call a rule having the first kind of relation to action an "enabling rule" and one having the second kind of relation to action a "restriction". The distinction is always "in relation to": the same expectations or rule may operate as a restriction on one kind of behavior (greeting a guest) and enable another (formal salutations to a visiting head of state).

Second distinction: Conformative behavior may or may not admit of alternative rules. There is only one way of spelling "girl" in English; by way of contrast, there are any number of different ways of saying *girl*, e.g., Maedchen, "muchacha", or "girl". The rules of English are appointed for speaking English and cannot be replaced for that purpose by the rules of Spanish. On the other hand, the rules of Spanish may very well subrogate the rules of English for saying that a stone is hard, or, if you wish, for saying que una piedra esta dura. I shall say that rules which do not admit of substitutions relative to the practice they underlie are "appointed" to that practice, and rules which do admit of alternatives are "subrogative", relative to that practice. *"The"'s being a definite article* is a rule appointed to English and subrogative for selecting referents.

The *last distinction* in conformative behavior which we shall need is this: It is not generally true that conformity to rule assures success in action where all the conditions of success obtain; that is sometimes so, however. Thus I may persistently fail to get food into my mouth no matter how mannerly my actions; contrastingly, if I write down the word "girl" in the right way, I shall have correctly spelled that word. I call rules of this latter kind, e.g. rules of spelling, "rules for succeeding". (It seems to me likely true that all rules of succeeding are also enabling in respect of what they are "for", but not conversely, e.g. the rules that enable players to bat in baseball are not also rules for succeeding, say in getting hits.)

The definitions of *proficient* and *conformative* action, slightly altered from what I previously developed, are as follows. (The commentary needed to make these definitions plausible defies summary.)

Proficient action: K is a (perhaps non-generic) type of proficient action if and only if (l) K is a type of action, where (2) a condition for doing K is that the agent should be able to succeed at an act of another type K' whose measure of success is that the agent should come to believe a condition of kind C, where C is a condition of success for whatever the subject generically does in doing K. (If K is a genus of proficient action, C is a condition of success for K). K' is some kind of what ethologists call investigative behavior.

The "rules" of conformative action may be either "communal" or "private".

Communal Conformative Action: A is a communal conformative act just when the subject does A in the belief or with the reason that there are or were others belonging to a specifiable community who, were they present, would believe that he would act from the belief that they would expect him to do act of kind K, where A may or may not be a K.

Private Conformative Action: A is an instance of a private conformative action iff subject does A in the belief that he has determined to act in a certain manner, which determination he may publicly declare, thus giving others reason to believe that he would act from the belief that they expected him to do an act of kind K, where A may or may not be K.

9. CONVENTIONAL ACTION AND THE USE OF LANGUAGE.

It is natural to say that *the use of money* (any use of money as money in exchange, by anyone, anywhere) is conventional, or that religious *ritual* is (intrinsically, essentially, inescapably) conventional, or that *the use of language is conventional*[7]. People, paintings, armaments and wisdoms as well as deeds mays also be called "conventional", but with quite different understandings. Also, almost any kind of performance, and not only acts supposedly conventional, may be said to be done in "conventional ways", e.g. when we say that the Gogolac brothers did not kick field goals in the then conventional way, and here something else again is meant. Still again, in speaking of "conventional greetings" (as opposed to unconventional ones), one would mean something different from what one would have in mind in thinking of greetings as something which can be only "conventional". This latter notion is our quarry; it is elusive and "problematic".

"Conventional" in this understanding does not signalize, as it may otherwise do, a contrast with what is universal, novel, idiosyncratic, spontaneous, inspired, untraditional or even natural.

Our first question is over whether utterance is distinctively "conventional".

Taking it now that utterance, ritual and the use of money in exchange are "conventional", in the same understanding, we are back to our first guiding question (pp. 11f.), now reformulated to ask whether utterance and its cousins are conventional in some behaviorally distinctive way. The central task of this section is to vindicate my working assumption that it is.

Most, but not all, traditional authorities would have denied it.

I don't think the question was ever explicitly asked before this century. Nonetheless, I do certainly have the idea that most of the traditional writers on language, including Plato, Aristotle, Leibniz and Russell, if queried on this, would have denied the assumption. Language, for these writers, is an expression of the mind--of judgement, want, will, emotion or the like--conventionally set-out in words, marks or other arbitrarily selected "expressions". For them, the conventionality of utterance is merely adventitious. Utterance, in this view of the matter, is distinguished from other "expression" only more-or-less, by some degree of apperceptivity. No doubt, there are also in the tradition some opposing tendencies, and perhaps even some authors who, if queried, would have voted on my side of the question.

Frege, I believe, implicitly stood opposed to the traditional assumption: so far as I can determine, his *Sinne* are needed only to give sense to language and as constituents of propositions. Locke also, in his bewildering appeal to *abstraction*, sometimes seemed to lean in my direction, for (sometimes anyway) Locke treated abstraction as a mechanism by which ideas, first vented in brute experience, are fitted out to serve as constituents of propositions. Wittgenstein's way of talking about *Ausdruecken* in his *Tractatus* (3.31-3.318) also brings in something like fregean *Sinne* and, in his criticism of Russell, Wittgenstein notably held that judgement could not be nonsense (5.5422); a seeming commitment to the distinctiveness of language is most clearly visible, in the *Tractatus*, through the way in which representing elements are distinguished from other merely represented objects by the requirement that they reach out to reality and share among themselves a common *Form der Darstellen* (Nb, NOT *Form der Abbildung*, 2.1511,2.173). Neither Frege, Wittgenstein nor Locke is serviceable to us, however. Grice's analysis of "non-natural meaning" ("Meaning", *Phil. Rev.*, 1956, pp. 377-88) certainly is, and we shall take notice of it as we go on. Austin, who should have raised the question, was only "suggestive". In *How To Do Things With Words* he listed six kinds of conditions bearing on the use of language in general, but all of them seem to carry over to other kinds of conformative action too. What I find suggestive is his twice repeated remark, made in relation to the use of language specifically, that a "judge could know" (*op. cit.*, pp. 121, 127). I take this to mean that an observer could know what a speaker is doing from the speaker's words alone.

We seek a "vernacular index" to specimens of conventional action.

The two competing assumptions--the usually inexplicit traditional one, that there is nothing all that special about language and mine, that language, in company with ritual and some other things, is behaviorally distinctive--pose a crucial issue for theories of language. Here it would be well to have a collection of everyday specimens--a data base, if you wish--to fix the course and, as we go on, to check our position against. To that end, it would be useful to have a "vernacular index" to examples of conventional action narrower than what we found for action.

The "index" is that there be a positive answer to the question of what the subject "meant", where that question is in some manner distinct from the question of what the subject meant to do. The answer to both answers may be more or less full.

The wanted index, evident perhaps, but not unproblematic, is that there be a positive answer to the question of what a subject "meant". The "problem" in this question is whether it is distinct from and additional to our action-indexical question of what a subject "meant to do". I shall argue that it is. A child kicks a chair. An answer to the action-index question of what he meant to do is "to kick the chair". A psychologist could be curious about the *significance* of the deed, and might well ask what "it meant": Was it petulance, spite, misapprehension or just exuberance? Most of us, however, would be puzzled by any further inquiry into what the brat himself "meant". Kicking by kids is not a kind of action which ordinarily involves such further wonderings. Contrastingly, a visitor in Britain might wonder what this english child meant in saying of the family car that the bonnet was open. Is it a convertible? We tell him that the child meant that the hood was up. This question of what a subject meant is always appropriate to language--spoken, written or gestural--and also to those adult kicks that are as good (or as bad) as language, e.g. under the bridge table. Again, looking to other orders of conventional action, a shopper might be said to have meant the candy when he offered the puzzled clerk money insufficient to cover the cost of a cigar, or that an arm-raising priest meant it's time to pray.

We know language and ritual for what they are, of course; my claim is that our sense of what a subject meant goes with that so-far unspecified knowledge and guides the classification. If the question of what a subject meant is not appropriate, then, *prima facie*, subject's deed was not linguistic or otherwise "conventional" in the distinctive sense we wish to define. If the question is appropriate, then (unsurprisingly) the deed so far qualifies.

Answers to the "What did he mean?" question, when appropriate may be more or less full. A Spaniard says, "Los pantalones son de lana". A non-hispanic may be told that the speaker meant that *he thought that the pants were made of wool*, or that *the pants are made of wool* or that *by* "lana" he meant *wool* or that *by* "pantalones" he meant *pants*. Any of these could stand as a more or less foreshortened answer to the indexical question.

The "means" and "means to do" questions, though answered together, are distinct in that only the latter invites further questions over whether the subject succeeded in his action.

I once indeed did think that any question of what a subject meant was but a special case of the more broadly applying question of what a subject meant to do and accordingly unserviceable as an index to our sense of distinctively conventional action. Isn't it so? If the Spaniard meant pants, then didn't he mean to refer to those pants; and if he meant that the pants are made of wool, didn't he mean to say just that; again, if he meant that he thought that the pants were made of wool, didn't he mean to express that thought? Conversely, if he meant to say (e.g.) that the pants were made of wool, it seems he meant that the pants were made of wool. An answer to either question--what subject meant to do and what he meant--, when both are appropriate, implicates an equally full answer to the other. If (as is plausible) questions were defined by their ranges of answers, it would seem our two indexical questions must be the same, and we would have lost our everyday evidence for the distinctiveness of language.

Still, the two questions are different, since the "meant to do" question invites a further inquiry into whether the subject succeeded, whereas the simple "means" question doesn't. The subject would, for example, have failed in his utterance if there had been no pants on the hook, perhaps because the cloth hanging there hadn't yet been fashioned into a pair of tubes. Yet subject still unfailingly meant pants. Meaning unlike meaning-to-do is not the kind of thing one can fail to bring off. Further, we could not determine whether a speaker's utterance succeeded or failed unless we knew what he meant.

That subject should have "meant" something is, not a condition of success for what it meant to do, but rather a "condition for doing" for conventional action; such "meaning" is "indicating" what the subject meant to do; otherwise, indicating the conditions of success definitive of that genus of action.

That subject meant something cannot, as we may now see, itself be a condition of success for conventional action, for if that condition lapsed there wouldn't be an utterance or other conventional deed in the offing. That subject meant something is, rather, a condition of doing characteristic of language and that subject meant what he did a condition of doing (not of success) for his particular utterance or other conventional act. It is a "truth condition" for his having done that kind of act.

That condition would seem to be that the subject, in his act, have "shown"[8] or "given indications" of what he meant to do. That hypothesis explains what otherwise is "problematic", that an answer to either of our indexical questions should implicate an equally full answer to the other.

Non-conventional sorts of action do not thus require the "giving of indications," and my thesis is that a distinction of questions pertains exclusively to conventional action, toward whose desiderated definition we now are verging. But first:

"Giving indications" or, more briefly, "indicating" cannot in this understanding be taken as the name of a genus of action. If meaning is indicating, then *indicating* cannot fail as generic kinds of action may. Furthermore, the indicating that goes on in an utterance does not itself have conditions of success, for (jumping ahead) it is a presentation of the conditions of success for the utterance. "Indicate" and "show" are of course also and originally used as names for proper genera of action. I fret about that now less than I used to, since the force of the "What did subject mean?" question establishes a pretty concrete sense that naturally attaches to such words as "indicate" and "show", and (as we have shown) the sense in question is not that of an action name.

> This notion of *indication* may be somewhat demystified even "naturalized" by bringing it into relation to the behavior of non-humans. Dogs claim territory and warn other dogs off by barking. Other dogs usually take proper notice of these "expressions of

intention" and dutifully keep off. They also develop habits of dutiful response to human scolds and threats. Dog keepers will have no trouble distinguishing that kind of response from my dog's response upon seeing me display his walking leash: here he not only takes in my "expression of intention" (to go for a walk), but also sees that I intend that he should take in that intention, *viz* he sees that I am indicating what I want to do. He understands this language. He can't yet do this sort of thing himself; but I do rather expect him any day now to come and leave the leash at my feet[9].

Action-classifications selected by the "means" question are "distinctively conventional". That answers our first question. Such classifications of action are "non-generic".

We have now effectively answered our first question: A kind of action is "distinctively conventional" when it can be given as a positive answer to a "what did he mean?" question asked of an acting subject. The answer is appropriate and correct just when the subject has "indicated" or "shown" what it means to do.

In gist, action is distinctively conventional when it is governed by the condition that subject should indicate what it means to do. It comes out that the classification of action as conventional is not "generic," at least not within that theory of action-classification we worked out in #6 above. A generic classification would be supplied in any answer to the question of what it was that subject meant to do. Such a characterization would be defined by a list of conditions of success. The condition that fixes action as distinctively conventional is not that, but a condition for doing.

This condition, that a subject engaged in conventional action should indicate what it means to do, can be approximated in two alternative formulations that may serve to enhance the plausibility of my thesis.

Conventional action "speaks for itself" and requires that the agent should adopt means calculated to show what it means to do.

First Alternative Formulation: Action is distinctively conventional only if "it speaks for itself". A patient is wheeled into a room and made comfortable, that is one thing; but an usher who *greets* us must somehow show by his action that that is what he means to do. In saying what he does, a speaker means something, and that meaning identifies what he means to say. Ritual performance, politeness, and the use of money similarly "speak for themselves".

Second Alternative Formulation. A subject who does a conventional act must adopt means calculated to show what it means to do. It utters a certain sequence of words, or hands over a coin, or makes a certain gesture, and thus makes clear to those who are familiar with its conventions that it wishes to make a request or make a purchase or invoke the presence of God. I say "Hello" to greet you. My words are calculated to show that that is what I mean to do. Similarly when I hand over money to make a purchase. The rower's skipping oar (see p. 13) is not in this way calculated to indicate what he means to do--inhibit progress through the water perhaps or irritate his mate.

A subject indicates what it means to do only by conforming to conventions, and that the subject's action be "conformative" in this way is another condition for doing conventional action. That there exist appropriate "enabling rules" is another such condition.

It seems altogether plausible to maintain, though I can't prove it, that a subject indicates what it means to do only by conforming to conventions, e.g. those governing English usage. Since subject's showing what it means to do is a "condition for doing" (a truth condition) characteristic of conventional action, the existence of conventions is another condition for doing characteristic of conventional action.

We have (admittedly without proof) now given a gloss on why conventional action is conformative. Since a subject does a conventional act only if it shows what it means to do and it manages that only by conforming to conventions, the conventions are enabling in relation to conventional action. The conventions can be for the employment of words, gestures, symbols or whatever and may be communal or private, slang, vernacular or technical, figurative, literal, neologistic, Swahili, English or non-English, and so on. Conventions are "rules" or "understandings", and, in utterance, a subject employs "expressions" in conformity with those understandings to show what it means to do. The conventions are commonly the rules of a language, e.g. English. Since, as we noticed, these rules may be various, the rules are what we called "subrogative" in relation to language (p. 41). One may say the "same thing" in different languages. This has abetted a suggestion I have found in some recent writings, that in addition to rules governing usage in various languages we need additional rules for language itself. Not true. Rules for languages are the only rules for language that we need.

The means a subject adopts to show what it means to do are, in ultimate reduction, the subject's movements, and the purpose of the conventional act is made identifiable in those movements. Our theory of action directs that we define the purpose by listing sorts of conditions of success for the act. The conclusion is that subject in doing a conventional act must make movements which, by the conventions, indicate the conditions of success of that act. So, in summary of what has so far been said: A condition for doing a conventional act is that the subject should make movements that indicate the conditions of success of its act. We further hold that a subject could not indicate those conditions except by falling in with rules or conventions of some kind of other.

A condition for doing conventional action is that the subject should make movements that indicate the conditions of success of its act.

"Ultimate reduction" is meant to fend off those who may wish to protest that, while gestures may be movements, words are not. Telecommunications allows for the identification of the "ultimate movements" at a great remove in evidences of many kinds. I might perhaps add that the now widespread use of recorded messages, while it does presuppose language in the background, is itself no more to be reckoned as language strictly taken than is the ringing bell that calls you to the phone.

An agent doing a conventional act may dissemble its purpose.

I now proceed to some observations and clarifications. Our characterization of conventional action, which says that the subject should indicate what it means to do, allows for the deliberate deception of interlocutors. A subject doing a conventional act may dissemble his purpose by making as if to follow one set of conventions when in fact he follows others. He doesn't wish his interlocutors to identify what he really means. But even here conventions are needed, and others who had divined them could know what he meant to do from the movements he made.

Many questions arise over the relation between the selection of conventions and conventional action, questions which I should like to put down without really answering. First, I speak of conventions as "rules", but of course they are not literally "rules", but either systems of community expectation or determinations of mind that would, if made public, create such systems of expectation. Linguists are much concerned with how these systems work.

Subjects may conform to deviant and idiosyncratic conventions. A man may deliberately or mistakenly fail to conform to the conventions which regularly operate within the community of his interlocutors. They may then get a wrong idea of what he means. In such instances I hold that the speaker does indeed conform to some conventions, but not to the ones expected by his communicants. This may be mildly mirthful as when The American in Paris says "Sein" meaning Seine, or devious, as when one makes as if to assert what he disbelieves, or is uncourageously sarcastic. Again a man may speak his "own language", and all of us sometimes do; only one wise to that language could be expected to understand what the speaker means. In all these familiar if deviant cases, I hold that the speaker would still be "making his purpose identifiable" even though no one else in the world could identify what he meant. That is a second "put-down".

Conventional action may fail without being "nonsense".

How could a completed conventional act ever actually fail? If the words get out, the deed is done; an observer could identify what was meant, and nothing more is needed to assure success. How can one fail to say what he means if he actually gets it said? Well, he won't fail to say what he means; still, conventional action may fail simply because certain of the indicated conditions might not be satisfied. Examples may suffice to break the doubt. Consider first our example of the pants said to be made of wool (pp. 45f.): that utterance would have failed if there had been no pants on the hook. Second: I ask a stone, which I mistake for a man, the way to town; my act indicates, as a condition of success, that I am addressing myself to a fellow-human-being, and that is not so. I fail in my effort to make the request. Third: "blue", "green" and "bluegreen" are "meaningful" as also are "orange" and "blueorange". I may successfully order the painter to paint the door bluegreen, but not blueorange. (A disputable case: A person innocent of genetics, tells me that he is one-third Irish. He does use meaningful words in a meaningful way, but there is a question whether such an utterance could succeed.)

It is evident from what has been said that unsuccessful utterance isn't "nonsense".

A speaker's "meaning" is his indication of the conditions of success. Listings of those conditions that serve to define the speaker's purpose also define his meaning. Such listings can never be finally completed. This is a welcome instance of the rule that generic sorts cannot be defined as particulars. Meanings are not particulars, anymore than species are. It may seem from this--and it has so seemed to me--that our account of conventional behavior yields a doctrine of the "indeterminacy of meaning". However, this indetermination, if such it is, holds as much for "natural kinds" as it does for meanings. A creature is not an indeterminate equine because we cannot finally define that genus *per genus et differentiam*; similarly the meaning of an utterance, which is anticipated in its expression, is not indeterminate because we cannot complete its definition. Every utterance has the meaning it does in despite of there being no finally determinate definition for that meaning (Nb: this "indetermination of meaning", if such it be, is merely generic and not "categorial" a-la-Quine.)

Two consequences: Conventional action could not be identified for what it is as conventional action except by subjects who know how to do conventional action of the kind in question. Conventional action is "apperceptive".

The notion that conventional action "speaks for itself" and is "self-identifying" is confirmed by two consequences of our definition; we shall later add a third (pp. 56f.).

First: The identification of a conventional act as conventional action by an observer requires that he should also know how to do that kind of action. Proficiency in the practice is a necessary, but not a sufficient condition for making a correct identification. The argument, in gist, is that if the observer identifies the act both as being of its generic sort AND as being conventional--if he knows both that the act is conventional and what the subject means to do--then he must know that the conditions of success could be indicated in the subject's movements. He must therefore know that the act could be undertaken by making those movements. The observer may not know how to make those movements. (I have embarrassing problems with Indian and Polish names.) Still observer knows that these conditions could be indicated by making those movements. That is enough to conclude that he knows how to do that kind of act, e.g. make that request or issue this greeting. Contrastingly, an observer could tell that the football player threw a pass--a kind of conformative action underlain by enabling rules--without knowing how to throw passes. Here also lies the contrast between conventional acts, e.g. making a request, and nonconventional acts of mouthing certain sentences, e.g. "Je voudrais des pommes frites". I may recognize and identify the language in which an utterance is enacted even though I do not know how to speak that language; but I could not identify a request without knowing how to make requests. On the other side of the story, I may see that a speaker is making a request or issuing a greeting even though I do not as yet speak its language, provided I know how to make requests or issue greetings in some language or other. That may happen when, for the first time, I hear an utterance in a tongue I had no previous familiarity with, e.g. "Jambo" in Swahili. I add, finally, that an observer's know-how is not sufficient for it to be successful in a like performance: I can tell a curtsy, a kind of conventional action, for what it is, and I know how to do it, too; but I also know that I am barred by sex from success in such endeavors.

> *Objection* (from T. Griffin): When you show your dog his leash, he sees what you mean, yet surely he doesn't know how to offer you an invitation to go for a walk. I do indeed allow that old Pan identifies what I mean to do, but not that he also sees that I am engaged in conventional action. I would be readier to credit him with the latter capacity if he were to bring me the leash on his own, as I rather suspect he'll be doing almost any day now.

Second: A subject who does a conventional act must be capable of "reflective knowledge" of what it means to do. Conventional action is "apperceptive" in a way in which bone-burying by a dog or infantile nourishment-taking needn't be. The subject of conventional action must be familiar with those rules or conventions by which it indicates what it means to do. So merely by looking in on its own words, gestures or other expressions, it can know what it means to do.

One may also plausibly, plausible to conjecture that a subject who does a conventional act must know that its act has that quality. The action is underlain by enabling rules and would not exist without those rules. A subject could not be responsive to such rules, as our imagined subject must be, without knowing what they are "for". But they may be for more than one thing. So, for example, the rules which govern my use of "Greetings!" enables me both to speak English and to offer greetings. Now in thus offering greetings I surely needn't know that I am speaking English, as it may seem that I must know that I am offering greetings. It also seems to me that if speaking English were, not my "means" but what I meant to do, then I would indeed have to know that that is what I was meaning to do. I think that is because that kind of deed--meaning to speak English--requires the presently conjectured self-consciousness that goes with the use of language plus the knowledge that different languages are available for that purpose. The conjecture, explicitly formulated, is that a subject must be capable of reflective knowledge that it means to indicate what it means to do when its reflective knowledge of what it means to do is meditated by its familiarity with enabling rules for indicating what it means to do. That, as I said, seems plausible; I do not see how to prove it.

Why Introspection Works. T. H. Green once charged John Locke with being an "introspectionist", and thought that Locke's "empiricism" was incoherent on that account. That subjects gain their ideas by experience must, for Locke, surely be a matter of observation, not introspection; moreover--a more important objection--, the features assigned to those ideas by introspective reflection, e.g. that *white* is an idea of sensation, are not features that subject experiences those ideas to have. Locke anticipated a defense by maintaining that others certainly had the same ideas he did; so he must have theirs, as objects of his introspective reflection. How could he be sure? Holding to something like our position of conceptual epiphenomenalism, he could maintain that one may observe the possession of ideas in the behavior of other subjects. Which ideas, you may wonder? And mustn't the observer already have them and know them--not by experience--but by "introspection"? Well, these ideas have, in the more advanced experience of the conceptualizing observer, already come to figure as constituents of propositions and as "meanings". This observer can always realize and indicate those ideas in his own linguistic behavior and know them in his own reflections. The argument of the main text just above shows that Locke's introspective analysis of ideas is compatible with his claim that we also observe creatures to attain ideas by experience. Similarly, while our "behavioristic" conceptual epiphenomenalism seems opposed to "subjective" Phenomenology, the difference is vanishing in actual practice: What Husserl called "bracketing" ("epoche"), is an abstraction of sense from its referent or other "extension"; a reflection on sense must, if the argument just above is right, work as well for the conceptual epiphenomenalist as for the phenomenologist.

Conventional action need not be and probably never is "generically conventional".

A condition for doing a conventional act is that subject should indicate the purpose of that act. The purpose of that act is itself to be defined by listing kinds of conditions of success. Those conditions are "indicated" in the agent's movements. Our second guiding question (p. 12) asks whether that purpose is or may be in any way *generically conventional*. That could be so only if the stipulated condition *for doing* conventional action somewhere constrained the indicatable conditions of success, as may happen for proficient action (see p. 37). The evidence is negative. *Some* "No"-sayings are *refusals*--a sometime non-conventional kind of action; moreover, there does not seem to be anything that cannot be linguistically refused, and, furthermore, anything that could be refused by saying "No" could have also been non-conventionally refused. There seems to be no kind of refusal which, generically speaking, could be only conventional. It's the same with "expression". I could express disdain by "avowing" it--that is "conventional"; but both Fido and I could do the job "non-conventionally" by turning up the nose. If the examples hold up--the issue will be resumed in connection with the definition of assertion at pp. 92f. below--, then the conventional character of an act does not affect its generic determination as action.

Of course when we characterize an act as an "avowal", "conjecture" or "assertion" we do imply that it is a conventional act. What is not required is that the implicated "preferred" generic classifications of the act, defined by lists of conditions of success, be in any way essentially conventional. The suggested general conclusion is that the main classifications of essentially conventional action are not themselves essentially conventional.

Because of what has now been noticed, I will, when (and pretty much only when) it matters, speak, not of "conventional purpose", but more tediously of "the purpose of conventional action". Refusing a pill may be the purpose of both conventional and non-conventional action; if one does refuse, e.g. by saying "No", then that must indicate that subject means to refuse. It's otherwise with the dog who, after sniffing, refuses the offered pill by withdrawing.

I have now answered the first two questions posed in #3 above and again on pp. 43 & *supra*: Action may be "distinctively conventional", but it is doubtful whether it is ever generically so.

The third question was whether there is anything that cannot be done conventionally in our now established distinctive sense of "conventional". Consider: I pluck a shell from the shore and that makes it mine; I give it to you, and now it's yours. The first act is not conventional in present understanding; the second is. The same kind of effect is brought about in both cases. Again, the rules of baseball enable fielders to put hitters out by catching flys; but the same result may be gained by an umpire's "declaration", as in cases of "infield fly". Catching is not conventional, declaring is; but the result in either case is the same[10].

In answer to our third question over whether there is anything that cannot be conventionally secured: Results can be conventionally gained just when all possible circumstances of-failure can be anticipated.

The rules enabling conventional action are rules for succeeding.

Is there anything that can't be conventionally achieved? Well, I won't get that book read by any mortal declaration that it be done or assure your understanding or "uptake" of what I am expounding (let alone your acceptance) by simply saying my piece or by doing any other conventional deed. It appears from these examples that, while conventional and non-conventional acts sometimes share measures of success, that is not always so, not at least for our human selves. Our own powers of thought and language are more limited than the reach of our hands and feet and voices.

Our third question now yields to another: What *are* the limitations of our thought and language? Consider again my giving you the shell. I won't succeed in making it yours if you won't have it. I know that, however; and my words have indicated your willingness to accept as a condition of success. I cannot in this way anticipate all the circumstances that might keep me from reading a book or from securing uptake, as perhaps God could. Again: The rules of baseball enable an umpire but not a fielder to declare a runner out. The umpire's conventional act abstracts from unpredictable circumstances of failure--strong winds and unaccountable miffs--as a player's fielding act cannot. The upshot of this, as it seems to me, is that language rules for indicating conditions of success must anticipate all possible circumstances of failure. The enabling rules underlying conventional action are what we called "rules for succeeding" (p. 41).

The One God, as we conceive Him, can do anything by declaration. I am stunned to think that we ever do *anything* in this way and am perplexed to say by what devices we manage to bring it off. Conventional performance requires an abstraction from circumstance and the "denaturalization" of success. The products and effects of successful conventional action, while they manifestly depend for their production upon material conditions, are not themselves material.

Products of conventional action are produced in their "productions", viz in the very performances in which they are identified.

The third consequence of our characterization of conventional action has to do particularly with the products of conventional action, and was anticipated as an observation on p. 7. We now "rationalize" that observation. A non-material product is always individuated and identified in a medium less abstract than itself. Thus an issue of a newspaper is identified in a material copy, and a piece of music in performance. Call the instantiating medium a *production of the product.* Our examples show that the production may in general be different from the act or acts by which the product is produced, e.g. the acts done in editing the paper or composing the tune. That is not so with productive, *conventional* action. The product is produced and instantiated in the same performances. A request, for example, is individuated and identified in a speaker's very act of making the request. Put formally, a conventional product is produced in its productions. Our explanation of this is simple: Since the act is "self-identifying" and since the conventions for indicating the conditions of success are rules for succeeding, we can tell from our understanding of the act what product it produces if it succeeds at all.

I say that this third consequence, that conventional products are produced in their productions, confirms our intuitive sense of what conventional action is; it also raises a serious difficulty. I have just argued that conventional action is not generically distinguishable from non-conventional action (pp. 54f.). The above argument suggests that conventional products are, so to speak, free-standing upshots of conventional action *only*. That conclusion, taken with the point about conventionality not being generically distinctive, conflicts with a notion that productive action is generically so, because measures of success for action are tied in with conditions of success for action. A sense of difficulty deepens with the plausible thought that the generically indistinguishable non-conventional counterparts to such conventional productive deeds as promising and asserting are probably not productive at all (see below). I believe that the escape from this lies in a simple allowance that measures of success need not at all points always be logically tied to conditions of success, and indeed are often not so for non-conventional deeds (see p.30). The wondrous thing about conventional products is just that they are "free-standing" and produced in detachment from the further consequences of action, this allowing conventional action to succeed without "uptake" or other effects.

The Great Mystery of Language. This freestandingness of statements and other products of language has struck me as being a mystery of theological magnitude. First Causes are credited with cosmic powers of creative intellection: Is there a place for any such creativity *ex nihilo* within the frame of our natural history? I have just above had occasion to speak of the "denaturalization of success". Still, one may well wonder how we bring it off. It may be useful as a first observation to recall that language may, unenigmatically, stand in for or replace natural expressions, e.g. of pain, delight and the like. Here the only relevant condition of success is that the subject should actually be in the state at question. But conventional products such as statements and promises may continue to impress one as having a more supernatural mold. In discussions with Melnick, I have come to think that "standing-in-for" can be extended to these cases too. Consider that a *credit*, which is a conventional product, may stand-in for goods still undelivered. So too, assertions of statements, which, as I shall urge, indicate as conditions of success that procedures of verification and falsification could be applied, may be taken as standing-in-for the verifications or falsifications they anticipate. Stand-in's are of course distinct from what they stand-in for. Credits may, after all, be voided and putatively asserted statements be unsecurable by test. Problems there too. But now perhaps I've said enough to illustrate how conventional products are to be placed in relation to the palpabilities of everyday life.

Now, altogether, finally, in brief definition: Conventional action is conformative action underlain by enabling rules for indicating all conditions of success for the action kind in question, these conditions to exclude all possible circumstances of failure.

My "conventional action" is very like Grice's "non-natural meaning". Our definition notably differs from his by severing definitional connections between utterance and communication. However, conventional action is most easily spotted in contexts of communication. To illustrate the definition, consider the controverted case of communicating dogs, which has already come up. Fido lets me know he wants out by barking at the door. His deed may or may not be conventional. Now consider the other case where he moves ingratiatingly between me and the door, perhaps taking me by the pant leg and leading me there; he wants to let me know something, no doubt about that; but now, he is letting me know that he wants to let me know something, and what he wants to let me know is indicated in his movements. This last case, unlike the first, is as good an example of conventional action as would be a child's request to go out.

The foregoing account of conventional behavior is serviceable as a "universal grammar" along lines I have anticipated in other writings[11] and as adjuvant for resolving the snaggy distinction between *meaning* and *saying* (*Foundations of Language*, 1972, pp. 66-97) and the heavily controverted issue of figurative usage ("A Semantics of Utterance", pp. 358 f). I have lightened this text of those applications and all others save those that play into succeeding chapters. For now, in the balance of this chapter, I shall be concerned to fix the boundaries around *utterance*, which, adopting what was proposed in #4 above, I take to be conventional action for which inscriptional conventions could be supplied.

Caveat: In what now follows I indulge myself and the reader by speaking of "conventional" and "non-conventional purposes" and of "conventional means" instead, more carefully, of the "purposes of conventional acts" (See p. 54 above). Doubts about this may, I believe, always be allayed by deep breaths and long exhalations; e.g., instead of speaking of doing a non-conventional P by doing a conventional P', I could say, "An act, A, with purpose P is done collaterally by P'-ing, where the movements of A indicate all of the conditions of success for P' but not for P." The execution of such "accuracy" is usually more painful than exhilarating.

Special caution must be taken with one danger posed by this abbreviated manner of expression. Conditions for doing never factor into the definitions of collateral and hybrid action. Conventional action, as we have explained it, is featured by a distinctive condition for doing and not so far by any distinctive condition of success. That means that our succeeding talk about conventional hybrids and collateral conventional action is only incidentally about conventional action. An important case in point: Expressing what one thinks one knows to be so is *assertion* only when conventional and featured by a condition for doing of indicating all those conditions of success that might be listed for purposes of defining expressing what one thinks one knows to be so. But then *asserting* is not to be explained as a hybrid of expressing what one thinks one knows to be so into the use of language, for the use of language cannot be explained via a list of conditions of success. In contrast, asserting in French may perhaps be explained as a hybrid of expressing and asserting what one thinks one knows to be so into mouthing French.

Some "hybridizations" of utterance.

A principle of my theory of action is that any kind of act can be done "on its own", and not inevitably as the collateral means of attaining some further purpose. That goes for utterance too. "Pure locutions" are possible if rare. Usually we speak to some further purpose, e.g. to convey information or to get someone to do something for us. However, a sylvan ambler may just remark to himself that this tree is an aspen and that a lodgepole pine, with no interest in recording facts or conveying information. While one certainly needs a respondent to give an order to and normally seeks a compliant response, a speaker may instead wish to goad his subordinate into disobedience, or (more to the point) be simply indifferent about what follows, even to whether respondent understands. "Securing uptake", in Austin's phrase (but in opposition to his doctrine), is never any part of the purpose of conventional action *per se*.

In the usual case, when a subject does speak to some further purpose not attainable in conventional action taken pure, I shall, following Austin, describe the act as "perlocutionary". Otherwise: non-conventional action done collaterally by way of utterance is "perlocutionary". "Perlocutionary" names no kind of action, but is rather a word used to describe how an act is done, *viz* by utterance.

Purposes may be fused into hybrids, and the result defined by a consolidation of the respective lists of conditions of success (pp. 27 & 32). The purpose of a conventional act may be hybridized either way, as the "original purpose or as the "original way". Ordering a meal is conventional action; mouthing French is not. When travelling in France, I normally order meals by using my little French in the best way I know. If those words fail me, I may be grateful to an English-speaking waiter. That shows that *what* I was meaning to do was to order the meal. On another better-fed day, trying to show off, I may insist upon my crippled French, perhaps to the amused contempt of this french waiter. His assistance would only frustrate my effort. Here my action fails when my French does: what I mean to do is *to order the meal in French.* That act would be a *hybrid* of a non-conventional mouthing French into the purpose of a conventional meal-ordering. On the other side, purposes realized in language may be hybridized into other kinds of action. For example, one may try to deceive another person *by* telling that person something; but if words don't succeed our agent might seek to produce a false thought in some other way; in perhaps subtle contrast, one may *in* telling someone something try to deceive that person. *Lies* are one such sort of deed. Other hybrids of this *genre* are *informing, convincing*, and perhaps *persuading*. (These latter kinds of hybrids, when actually distinguished as language, are partly conventional and not perlocutionary, as we have defined that idea.)[12]

The examples used in the above discussion illustrate one kind of "doing something in a conventional way". Some conventional *means* are not initially forms of language. "Politeness" covers one kind of case, and "deference" another. One can do almost any kind of thing politely or deferentially. Here one employs conventional means, where "conventional" has the force we have defined: Subject adopts means calculated to show that he wishes to act politely or deferentially; conventions are at hand to enable him to do that; conditions are indicated, e.g. that there is a respondent who owns some kind of rank or distinction that subject lacks. Actual words may be employed for that purpose. These words would naturally occur in the full flow of utterance. Take a case where *uncertainty* is conventionally indicated. A speaker says, "He will be here by four, I believe"; he could just as well have said, "I believe he will be here by four" or "He will, I believe, be here by four". Here the "I believe" occurs merely "parenthetically" (Urmson). *What* the speaker tells us is that the person in question will be here by four (as speaker thinks). Speaker's "I believe" conventionally modifies, not what he relates to be the case (as he thinks), but rather the manner in which he issues his opinion. "I believe", in such cases, has what I call "act-adverbial usage".

Finally, the purposes of two autonomous *conventional* acts may be hybridized, one into the other. *Bidding in bridge* usually has that character: one's act of bidding not only arrives at a position in the auction but may also issue a report about one's hand and request a response from partner.

We can readily multiply distinctions in the characterization of action and of conventional action. All those we have now introduced--among conventional and perlocutionary action and various hybrids of non-conventional into conventional action and of conventional action into both non-conventional and conventional action--will be used in what follows.

To illustrate the foregoing, I would like to consider the two hard cases of *asking questions* and *telling*. Here, most especially, the presentation opts for laxity over prolixity in the use of "conventional purpose".

It is commonly taken that *asking questions* is a sort of utterance, quite on a level with such others as *requesting, ordering, promising,* and *conjecturing*. That assumption, however, proves hard to square with one's sense that *questioning*, if successful, produces questions that have answers to be provided by other utterances. The problem is that these responses may be of different kinds and are not generically classifiable together in any obvious way. Answers may provide information, or solutions to problems, or expressions of intention, or promises, or even orders ("What do you want me to do?"; "I want you to hire another foreman.") Surely these different kinds of answers are given to different kinds of question; and there must be as many kinds of questioning as there are kinds of question. One would then naturally conclude that questioning is not a single kind of utterance or use of language, but many.

There are also difficulties in that. First, it is unlikely that we shall ever know what all these different kinds of questioning are, because almost any kind of utterance may occur in response to a question, and it is most unlikely that we shall ever have anything like a complete list of utterance-kinds apt to provide answers to questions. Second, there remains a problem of saying what makes all these different kinds of things kinds of *questioning*. Third--and this is connected with the other two--, if every kind of answer has a corresponding kind of questioning, the supposed general order of questioning would simply duplicate with a question mark the many different uses of language within which answers can be given. Seen in this way, the linguistic order of questioning seems superfluous.

The second difficulty is the most easily attacked. Different kinds of "questions" are so-called because they commonly seek to elicit answers. An answer is nothing if not a conventional response. That sets *questions* apart from *requests*, which may be satisfied by non-conventional responses. Also, *questioning*, unlike *requesting*, need not be directed toward particular interlocutors. A lecturer in mathematics who has forgotten the explicit form of a certain integral asks the class at large what it is, simply because he wants to know. He would be glad to hear a voice from the corridor, or perhaps he finally comes up with the answer himself. This is a different kind of thing from requesting a response, which he might have done because he wished to know whether one or another of his students knew. So the common measure of success of questioning of all kinds is the occurrence of a conventional response: It may be a promise, an order, the expression of an opinion, guess, or an "assertion". (The publisher's reader suggests that bidding in bridge is another non-questioning kind of request for a conventional response. I don't think that is so. The bidding may be taken "pure", and still establishes a position in the auction whether partner responds or not. In actual play, it may also request a response; but that is a matter of asking partner a question. The reader is, nonetheless, right to challenge me here: I have not and cannot establish the sufficiency of any formula for *questioning*.)

Now the desired response cannot be assured by the mere fact that the questioner utters the words he does. By this, then, asking questions cannot be something which is purely conventional or merely linguistic.

"But surely questions occur in words. There is no questioning without utterance!"

The two points are easily reconciled by casting *questioning* as a *hybrid* of utterance into the soliciting of a conventional response. *Asking a question* is in a class with *trying to convince, persuade* or *explain* and *lying*.

The difficulty over determining what sorts of questioning there are is now transformed into a question over what kinds of utterance can be hybridized into acts of questioning. Well, "lots and lots", without limit. One may, for example, *wonder whether* something is so and then go on to ask. The use of language incorporated into such a questioning would be that of *expressing one's wonder whether*. Again, one may *pose a problem, formulate alternative policies*, and so on.

It emerges from this "analysis" that, had we treated *questioning* as a kind of utterance on its own, we would indeed have gained a gratuitous duplication. The conventional component is language already named and classified, e.g. as an *avowal* (of uncertainty over what is so and what to do, disbelief, lack of information) or as a *formulation of alternatives, possibilities, etc*. Nothing more in the way of language is needed or wanted. What was a difficulty for the view that *questioning* is a kind of utterance is an unproblematic consequence of our analysis of *questioning* as a hybrid of conventional action into the soliciting of a conventional response.

Tellings[13] are told to respondents. Under that general condition, *telling* may seem to be something very like *questioning*. Almost any kind of conventional product may be *told*, just as almost any kind may be *asked*. Thus,

"I tell you, John has come." (Assertion)

"I tell you, I promise to call her." (Promise)

"I tell you, I need your help." (Request)

"I tell you, shut up!" (Order)

"I told you, that would be unwise." (Advice)

"I told you, go to Hell!" (Imprecation)

In all these cases I mean to let my respondent know what I'm doing, and it is reasonable to see *telling* as a hybrid of almost any kind of utterance into letting someone know. That impression does not survive examination. I may succeed in telling an uncomprehending respondent even though he doesn't come to know: "He didn't come. Did you tell him? Yes, I did. Perhaps he wasn't listening."

Telling surely isn't simply a matter of saying anything to anyone, because then we could not distinguish telling one's promise from merely promising, and one must suppose that telling a promise differs from promising in the same way that telling what one thinks differs from expressing an opinion. But if *telling a promise* is not the same as *promising* and also does not reach beyond the use of language, what greater or lesser or other thing can it be?

To tell a promise or to tell a person to shut up are, *at least*, to promise and to order. So it looks as if we need "something more". Is there any special kind of utterance of which *telling* is always a hybrid?

Telling involves a *display* of a promise, order, statement or whatever. There's the track we seek. Almost anything can be displayed, and displays of almost anything may be *conventional* in our sense. In making a conventional display, the subject indicates that he seeks to make a display. This covers a variety of kinds. *Exemplification*, for example, is one kind of conventional display; *demonstration*, whether in mathematics or physics, is another; and *telling*, I submit, is still another. *Telling*, however, is special. One may give an example of a promise without making the promise displayed as an example. In contrast, one cannot tell a promise without making the promise: If I tell you that I promise to see your sister, then I have promised to see your sister. One tells a promise, order, statement or whatever only in making it.

Telling is very like what is conveyed in English by "I say" (not in reporting an utterance but as part of utterance; see fn.. 13). There are two differences. First, "I say" usually affects an "act-adverbial" modification on the manner of utterance, whereas that factor, in *telling*, is hybridized into what the subject means to do. Second, *telling* is always *to* a respondent. The respondent may not comprehend what the speaker displays, but if he does, then he will also "come to know" what the speaker meant to do.

If *telling* is, as I say, a hybrid of almost any kind of productive utterance into a conventional display of that product to a respondent, then *telling* is a doubly hybridized kind of conventional action. The conditions of success indicated by what one tells carry over; we have also noticed that *telling* additionally indicates the existence of a respondent and other conditions of success having specifically to do with *display*. (A first condition of success for *display* is that other conditions of success should also be indicated and that secures the semantic linkage between *telling* and *what* is told.)

Telling by this analysis, while a hybrid, is, unlike *questioning*, nothing other than conventional.

In *summary*: We have embedded a theory of conventional behavior and of utterance within a general theory of action. That theory of action is featured by the use of two technical notions, that of a condition of success for action and that of a condition for doing action. An action must fail if any of its conditions of success is not satisfied. Genera of action may be defined by listings of conditions of success. No action of any specifiable kind is done failing any condition for doing that kind of action. A condition for doing any generic kind of action is that the subject should believe that conditions of success defining that kind of action are satisfied. A condition for doing *conventional action*, no matter the conditions of success, is that the subject should, by conforming to enabling rules, indicate conditions of success for the particular act in question that exclude any possible circumstance of failure for the act. We now proceed to use this theory for purposes of defining *assertion* as a kind of conventional action.

NOTES

[1]Aristotle repeatedly noticed some such point of distinction, e.g. in the *Nichomachean Ethics*, I.1. He usually marked the distinction as one between *energeia* and *praxis*. I tried to sort out the differences between "productive", "effective" and "endotychistic" action at #19 of Part II of my book *The Stratification of Behavior*.

[2]In Aristotle's thinking, everything categorially sayable of substances is genus-sayable of something, but not of everything of what it is sayable. This sheet of paper is white; the sheet's coloration but not the sheet itself may be genus characterized as a white (See *Topics*, I.9)

[3]Aristotle was the first to observe that KINDS are "intentional objects", grammatically posited as objects of thought and knowledge. They are, I hazard, our most basic and primitive such "objects", for they owe their possibly contingent existence to the existence of their instances.

[4]On the occasion of my final revisions, I find that I have, upon reflection, become more satisfied with the "spinozism" I previously regretted. Beyond what is argued just above, it is also a riposte to that pack of contemporary writers who, in their pursuit of the "mind-body problem", howl a preference for deep physiology. It now seems to me that, while various characterizations may be classified as being (e.g.) physical or spiritual, the thus characterizable is, so to speak, just "there". We shall in Chapt. 13 work up apparatus for characterizing those various characterizations, and shall then discover that certain everyday physical characterizations are, as I shall put it, less "dependent", than are other characterizations--physical and spiritual ones alike. That conclusion, however, bears no ontological freight.

[5]Intimated by the difficulty Aristotle noticed over acquiring virtues as habits of good choice--that there must but cannot be a time when the habit is established. I hope that my valve example from the main text does something to alleviate discomfort on that point. Another difficulty over dispositions was raised by Wittgenstein in his investigation of *reading* and then implicitly extended into a general doubt over the idea of *knowing-how-to-go-on*: There is no record of performance that cannot be fitted to any number of different patterns: So what are these skills? By my reading of Wittgenstein's *Investigations*, he believed that such skills lie at the foundations of thought and are not to be denied, and that is certainly my own opinion. Kripke, in his well-known interpretation of Wittgenstein, makes an objection of the difficulty, one that seems to argue that our various conceptual skills cannot be conditions present in us. The "contentions" above are meant to counter that conclusion.

[6]Roughly: formulated guides for the use of advisors and administrators in their tasks of managing and adjudicating the activities of specific populations such as drivers, school children, players, entries at exhibitions, and importers of dutiable goods.

[7]See Sausurre, *Cours de Linguistique Generale*, Paris, 1966, p. 33. He would call our "conventional action" "l'ensemble des faits semiologiques".

[8]Cf. Wittgenstein's pronouncement that a *Satz* "shows" its sense. *Tractatus*, 4.022.

[9]I came to appreciate the need for this kind of example in the course of a conversation with T. Griffin.

[10]Isn't the fielder's catch an indication of what he means to do? No, for he might make the same movements in conformity with the same rules and mean to

drop it. (He was bribed.) Here we need to go deeper than his movements to find out what he really meant to do.

[11]*Stratification of Behaviour*, pp. 390 et. seq.; "A Semantics of Utterance", in French et. al, *Contemporary Perspectives in the Philosophy of Language*, pp. 244-259.

[12]Austin distinguished them as "Ca". See *How To Do Things With Words*, pp. 101f. My "when distinguished as language" is calculated to avoid saying that the hybrid purpose is in itself either conventional or non-conventional (see pp. 58f).

I am glancing complications. One can, I concede, "Tell a lie" with no intent to deceive. (See F. A. Sigler, *Amer. Philos. Quart.*; 1966, pp. 128-36). But then I sense a difference between telling a lie and lying. The use of a theory of behavior must be as labyrinthine as the solicitudes of our usage.

[13]Nb. the ambiguity in "tell". I can report almost anything I said and meant by saying what I "told" or "was telling" someone, even myself. In this digression I'm interested only in cases where "tell" is used, not to report utterance, but as part of the utterance, as exemplified in "I tell ya', she's coming". There's a like ambiguity in "say".

CHAPTER 2: ASSERTION

1. ASSERTION IS SAYING AND MEANING WHAT ONE THINKS ONE KNOWS IS SO.

Statements are products of utterance, specifically of assertion successfully undertaken. That is a thesis and a proposal for whose elaboration I have tried to establish a foundation in the previous chapter. Something like a definition of *statement* is needed. Just as mice, guinea pigs and beavers, abundantly familiar in their own places to householders, juveniles and trappers, are[1] together technically classified as rodents, so statements constitute a technical, factitious classification of various items which are, in their places, abundantly familiar to specialists and non-specialists alike. For starters, we need some irrevocable specimens to illustrate what we want our technical idea of a statement to cover, and the following is a representative listing, with contrasting items in parentheses

Controlling specimens selected to illustrate various possibilities of human concern, truth-value, form and subject-matter.

(i) *Aristotle was twice married* (but not: *Plato was a homosexual*, an "opinion").

(ii) *Crete is an island in the Red Sea* (but not: *Maybe Atlantis was an island off the coast of Spain*, a supposition or "declaration of possibility").

(iii) *There will be a total eclipse of the sun visible from Central Europe on Aug. 11, 1999* (but not: *There will be a major earthquake in the Bay Area before 1985*, a onetime prediction).

(iv) $e=1.6020 \times 10^{-19}$ *columbs* (but not: *Heat is matter in motion*, an hypothesis).

(v) *All earthly mountains higher than 8000 meters are in Asia* (but not: *All birds are warm-blooded*, a generalization).

(vi) *Pampadour was once the favorite of Louis XV and Barry was later on* (but not: *If Philip II had not reigned for so long, Spain would not have languished*, a conditional).

(vii) *28 is a perfect number* (but not: *There is no odd perfect number*, a conjecture).

The listing also distinguishes statements from other comparable products of language.

Now our list of statement specimens was gained by aping their assertions and accordingly conveyed a specimen list of assertional kinds. No such a list of course can tell us what an assertion *is*, and (by our doctrine) something like that is required if we are ever to say what statements are. For starters in that direction, I suggest that an assertion "in other words" is an act of meaning by saying-- an "utterance"-- of what one thinks one knows to be so. Those "other words" are a partial paraphrase of "assertion", not its "analysis". The paraphrase does, however, give leads to a definition, while also making it evident that assertion is something we do a great deal of. It also raises questions.

Assertion is saying and meaning what one thinks one knows to be so.

The first question is prompted by "partial"[2]. Saying *All birds are warm-blooded* may be of what one thinks one knows, even though this case was deemed a non-assertion at (v) above. We won't get into position to define the distinction between generalizations and (possibly universal) statements until, at #6 below, we'll have listed some "conditions of success" for *assertion*. Suffice it for here to observe that *All birds are warm-blooded* certainly cannot be verified or falsified by test as may be the statemental products of assertion. Anyway, that's one question left unsettled.

The second question is over *what* one thinks one knows in asserting a statement. I believe that it is always the *truth* of the produced statement itself and not just the fact formulated in that statement. I leave that still open, however, if only to avoid what I think would be an unnecessary suggestion of circularity.

Contrasts with some other "constatives".

Assertion is to be contrasted with such other if related "fact-stating moods" or "illocutions" as *expressing opinion* (simply saying what one thinks), *predicting* (saying what one thinks will be known), *hypothesizing*, *postulating*, *conjecturing*, *defining* and the like. If *generalizations* are something other than *statements*, then *issuing generalizations* must, as a kind of utterance, be something other than *assertion*. We should in principle be able to define these[3], and assertion would, I hope, then be seen to occupy a central position of reference within this broader field of "constatives" (Austin's term). Arguments can be marshalled to establish the need for these various contrasts. Consider, for example, that, while it makes good sense to say "Maybe Atlantis was an island off the coast of Spain, but I don't think it was", it doesn't make evident sense to say "Maybe Atlantis was an island off the coast of Spain, but it wasn't". The easy explanation of this is that *assertion* ("It wasn't"), unlike *declaring a possibility*, makes a "claim to know" that is incompatible with leaving opposed possibilities open.

This conception of assertion allows for a doctrine that is consistent with philosophical skepticism.

Objection (from Fred Schmitt): According to this paraphrase of assertion as "saying what one thinks one knows", a skeptic who believes (not knows) that no one ever really knows anything could never assert anything, although, certainly, such a skeptic would be abundantly able to say what he thinks.

Answer: I agree, especially if (as I believe) the knowledge an assertor thinks he has is of truths; but this agreement is with reservations. I once found the objection troubling, because I believe that *assertion* occupies a central region within the field of "constatives", a region the skeptic believes should be evacuated. If that evacuation were enforced, then my picture of the relationships between *assertion* and other sorts of saying what one thinks would go to pieces and the motivation for any interest in assertion be destroyed. My misgivings here are allayed by the thought that a skeptic who doubts our knowledge of truths must anyway know what it is to say what one thinks one knows and know how to do it, as do we all. Indeed, he must be supposed to have done a good bit of asserting in the days of his innocence. Skeptical doubtings do not work to evacuate the central region of assertion but to discourage visitations by philosophers.

I should like, finally, to give advance notice that what follows in this treatise is meant to be consistent with the skeptical holding (which I disbelieve), that we never actually have genuine knowledge of truths.

2. THE SEMANTICS OF MOOD: AN HYPOTHESIS.

We ask why all utterances should be distinguished by "mood" ("illocutionary force").

We called *assertion* a "mood" or (in Austin's terminology) an "illocutionary force". There seems to be no utterance not classifiable in "this way", by mood--as an act of promising, vowing, ordering, commanding, saluting, greeting, requesting, voting, betting, predicting, conjecturing, formulating an hypothesis, naming, defining, expressing an opinion, expressing an intention, depositing money, cheering, and so on. Why is that so, and why is our list of moods open-ended and so unsystematic? Mood, after all, isn't all there is to utterance. Determinations of predication, attribution, reference and "form" cut across determinations of mood. Why, as it seems, should only distinctions in mood be mandatory? The answer is elusive partly because the pattern of mood determination is unclear.

We hypothesize that those conditions whose indications contribute to the determination of mood are one's that facilitate the perlocutionary connection of utterance to other activities,

An utterance's mood, I submit, is part of its "meaning" and, accordingly, by our theory of conventional action, something fixed by indications of conditions of success, this often secured by the employment of an expression that incorporates a name for the mood (e.g. "I vote", "We conjecture", "Let us define") but also with the use of copulas and by features of word-order and intonation.

An hypothesis for mood: The conditions of success indicated in an utterance that contribute to the determination of its mood are ones that facilitate the "perlocutionary" connection of utterance to our other activities. The conditions that contribute to the determination of predication, etc. do not do that. By this hypothesis, the need for mood is corollary to the plausible thesis that no form of language could exist unless sometime brought into connection with the larger affairs of life. We allow, indeed insist, that any kind of utterance may, in splendid, unsubordinated isolation, be executed in disconnection from these other matters (see pp. 59f.). But surely the isolated occurrence of utterance is the exception from and an attenuation of the richer case.

The evident variety of these "connecting conditions" accounts for the open-ended and unsystematic appearance of our list of moods.

and further that these conditions have to do mainly with the state of the subject, its respondents, institutional arrangements and knowledge.

I once thought that these conditions can be exhaustively if vaguely classified under four headings; as having to do with: (i) the situation, competencies and knowledgeability of the subject; (ii) the existence, situation, identity and competencies of addressees or respondents; (iii) the existence of "institutional" arrangements underlain by enabling rules (e.g. games, banking, elections, "property", "justice", courses of instruction) and (iv) the acquisition, organization and transmission of knowledge. Language, when available, is a natural stand-by in the service of other non-conventional usages underlain by enabling rules and an obvious instrument for organizing and transmitting information and for coping with its absence. No surprise, then, that utterance should be answerable for its success to conditions of these several sorts. All of this is distressingly "soft" and I have no proper proof, which makes me less sanguine than I used to be; I shall nonetheless continue to use the classification as guide, but warily, in the hope that what follows will confirm the track.

There seem to be no defining conditions of success characteristic of all "constative" moods.

3. CONSTATIVE MOODS, EXTERNALLY DELIMITED.

Our policy of concentrating upon the factitious mood of assertion is built on the unproven assumption that it occupies the central position in the still undefined field of "constatives". I do not see how to secure the desired delimitation of the field of constatives in a manner conformable to my theory of utterance, solely in terms of characteristic conditions of success. I propose instead to consider those subordinations to the larger affairs of life that occasion constative utterance. We have hypothesized that the connections between language and "what it's for" are facilitated by mood determining conditions on language itself, and that same hypothesis predicts that there will be different constative moods featured by different ones of these conditions.

My assumption, which may be questioned, that there is indeed this field of constative language is like but more modest than the hypothesis that language may be exhaustively and neatly divided into two parts, exemplified on the one side by expressions of belief and, on the other, by expressions of intention, the former to be generically classified perhaps as "theoretical", "propositional" or "constative" and the latter perhaps as "practical", "active", "performative", or "volitive".

We propose to delimit the field broadly by appeal to the normal perlocutionary subordination of them all to epistemic enterprise or the "life of knowledge".

Why shouldn't there be at least some conditions of success, characteristic of all constative kinds, laid like a dike across the whole field? Looking at our classification of mood-determining conditions, inspection suggests that the desiderated conditions would have nothing to do with respondents or institutional arrangements, for all of us sometimes speak "constatively" at home alone. The epistemic competencies of the subject and the general state of knowledge are obviously and unsuperficially pertinent to the fact of constative language, but doubtfully in any "common way". That is because constative utterance trades so variously on the state of knowledge and contributes to the conduct of theorizing in ways so variously complementary as to extrude any candidate common condition of success. Constative utterance, no doubt, typically engenders commitments to withdraw in case of insufficient evidence or proven falsehood; still, the general applicability of these notions of evidence and truth-value does not supply a common field-bounding condition of success, for (e.g.) *considerations of possibility* avoid such commitments. We similarly cannot appeal to any general condition regarding the availability of knowledge without jeopardizing the good standing of *conjecture*. What qualifies *considerations of possibility* and *conjecture* as stand-in's for assertion also rules out any unified definition of constatives in terms of conditions on knowledge or belief, truth and evidence. Constatives come together through their interrelated contributions to the "life of knowledge", and that is the external point of perspective from which they may be comprehended as a group. They are back-up's for and continuations upon each other. So, if the usual perlocutionary purpose of assertion were to transmit information to others, then, on an occasion apt for such a deed, a speaker, knowing that he did not know that such and such was so and therefore also knowing that he could not assert as much, might still wish to convey his thought that it "might be". An accumulation of data reported in flat assertions of fact might lead one on to a generalization, thence to a law, and then back to more reports of data.

> I am encouraged in this policy of seeking an external delimitation by an analogy from the theory of practical reason. The direct assessment of conduct and character is most usually secured without any immediate appeal to resulting benefits or harms. Yet surely this elaborately variegated field of assessment, when taken all together, must have some relation to biological well-being. The fact that behavior of this kind is beneficial or harmful to the assessed individual or to the groups to which he belongs *explains* our having the norms we do. The immediate vindication of character and the justification of action is secured by appeal to those norms themselves and not to the biology that underlies them. The biology, however, demarcates this field of justification. Due notice of the natural history and setting of conduct and character lets us better understand what sorts of considerations contribute to their assessment. So too (I suggest), while a consideration of the natural history of constative utterance and of what it contributes to the larger affairs of life--those matters which explain its existence--would not define it as a kind of utterance, it might still serve to delimit the field and sensitize us to factors that could be legitimately cited for purposes of defining different constative moods.

It is evident that constative language variously serves the interests of acquiring, preserving, organizing and transmitting knowledge of truths. Our sense of something common can be roughly and externally demarcated by the usual subordination of this language to those interests. The flat assertion of fact is commonly enacted to let others know, but also often to record information gained. Predictions, conjectures, assumptions and hypotheses are commonly issued to show the way to what may later come to be known. Definitions and postulates are characteristically laid down with an eye to the systematic organization of circumscribed bodies of knowledge. Many moods are needed in order to get on with the business of knowing things better in the absence of established information. The regular subordination of constative moods to the ends of knowledge, if I may now put it so succinctly, explains there being different characteristic conditions for different ones of these moods. If subject does not know and thinks that no one does, he may still "predict", saying what he thinks will be known, in which event he must indicate that (he thinks that) what he thinks is not yet known.

(Nb: *What* he predicts in such cases is not the knowledge, of course, but what he thinks will be known.)

4. ASSERTING STATEMENTS AND ASSERTING FACTS.

In defense of statements.

Distinctions between the sentences and other "expressions" we employ, the "speech acts" or utterances we perform and the propositions we express are familiar and widely accepted. There are unsettled questions, no doubt, over what these items are and how they are related (see W. E. Johnson, *ibid.*). My distinction between assertional utterances and their statemental products is more elusive and less observed. Statements like other conventional products are identified in the acts which produce them, *viz* assertions. It may then seem that we are trading in a distinction without a difference and that statements are mere "aspects" or shadows of assertions or are assertions masquerading as something else, or are "logical constructions" from classes of assertions, or mere "internal accusatives", creatures of grammar alone and spuriously taken as objects on their own. I do not know how to refute disparagement framed in terms so uncertain as these. Professor Church once remarked that only a misogynist could take the fact that women are what men marry when they do as an argument for the "logical reduction" of women to men. "I saw a sight": Why can't the "internal accusative", "sight", be referential on its own? It might even seem that it must be so if what I said was true. Statements are tightly bound to assertions, as their products; still, meaningful but unsuccessful assertions may occur and no statements be produced and different assertions may produce the same statement; also unasserted statements may exist so long as they could be asserted.

Comparative "Categorization" of "assertion" and "statement".

Our sense of distinction between assertions and their statemental products is sustained by amplifying the list of the different things that "can be said" about these presumptively distinct articles. Such consideration of "things said" does not of itself prove a distinction in fact; it creates a presumption for a distinction.

An assertion is an utterance or conventional act which succeeds or fails and is an episode enacted over a definite stretch of time.

We must be able to say of it where, when and by whom done. Derivatively, we can say that it was long, opportune, relevant, or justified. Assertions are always performed by use of definite expressions, and because of this will have various kinematic, phonetic and lexical properties; they may be euphonious, loud, in English or dialect. They will also have certain grammatical properties, e.g. they may be ungrammatical. Finally, assertions may have semantical properties assignable in consideration of the manner in which conditions of success are indicated. An assertion may be tentative, terse, or unclear. Statements are none of these things.

Statements, in contrast, may be true or false, provable, accurate, consequences of other conventional products, e.g. of generalizations or guesses, contain elements like references to Churchill, and have various properties of "form", such as being singular, universal, existential. Statements have or lack these features regardless of when they are made or by whom.

Assertions may indeed have features which borrow from these last mentioned properties of statements, but then only under special conditions and always in consideration of the statement that would be produced if the assertion were successful. Thus we may say in borrowed senses that an assertion is "true" or "false", but then only if the assertion is successful. Properties of form may carry over to assertions, but then only if the assertions contain elements corresponding to the form determining elements of the statement: "All of these books are paperbacks" schematizes a universal form of assertion, but if one produces a statement to the effect that all of a certain group of books are paperback by saying "Yes", then that assertion is not of universal form.

Statements and assertions may also have or lack a variety of other properties of possibly disputable application. There is, I believe, a sense in which statements may be *long* or *short*, not in duration of course, but in content, independently of how tersely or long-windedly they are made. I shall argue in Chapter 3 (#6) that the philosopher's notion of *necessity* has its primary applications to assertions and other utterances and only a derivative application to statements; on this score, *necessity* contrasts with the everyday notion of *truth*, which has its primary application to statements and other products. (See Appendix A for opposed opinions.)

Assertions are "episodes"; statements are no kind of "particular".

Assertions are episodic particulars, localized in space and time. Statements may perhaps have dates, but not places; and, putting it fancily, statements are not "particulars". A statement exists only if all the conditions of success that would be indicated in a producing assertion are satisfied. But there is no point at which a list of such conditions must terminate. Here statements coincide with assertional kinds, for both are together defined by the same non-terminating lists of conditions. Statements, like such continuously ordered qualities and magnitudes as color and graduated weight, are defined as bands, not as sharp lines. Again, statements are like animal species for which no final list of "essential" features can be given, and not like actual cageable brutes.

The co-definability of statements and assertion-kinds is a factor that operates to encourage the idea that statements are nothing but "logical constructions" from assertions. The equation remains suspect, nonetheless, for there exists no statement answering to an assertional kind if any of the listed conditions is not met. That side of the case for a distinction between assertions and statements is most powerfully argued from the consideration that existing, completed, meaningful assertions, unlike statements, may fail, in which event no corresponding statement exists. Statements are contingent upon circumstances about which speakers may have false beliefs. A politically uninformed speaker may meaningfully declare that the president of Spain during the 1960's was from El Ferrol or, in all arithmetical innocence, aver that 5/0 is not a perfect square (because, e.g., it is the undivided prime number 5).

Assertions may be faulted for failure to produce statements or, differently, for the qualities of the statements they do produce. The falsity of a produced statement is one such fault.

Assertional failure is a fault. Assertions also suffer other faults, e.g. unclarity. More to be remarked, assertions may be faulted for producing false statements. Then there must be that statement, and the assertion was a success. Where's the fault, then? One who asserts a statement must be taken to "endorse" the product for truth and be willing to withdraw it from circulation should reasons appear for thinking it to be false. *Truth* is always desired from assertion, and that is reflected by its being a condition for assertion that the speaker cannot think that what he says is even apt to be false. Yet it may turn out to be so, as the speaker must also know. This ungainsayable possibility of asserting falsehoods has been a difficult axiom for philosophers since Plato. Our reading of the principle is that the truth of the product is not a condition for its existence.

The assertion of a false statement is faulted by cause of the non-existence of something of what I believe we really do commonly call "facts". For every statement there will be an equivalent statement of the existence of a fact, and the two statements are true together. Some readers will find me duplicitous in thus appealing to both *truth* and *fact.* I protest innocence, if only because other varieties of sometime true "constatives" do not seem to be "of fact" in the way that statements may be; so, for example, while we may conjecture true conjectures we do not conjecture facts in the way in which we may be said to assert facts

A subject who asserts a statement also "endorses" that statement for truth. An assertion that produces a true statement may also, in a related understanding, be said to be an assertion of fact.

With that said, let's move back to the assertion of facts: Just as we speak of shooting targets as well as shots, so too we may and do speak of asserting facts as well as statements. A missed target is not a misfire; a false assertion is competent to miss the facts only because it succeeds in producing a statement. When and as it is convenient or necessary to honor the distinctions flagged in the foregoing remarks, I shall say that assertions "assert facts" and "produce statements" and that statements "formulate facts". Where conjectures and generalizations, for example, are "responsible to the facts" for their truth, it is (as we have remarked) doubtful whether they can be said to formulate facts.

On Being Fair to *Facts.* (With acknowledgments to J. L. Austin.) *Fact* of course poses hard questions over whether facts exist and what they are if they do. Even though the issue appears repeatedly in this treatise, I shall never produce a satisfactory analysis of speaking of facts. Here in gist are my present thoughts on the matter. Facts are said to exist and are established with reference to the "state of the world", and in that vague sense may be said to be "in the world" and not merely "in language". (Of course, the *existence* of an assertion or statement, like anything else, may also be a fact: "His assertion that she would come is a fact". An assertion is an episode like her arrival and a fact, but different from the asserted fact that she would come.) A sense of "fact" is gained by taking note of the kinds of things facts may be said to be, e.g. "known", "noticed", "observed", "overlooked", "forgotten", "discovered", "surprising", and people may be *informed* or *apprised* of facts. They figure preeminently as "objects of knowledge".

What are facts? I'm inclined at this unprepared stage to say that *anything* may be a fact: Socrates was a fact and so too is Iran. The hardness of a rock may be a fact, as also may be John's being taller than his sister, or the later arrival of a plane, or the shot's having been fired by the accused. The death of Hitler, stormy weather, Spartan heroism and the heroism of every Spartan at Thermopylae--all these are facts. It may seem from this that facts are to be identified with things and even with non-things (like stormy weather) and with every manner of quality, magnitude, happening, relationship, collection, state, circumstance, situation and condition that may be located in or among things. It would seem from this that nothing is not (a) fact. There are differences, however, in *how we talk about* things *called* (e.g.) "qualities" and things *called* "facts". The hardness of the rock, taken as quality, may be said to be *felt*, or *penetrated*, but not the fact of the rock's hardness, which may instead be *observed* or *taken note of*. So far as I can see, there is nothing not "intentional" said of things when they are called "facts". I believe that that is because facts are said to exist or not to exist only as assertible objects of knowledge. Facts are "located in" assertions and shown to exist with the truth of these assertions. Facts are not therefore to be equated with those truths. After all, different assertions and different statements may be of the same fact. *My oldest brother was stout* and *My brother Herschel was stout* are distinct, contingently related statements of the same fact. A condition sufficient but not necessary[4] for the identity of the facts conveyed by distinct assertions is that both assertions, while producing different statements, say "the same thing" about the same object, e.g. that *Denver* or *My home town* or *The Capital of Colorado is the capital of Colorado*. A formula--vague and much in need of trimming--that may be serviceable in what follows, is that a fact is just anything you wish *qua* assertible object of knowledge.

5. REPORTING AND PROPOUNDING.

Constative utterances may characteristically reportorially be "fitted to the world" or propoundingly be issued to be "fitted by the world" " or neither or both of these things.

One can *report* facts or *describe* what one sees, but not *postulate* facts or *conjecture* what one sees. Contrastingly, one can advance or propound a postulate or a conjecture, but not a report or a description. One can, however, on the basis of the observed facts, advance or propound a *generalization*, at once recording what has been and looking ahead to what is yet to be. The differences are "characteristic" of the constative moods we mentioned, and are conveyed by indicated conditions of success. Reporting utterances characteristically indicate that they are performed "after the fact", other propounding utterances indicate that they are performed "before the fact", and still others indicate that they are performed "after" one range of facts and "before" another. "After" and "before the fact" are of course figures of speech: definitions are "propounded" but not exactly as facts. "Fitting" is perhaps a more suitable idiom: Reportings are "fitted to" the world, and the world is "fitted to" propoundings. Generalizations are fitted to one range of facts and fitted by another.

In comparison with Austin. My distinction between *reporting* and *propounding* resembles the distinction Austin employed under the title "direction of fit" in his paper "How to Talk" (Reprinted in *Philosophical Papers*). He observed, in reference to a simple kind of speech situation, that we might, in asserting that S is a P, be responding to different requests roughly schematizable as "What kind of thing is S?" "What about S?", "Give me an instance of P", "Which is the P?". Responses to the first two requests would record that we have found and fitted a sample P to the item S. The responses, in the latter two cases, record that we have found and fitted an item S to the sample P. The first case ("cap-fitting identification") differs from the second ("statement") in respect of "onus of match". With sample P in hand to fit the item S, we identify the S by matching P against it; but the "statement" records that we have successfully matched the S against the P. The same difference marks the distinction between "instancing" (S as a P) and "bill-fitting identifying" or "casting' (S as the P).

Verbal records of successful cases of finding a pattern to fit an item are like (but not the same as) reportings; verbal records of finding an item to fit a pattern are like (but not the same as) propoundings My one-parameter distinction will do for our purposes.

Reportings and propoundings both cover a variety of cases in respect of what is fitted to what. We may fit words (patterns) to things described (e.g. a face) or to events reported (e.g. the arrival of John). On the other side, we fit applications to postulates ("By postulate, we can construct a line parallel to *l* through point *p*.") cases to hypotheses, and knowledge since gained to predictions previously made. We might also speak of fitting later observed facts to earlier observed facts through the mediation of generalization.

Reportings and propoundings, as classes of utterance, are formally distinguished by indicated conditions of success. Among the conditions of success for reportings are that specifiable facts have been or are being observed, and among the conditions of success for propounding kinds are that specifiable facts have not been observed. Further distinctions can be drawn in reference to what facts have or have not been observed, in what manner and by whom, and still other related considerations may come into play.

Austin observed that although we can state that *S is not P* or give *S* as an instance of something which is *not* P, we cannot identify an *S* either as *not-P* or as *not the* P. He intriguingly conjectured that negative assertion is restricted to cases where the directions of fit and match are opposed. One can sense a reason why that should be so: If A is both fitted and matched to B, then we cannot effect the contrast available when A is fitted to B and B matched to something else. This may explain the unnaturalness of "All S's are not P", which (unlike "No S is P"), would look backwards to fit observed facts and forwards to find fitting facts, which in both cases are matched to our words. (The "All S's are not P" has become fairly common in american English, but with the "wrong" understanding of "Not All S's are P". The force of that "existential" is of a report on a fact that is matched to the word.)

We shall define assertion so that it is characteristically neither reportorial nor propounding.

We now have in hand a broad classification of constative kinds into reportings, propoundings, mixed cases (e.g. generalizations), either's and neither's. Assertion in my usage (as in Austin's) is an "either" or "neither". We may assert the same statement before or after the fact, with or without indications either way. In saying what one thinks one knows to be so, one need not indicate as a condition of success that he does in fact know or that he does not, and certainly not that he either has or has not observed a "relevant" fact. "Thinks one knows" is a license, not a restriction. In asserting, one may but need not give indications that certain things have or have not been observed. Since either "direction of fit" is as good as the other, this distinction does not affect the essentials of assertion, and we are free to use examples of either or neither kind. Now occasions of assertion matter little in what follows, whereas occasions for verifying and for falsifying produced statements are of the greatest moment. For that reason, I shall prefer "neither" specimens, since they conveniently unfetter occasions of assertion from occasions for testing.

6. CONDITIONS FOR ASSERTION AND THE EXISTENCE OF STATEMENTS.

A statement exists at such times as conditions are satisfied for doing an assertion of a certain kind and all the conditions of success that would be indicated by such an assertion are satisfied.

In order at once to retain the conception of a statement as a product of assertion and to allow that statements may exist unasserted, I want to say that a statement exists when it could be successfully asserted. That dictum , as formulated, is full of uncertainty because of "when", "it" and "could".

"when". It is entirely concilliant with my views about statements (and also other conventional products, such as orders and promises) that they exist only at such times as all the conditions for their successful assertion obtain. No one before 1850 could have successfully asserted that Bertrand Russell used(s) the Vicious Circle Principle, so that statement did not then exist. Statements are "temporal" things.

"it". The presence of "it" suggests that we are saying only that the statement exists if it does, which is hardly news. We must seek a more circumspect and less circular formulation of the rule.

"could" shows that we are alluding to conditions for successful assertion without yet saying what they are. We must endeavor to be more explicit.

An improved version of our formula is this: A statement exists at such times as

(i) conditions are satisfied for doing a certain kind of assertion

(ii) all of the conditions of success that would be indicated in such an assertion are satisfied.

We would *prove* that a statement exists by observing the actual occurrence of an assertion of the kind in question and by showing that that assertion succeeds. By this, a statement could exist unproven and unasserted provided it could be successfully asserted at the time. (See pp. 8f)

This is all exceedingly abstract and will become less so only after we have elicited a short list of conditions of success for *assertion*. Actually, there can be no effective procedure for proving the existence of a statement, because the contemplated list of conditions cannot be completed. Disproof, on the other hand, is possible and never impossible. We are able to operate in the field of assertion because we are normally satisfied to take it that we understand what a speaker or writer says and leave it at that. Something may go wrong or raise questions, as when we stop to reflect what a person might have meant when he declared himself to be one-third irish: "Was his russian mother half-sister to his half-irish father? Wouldn't he be quarter-irish in that case? Perhaps he's just confused." Reports claiming or implying the success of assertion and the existence of statements are not to be indelibly registered or put beyond recision. Ineffective determination and interminable revisability are not fatal for statements, however; no more so than for roses: There may be no effective list of essential conditions for being a rose, and one need not be botanically expert in order to recognize a rose for what it is. "But surely someone must be able to say something about what it is to be a rose!" Similarly, I must be able to say something about what conditions must be satisfied if a statement is to exist.

The dictum is destined to remain "ineffective" but can be usefully amplified with the specification of conditions.

Working from the sense of "meaning by saying what one thinks one knows to be so", we select a "non-generic" set of conditions for doing for assertion and a short list of "generic" conditions of success schemata for assertion. Although these lists cannot cover everything that can be said about assertion, they will give me all that I shall need for what follows.

We shall seek the strongest conditions both for doing and for the success of assertions consistent with specimens and paraphrase and with the policies of minimizing peculiarities of subject-matter or peculiarities tracing to the evidential situation of the subject. We shall concentrate on "epistemic things said" in respect of the assertion of statements.

Constraints on the quest. Because *assertion*, undisguisedly, is a device of theory, we have some liberty in our selection of these conditions, but we are also subject to important constraints. First, our formulas are responsible to our list of specimens (pp. 68f.). Second, they must jibe with our (partial) paraphrase of "assertion" as saying what one thinks one knows. Third, they should reflect a regular subordination of assertion to the ends of knowledge. Fourth, they should locate a well-populated and stable position close to the center of the constative field. Fifth, they must not violate stipulations already entered, especially that *assertion* is not characteristically either propounding or reporting. These "prior restraints" of themselves provide resolutions for certain controverted issues. So, for example, that the speaker should actually believe and not merely not show disbelief is required by our paraphrase of assertion as saying what one thinks one knows to be so.

Within these restrictions, I shall adopt the discretionary policy of maximizing the concept of assertion by imposing the strongest reasonable conditions that suggest themselves. This policy is naturally prompted by our rules of regular perlocutionary subordination and that assertion should be central and familiar. Heavily packed moods are apt to be primitive, common and tightly connected with the larger affairs of life. Every constraint we put on assertion adds content to the claim that it is central and gives another hasp for tying assertion into connection with other constative moods.

Our concept-maximizing policy is limited by our incapacity to achieve a completed list of conditions. It is also, more consequentially, limited by (non-standard) peculiarities in regard to subject-matter or theoretical situation, which we wish to eliminate from our characterization of the general idea.

In mathematics, but not generally elsewhere, one cannot assert (as one may conjecture) what one thinks has not yet been demonstrated. Again, a statement--but not an estimate--of the value of some physical constant must wait upon controlled observation or the completion of an experiment. But surely, mathematicians and physicists do not have an exclusive on assertion and the peculiarities of mathematical and physical practice should not impose upon our general definition of *assertion*; neither should the definition put us out of position to accommodate those peculiarities.

The combined policy of maximizing content while minimizing peculiarities is served by schematic formulations of conditions--a method unavoidable in any case. I shall stipulate as a schematic condition of success for assertion that the sort of thing asserted could be known at the time of assertion Now it is part of the idea of mathematical knowledge that what has not yet been demonstrated could not be known. Similarly, it is impossible to know the value of a physical constant which has not been established through controlled observation or by experiment.

Normally, one who thinks he knows a fact is unaffected by reservations. He may, however, without relinquishing the thought that he knows, acknowledge gaps in the evidence, in which case he would do well to modify his assertion with an "I think", "if", "unless" or the like. The conditions for such "adverbial modification", if sufficiently recurrent and consequential, may come to be incorporated into alternative moods. This difference between adverbial and mood is definable (see pp. 60 and 100-103 below) and ought not to be obliterated. Certainly we must allow that assertion may be thus variously qualified and, accordingly, in our definitional efforts, we must seek to minimize peculiarities owing to the evidential situation of the subject. We should, in short, allow that assertion may be variously qualified.

Our policy for assertion is "minimally maximal". We narrow the definition until it would encroach upon the fact of express adverbial qualification in the manner of assertion or restrict assertion to limited theoretical practices or subject matters.

A variety of questions standardly arise in connection with what I shall take to be the assertion of statements; those questions are convertible into formulations of conditions for assertion. The questions flag the regular subordination of assertion to the ends of knowledge. The terms of the questions are also appropriate to other constative kinds but at different points and in different ways. These questions involve notions of *knowledge*, *belief*, *certainty*, *likelihood*, *evidence*, *confirmation*, *authority*, *truth*, *falsity*, *verification*, *falsification*, *proof*, *premise*, *conclusion* and *implication*. In observing how these quite ordinary ideas congregate, we should, by a progressive diminution of uncertainty, narrow onto the target, which is a maximal conception of *assertion*.

Epistemic categorization of assertion.

Assertion is an expression of what the subject *believes* (thinks) and of what might or might not be believed by others. An asserted statement is a formulation of what the speaker or others may or may not *know*. A statement formulates something of which the speaker or others may in different degrees be *certain* (sure) or uncertain. Differently, it may or may not formulate a "certainty", e.g. it is certain that Charlotte Corday killed Marat but not that Caesar Borgia killed his brother Juan. Again, some statements are of things known "more certainly" than are others: that 2+2=4 is "more certain" knowledge than that $169=13^2$. In asserting a statement a speaker normally does but need not always indicate that he knows; he may additionally, and somewhat differently, claim to know (*viz* say that he knows) what his statement formulates, and then be requested to establish his *authority* in regard to the fact or range of facts in question. The belief expressed in asserting a statement may be *confirmed* by citing supporting *evidence*, and the belief deemed *likely true* in degree that the evidence is strong. A statement is itself commonly either true or false and may be *proved* or disproved by being *verified* or *falsified*. A statement may stand as a *consequence* of or be implied by other constative products and *entail* other statements.

All this is, of course, "stipulative", but with real examples in hand. Surely these everyday ideas of knowledge, certainty, evidence, etc. do not, taken singly, apply exclusively to the assertion of statements; if they did, our explication would be shadowed by circularity. Generalizations like Balmer's Law about the distribution of the spectral lines of Hydrogen by wave-length formulate what can be thought and known, and may themselves be true, certain and confirmed. But generalizations of this kind cannot be proved by verification. Calculations, by contrast, are not formulations of what is known or unknown and are only dubiously called "true" or "false"; but calculations can be verified very much as can statements. Predictions, again, while they may be known to be certain, are certainly not formulations of what is (now) known; predictions are borne out or come true, but not verified. Descriptions (of a face, say) may imply statements but are themselves implied by nothing, because of their indefinite length.

Such reflections fortify me in the opinion that the mentioned ideas of *knowledge, belief, ... consequence*, while separately applicable to many things, collectively apply only to what I want to dignify as the assertion of statements.

Now, with an eye back to our hypothesis about the determination of mood (pp. 71f.), our foregoing survey of questions suggests that *institutional arrangements* will figure as conditions of success for *assertion* only if they have something to do with the scientific or other organization of knowledge.

Furthermore, since statements can be privately recorded, it is fairly evident that there are no conditions of success for assertion specifically on the existence or identity of respondents.

Further still, particular statements do not, as it were, "belong to" particular speakers as particular promises and orders do. Hence there are no condition-of-success schemata for assertions having to do with the competencies of the speaker.

The epistemic qualifications of an assertor are "non-generic" "conditions for doing" characteristic of assertion, as are various other matters

Yet surely the "epistemic" situation and competency of the speaker must come into play at some point. An infant cannot yet assert mathematical statements because he cannot yet know how mathematics is done. Again, a speaker must believe that he does know whereof he asserts. Now it looks as if these are all points on which the speaker, if he has any inkling of the matter at all, cannot be mistaken. That suggests that these are conditions for doing characteristic of asserting and not conditions of success. Here are a few conditions for doing implied by our idea that assertion is a kind of meaning by saying what one thinks one knows to be so. These are "truth-conditions" for saying that A asserts F.

(1) It is "meaning by saying", so, as we have hitherto quietly assumed, a condition for doing for assertion is that it is action and conventional action, and A indicates and believes all the so-far-undetermined conditions of success for assertion. Of course, the "generic", "preferred" classification implied by calling an utterance an assertion--the classification fixed by the conditions-of-success of the act--needn't be conventional (see pp. 54f. and 92f. below).

The next four conditions follow from the fact that, since assertion is conventional, subject must be capable of knowing that he is asserting (pp. 53f.), *viz* meaning by saying what he thinks he knows.

(2) A believes F.

(3) A assumes liability for others believing F on the grounds that A told them so.

(4) A undertakes a "commitment to the truth" of what he says in that he undertakes to withdraw from asserting F or to correct

his assertion in the face of proven falsity or strong contrary evidence, and must acknowledge as criticism that what he said was false, uncertain, unlikely, or did not follow from what was given.

(5) A could know facts of sort F (*viz* facts classified under F-formulations) in that sense in which I, who do not have absolute pitch, could not, unassisted with instruments or testimony, know the pitch of a note.

Now I repeat: These are entered as conditions for doing that take first steps toward a "non-generic" classification for assertion. Our preferred "generic" classifications of action are in terms of fallible conditions of success. (pp. 26f.)

Our hypothesis about mood-indication suggests that conditions of success for assertion must have to do with the objective availability of knowledge. This is a matter on which subject may be mistaken.

Three schemata for assertion.

I believe everything we need here may be fitted under three strong *schemata*.

I shall presently argue that these *schemata* are not only distinct but independent. One reader sensed that the second and third are mere amplifications upon the first. I think that's wrong, and that the reader may have failed to appreciate the thrust of my formulations being, as I put it, "schematic": different conditions may satisfy any one of them; and occasionally one such circumstance might concurrently satisfy all three, e.g. the actual verification of a statement by a subject is sufficient to show that what is said in the assertion of the statement could be known by authorities and that there are procedures of verification.

First schema: Any assertion indicates as a condition of success that a formulation-specific kind of thing could be known at the time.

First Elucidation: What is thus indicated as the kind of thing which could then be known would cover the asserted fact, were the assertion both successful and true. Nowadays we assert all sorts of truths and falsehoods about the values of physical constants, specific heats, the existence or non-existence of algorithms, and the constitution of tissue that could not have been even conceived in earlier times. On the other hand, there was once a time when all sorts of things could be known and asserted about the private life of Alexander the Great of which nowadays we can express only suspicion and opinion. So, to expand upon this first example: Were a gullible or careless reader to assert that Alexander was complicitous in the assassination of his father on the basis of a respected scholar's expressed opinion to that effect, the reader would fail in his assertion.

An assertion must convey as conditions of success that a certain kind of thing could be known at the time, that there exist authorities and that there exist sometime applicable procedures of verification and falsification.

Examples:

(ii) While I could successfully assert where my brother will be tomorrow, I could not, prior to 1984 successfully assert where my then as yet unborn oldest great great niece would be in fifty years.

(iii) Horse races being what they are, there is normally a presumption that the winner cannot be known in advance; a confident but honest better who knows all this would not assert, but guess.

(iv) Geologists do not assert but may conjecture regarding dates of impending earthquakes. An innocent layman--on the strength of what he reads in "People"--"Alignment of sun and moon makes earthquake likely next Tuesday."--may unsuccessfully assert that an earthquake will occur next week.

(Since this example was first set down, I had occasion to be in Gilgit, Pakistan at a time--1975--when Chinese engineers operating nearby had predicted an earthquake. My travelling companions at the time didn't know what to make of an announcement posted at the town center that an earthquake would occur that day. It wasn't exactly that they disbelieved it; they just didn't know whether the local officials were guilty of unsuccessful utterance or were to be complimented for alerting the local population to what, given the considerable success the Chinese had already enjoyed in predicting earthquakes, was a genuine likelihood.)

Further Elucidation. This condition has to do with the general state of knowledge. New discoveries no doubt bring in new conceptions, e.g. those of polarized light, of spectral lines or algebraic numbers, without which certain kinds of things couldn't be thought of at all. They also facilitate methods of investigation by which problems can be resolved and questions settled, e.g. for calculating the distances of celestial bodies. Our statement of the condition says that the appropriate classification of information is "formulation-specific". Known facts may be assertible under one such classification but not under another. Our observant ancestors, without yet having any idea of polarization, might have noticed and asserted that a hunk of calcite let through different amounts of light depending upon how it was held.

Second Schema: Any assertion indicates as a condition of success that there exist authorities.

Elucidation. For the kind of thing asserted, of course. Authorities may be qualified as trained historians, archaeologists, scientists, mathematicians, medical researchers, members of the family, eye-witnesses, etc. This condition schema also puts dates on statements. No assertion can succeed if all authorities have expired. Others may appear later. Our literary sources are bad on the exact sites of Alexander's battles in the Northwest Frontier area; archaeologists and scholars have been working on the matter and may succeed. This condition schema is independent of the first. That a certain pharaoh died young or old is a kind of thing that can now be known, even though it actually isn't known because no reputable archaeologist, as a prospective authority, had yet been given access to the tomb. The first schema is satisfied but the second isn't. On the other side, a mathematician might be heard truly to say "I could have known that" even though it could not have been known at the time because essential, preliminary lemmas had not yet been established.

Examples:

(i) Police investigators may "assume the worst" about a missing person when no witness has yet come forward. The whereabouts and condition of the person can't be known without some such authority to call upon.

(ii) Suppose that I have true thoughts about today's weather in central Greenland, whence no reports have been received for days; I think this because I mistakenly took a weatherman's forecast for the area as an actual report. I must fail in my assertion of the fact because then I would indicate as a condition of success that there was some such authority when there wasn't any.

Third schema: Any assertion indicates as a condition of success that there (now) exist specific, sometime applicable procedures of verification and falsification.

Elucidations. Since this schema will be the main subject of Chapter 3, a few preliminary elucidations will do for the nonce. First: Application of these procedures is but one however preferred way of coming to know asserted facts. People may know certain facts before or long after possible occasions for applying verification and falsification procedures. This condition-schema is independent of the first two. Second, although occasions of verification and falsification may fall within restricted intervals of time, those times may be indicated at other times and this condition schema does abstract from the occasion of assertion as the other two do not. Currently existing verification and falsification procedures may have been applicable long before anyone was around to apply them. Third, the indication of this condition in the utterance distinguishes assertions from generalizations (which cannot be verified), open existentials (which cannot be falsified), natural laws, and many other constative kinds. There may, however, be other constative kinds, such as computations, which also indicate this condition.

It may be objected at this point that I have defined "assertion" twice over, once in terms of conditions of doing and once in terms of conditions of success. I guard against this charge of duplicity with a formal distinction between "generic" and "non-generic" classifications of action. Assertion is non-generically classifiable as language. Particular assertions also have a "generic purpose" definable by a listing of conditions of success. There is a connection of course, for among the conditions for doing any kind of action are that the subject should believe that genus-defining conditions of success should obtain.

Problem: Do these conditions of success for assertion define a kind of action which could also be non-conventionally done? Examples suggest that that is so. The non-conventional attainment of such purposes would not eventuate in statements.

Now in calling an act an assertion we are sub-classifying it as a form of language and accordingly as being conventional, in our technical understanding of that term; it is, generically speaking, something less than that. In #1.6, at p. 54 we raised the question of whether such generic sub-classifications are in themselves "essentially conventional", and concluded that that is not always so. We must be especially concerned with whether the generic classification of *assertion* secured in our three schemata is of something that needn't be conventional. Otherwise: Could the purpose defined by our schematic conditions of success also be non-conventionally acted for, in the manner of acceptances and refusals (p. 34ff)? The evidence, to be collected just below, suggests to me that that is possible, at least for some orders of assertion. Of course (and to repeat), we wouldn't call an act an *assertion* unless it were conventionally done, just as we wouldn't call a horse a mare unless it were female; that, however, does not affect our question over whether the preferred "generic" classification of assertion, according to "purpose", defined by a list of conditions of success, is ever non-assertionally realizable in action. Two kinds of examples suggest non-conventional possibilities. First: I may let a betrayed wife know by an assertional say so; I may, however, non-linguistically and less bravely secure the same end by leaving another woman's handkerchief in a certain place for my friend to see. The "further purpose" is the same, and it would seem plausible to suppose that the two "means"--the one assertional and linguistic and the other not--are subject to the same conditions. I preface my second example with the question of whether only creatures having the use of language could ever be said to believe the rather fancy sorts of things we have schematized as conditions of success for assertion. There would certainly be that restriction if any of the indicated conditions involved the fact of language itself. (I thank S. Helmreich for this important proviso.)

Dated statements are an important instance, for it looks as if the verification procedures indicated in their assertions presuppose a (conventional) capacity for counting. But other assertions may not require so much in the way of underpinnings. We may still wonder whether conditions of success for assertion are *ever* doxastically accessible to non-linguistic creatures and possibly binding on their (non-linguistic) conduct. My second highly contrived example tries to fit the bill. Canine behavior might sometimes be conventional, but usually is not (see pp. 47f.). Take a non-conventional case. The household dog might give notice of the presence of strangers at the door by cocking an ear and growling. He might do this even when, as he knows, no one is home to hear him. One day he responds in this way to an alien presence. Here it seems to me that he could be said to believe that there may be something there of a knowable kind (strangers) for which there are authorities (his mistress: she could investigate were she home) and "procedures". He could be wrong on any or all of these things. We assume that his growls and ear-cockings do indeed give expression to his beliefs that these conditions are met but that they are not language strictly taken and that here there is no linguistic communication for him to fail at. Maybe it's a case. These two sorts of consideration--that non-conventional action sometimes has the same perlocutionary potentials as assertion and the existence of certain non-conventional expressions of belief--incline me to think that an assertional purpose may sometimes be non-assertionally acted for. The question, however, is still open and interesting.

If I am right in my best guess so far, that there are these non-assertional "expressions of belief", they will not be statement-producing when successful. We made provision for this contingency when we noticed (at p. 30) that measures of success are neither exclusive to nor distinctive of the purposes with which they are associated. Now I do want to hold that statements exist only as products (or, anyway, prospective products) of assertion, but must now allow that what one means to do in asserting may be non-productive and merely "expressive".

Of the conditions we have mentioned, the existence of procedures of verification and falsification will be "most central" for what follows.

The ideas of *truth, evidence, certainty, consequence*, etc., invoked for the formulation of these various conditions are interrelated in their application to statements. *Verification* and *falsification* are central for our purposes. I define statement-*truth* and *consequence* in terms of proof by verification, and other interesting properties of statements will be explained in terms of "criteria" for verification and falsification. We hold to *one* version of the old verificationalist formula, that to understand (identify?) a statement is to know under what conditions that statement would be verified and falsified. It remains that there would be no statement to be verified or falsified unless the other indicated conditions on the state of knowledge and the existence of authorities were satisfied.

We have now explained *assertion* as a kind of utterance, generically fixed by reference to the existing state of knowledge, the existence of authorities and (most centrally) by the existence of procedures of proof and disproof. The formulation of our three condition-schemata is the high-point of this chapter; what follows is downhill commentary.

7. ASSERTION AND TRUTH.

I wish in this section to stop over briefly to consider just one of the "implications of assertion" mentioned on pp. 87f. above, assisted with some back reference to thoughts about the "delimitation of constatives" brought to the surface in #3. We shall be engaged now in a preliminary skirmish over *truth*, with the main attack to come in #5 of Chapter 3.

In asserting a statement, the assertor "certifies' or "endorses" the statement he would produce for its truth.

Some writers (I think of Ramsey and Strawson) have urged and argued that "true", in the only use of that pesky word that would interest a philosopher or a logician, is an assertion sign. Others--and here I think of Frege (*vide* p. 364 fn #30) and of those who like to describe assertions as "truth-claims"--talk as if they thought that every assertion is "of" the truth of something, the "judgement" itself perhaps, or a "proposition" or maybe even a statement. These opposed but equally incorrect doctrines (arguments at pp .323ff.) owe their seductive attractiveness in considerable part to that assertional "commitment to the truth" of the produced statement we adduced on pp. 87f. An assertor who undertakes a commitment to the truth of that statement he would produce "certifies" the truth of that statement, rather as one who labels a can as a can of peas certifies the contents as peas. To change the figure, the speaker "endorses" the produced statement "for truth" or "as true". The "certification" or "endorsement" is binding on the speaker even when the asserted statement is itself false. To keep hold on this matter, we must respect the distinction between the utterance and the product: the utterance conveys a certification of the product; the product of itself certifies nothing; on the other hand, the statement-product is itself true or false, whereas the assertional utterance is true or false at best only in a derivative sense

Endorsement for falsity isn't assertion.

Why cannot one, in asserting a statement, endorse the statement for falsity or not endorse it at all? Wittgenstein's arcane response was, "For an assertion [*Satz*] is true if things are such as we say with the assertion. And if by "p" we mean not-p and things are as we mean, then in this new conception "p" is true and not false."(*Tractatus*, 4.062). But surely we do sometimes say what we think is false, covertly (as in cases of lying) or openly, as when setting a "multiple choice question", which vouches for the falsity of all but one option. I have argued elsewhere that such a "parasitical" utterance would actually fail if the corresponding statement were not false (*Stratification of Behaviour*, pp. 378ff). The short answer to the question, why, in asserting, we must endorse for truth, is simply that that is the kind of thing assertion *is*; other utterance kinds are different.

Wittgenstein, in the *Tractatus*, was concerned with assertion *only*; under that provision and with reservations over the equivalence of "not" and "false", we can agree with what he says, but still complain that he left out exactly what enables us to understand the matter aright, namely, a contrasting case.

Such stipulations about what assertion "is" give meagre satisfaction. Perhaps we can warrant the position as a conclusion from a consideration of the idea of a "commitment to truth" and (going back to the matter of #3) of why, within the repertory of our language, there should be utterance kinds that engender such commitments.

We use "commitment" in at least two related ways[5]: First, one may be committed to *do* something in certain circumstances. Thus one may assume a contractual commitment to deliver goods at a certain time and place. Again, one has a commitment to do as he promised to do. In a second sense, one may be said to have a commitment, if he is in some definite respect liable or answerable. Thus, just in being a parent, one has certain commitments to and on account of one's children. The "commitment to truth" engendered by a successful assertion is a mixture of both. In making a statement one becomes answerable to his interlocutors (if any) for misinformation. That responsibility may be discharged by a demonstration that the statement is true. Derivatively, one assumes a second commitment to support the statement with evidence or proof should his interlocutors demand assurances. And (third) one assumes a commitment to withdraw from making the statement and to acknowledge error in the face of disproof or strong counter-evidence.

The assertional commitment to truth traces to the natural fact that assertion normally subserves the purpose of communicating information to others.

Now an asserted statement is false when there is no asserted fact to be known or to be ignorant of either. One may suppose, as a matter of natural explanation, that, if there were utterance for transmitting knowledge of fact from one person to another, a speaker doing such a deed must stand ready to give assurances that there was indeed such a fact to be known. Assertion, I submit, is precisely such a kind of deed, and that suffices to account for its implicated endorsement of truth. When, as may be, assertion is disconnected from its usual communicative employments, the engendered commitments still stand, in the sense that one cannot be willing to disavow them and still be correctly held to be asserting.

8. SOME IMPLICATIONS OF ASSERTION

The implications of assertion comprise various items that would follow from a true report to the effect that an assertion was undertaken.

In the previous section, we spoke of the "assertional implication" of the assertor's endorsement and certification of the truth of the statement he would produce if his assertion were to succeed. I wish, in this section, to extend this way of speaking about assertional implications to some other, related matters.

Implicative relations of logical consequence among "things said" have long been the subject of fruitful study. I wish now to install a way of speaking about the implications of the *saying* itself, with special attention to assertion. We are concerned with what follows from the fact that an assertion was undertaken.

"Transcendental implications", of which Descartes' "Cogito" is the exemplar, are of this kind. Implications of assertions comprise most if not all that Wittgenstein in his *Tractatus* said was *shown* but not said by a *Satz*. Much of the literature widely read in the English speaking philosophical world of the 1950's and 1960's trafficked in other such implications, and here I mention Strawson's "presuppositions", Hintikka's "doxastic implications" and the "contextual implications" of Nowell-Smith and other writers. There are a largish number of papers on the general subject, some dating from the 1930's, of which a selection are reprinted in chapters four and five of *Philosophy and Analysis* (ed. M. McDonald). These papers concentrate on the implication of speaker-belief. The two most useful systematic attacks on the issue of the implications of assertion I know of are developed in Grice's doctrine of "conversational implicatures" (a special kind of implication of speech) and in Chapter IV of Austin's *How to Do Things With Words*. Grice's objective was to elicit the consequences of a list of plausible "conversational maxims". Austin, differently, provided a useful classification of conditions for utterances of every kind under six general headings, which, in my lingo, have to do with the availability of conventions (A-1 conditions), the satisfaction of conditions of success (A-2), actual and completed performance (B conditions) and correct intentions and beliefs (Gamma-1) and fulfillment of commitments (Gamma-2). The fact of assertion implies the fulfillment of conditions falling under A-1, B and Gamma-1 and furthermore implies that the agent believe that A-2 conditions are satisfied and does not in his act give reason to think that Gamma-2 conditions are not satisfied.

Implications of assertion seem to be a kind of implication of *truth* by *fact*. Confusion! Facts imply nothing in themselves. If they did, then we would have the unacceptable consequence that the same fact under different "expressions" or "descriptions" would have different implications. Any implication here must be by the "description" and not the fact described. It seems, then, that, to make sense of the implications of *saying*, we must repair to the more basic idea of the implications of *things-said*. We introduce the term "implication of assertion" as a *facon de parler* to mean what would be logically implied by a true report that a kind of assertion was done by a particular subject on a particular occasion.

To speak thus of "implications of assertion" affords a short way of saying various complicated things. We shall survey the field quickly and broadly, but first:

Take note that these implications do *not* normally include the asserted statement itself or any of its logical consequences. *Necessary* assertions and (I shall argue) *only* necessary assertions have the feature of assertionally implying the statements produced in them. (see p. 346, in #6 of Chapter 3).

Many "inspecific" implications of assertion, as we have now explained that idea, are presently of no interest to us because equally implied by any act of any kind, e.g. that a live and conscious agent made movements partly by cause of what he believed to be the case. Some implications of assertion are material to the general theory of language but lie outside the area of my immediate concern because having to do mainly with the perlocutionary constraints on conventional performance. (Grice's "conversational implicatures" fall here.) Narrowing it further, there are other truths, having to do with such things as the satisfaction of referential presuppositions, which are inspecifically implicated by the mere fact of utterance, whether specifically assertional or not. (These comprise the cases considered by Austin in *How to Do Things With Words*).

Notable among the specific or "characteristic" implications of assertion is that the assertor should believe what he asserts to be so.

Chief among what we may call the "characteristic implications of assertion" are that the conditions (for doing) of the assertional mood are satisfied. These all figure as "truth-conditions" of an observer's report that the speaker asserted what he did. Of these, the one that has received the most attention is that the agent believe (think) that what he asserts is so[6]. I have elsewhere argued (Shwayder, *ibid*, Part V, #7) that this is so, and that making as if to assert what one does not believe is not properly assertion at all. Assertion, properly taken, cannot be "insincere" in the way that successful acts of promising to do what one does not intend to do are insincere. Assertions on this point are more like conventionalized expressions of pain or acts of cursing someone. Deceptively making as if to endorse for truth is no more an assertion than deceptively making as if to curse is really cursing.

Other truths are implied by the fact that an assertion was *successful*, notably that all the indicated conditions of success are satisfied, in the understanding that no such condition is unsatisfied.

Among the "normal" implications of an assertion, implications the assertor may void by explicit disclaimers, are that he is certain of and would claim to know the fact he asserts

I mention, finally, a range of items which are only *normally* or *usually* or *probably* implied by the fact that speaker asserted what he did. These are implications which can be adverbially voided without tumbling into assertional failure and without transfiguring assertion into another constative sort. *Failing notice to the contrary* the speaker's interlocutors are entitled to take certain things to be true regarding him and his circumstances. One may, however, serve notice to the contrary without imperiling the success of the utterance. A speaker may limit or lessen his commitment while still certifying or endorsing the statement for truth; but, if he does not indicate such a limitation, the "normal commitment" is "assumed".

Among such "normal" implications are that an F-asserting speaker is certain of F and that he would claim to know F. Both want comment.

A speaker who asserts F and is also certain or sure of F need not and normally will not have made certain and he need have no strong feelings of certainty, possibly expressed by his affixing an "I feel (or "just feel") certain" (locutions that actually may suggest uncertainty). Prior certainty in the relevant sense is evidenced by the speaker's answer to the question "Are you certain?", "Yes, I was". Speaker is certain of what would be formulated by a statement *S* when he would assert *S* and knows no reason to doubt the truth of *S*. Since, in asserting, the speaker shows what he means to assert, it is additionally normally implied that he knows of no reason to doubt, unless he does something of a kind which would be elicited by a doubt-raising question. If subject does acknowledge such a challenge--"Oh, of course, he *could* recover, but almost certainly he'll die within the day"--subject still asserts the same statement, although now perhaps the utterance is tinged with uncertainty.

An F-asserting speaker also normally implies that he would claim to know F, *viz* we are entitled to take it, failing notice to the contrary, that he would, if pressed, say "I know F" and defend that claim. In making such a claim, one does not merely imply that he thinks he knows, for he already does that just in asserting F. In claiming to know he would additionally undertake to authorize his interlocutors to assert F (See "Appendix C", pp. 195f.). He may, in asserting F, refuse to convey such an authorization, although he may very well be quite certain that he knows, in the sense of having acquired and retained information. He may, that is, for one reason or another, wish to quash the implicit claim to know, and he can do so without forfeiting success in assertion. A challenge may cause him to flutter; he may, for one thing, without in any way confessing ignorance, wish to abjure on the ground that the credentials he would advance in support of the claim are less than those of his interlocutors. (I have blushed to be heard saying "I know" within earshot of superior authorities.)

The normal implications of certainty and of a claim to know are distinct. Doubts are raised and the speaker is for that reason a bit uncertain, but still he may correctly say, "Don't worry, I know that he'll be here tonight." Conversely, one who is perfectly certain of a fact, may still refrain from authorizing another to assert that fact. I am, for example, quite certain that protons are more massive than electrons, but I should never think to authorize an assertion of that fact by one whom I knew to be a trained physicist. "How do *you* know that? Well, *I* can't know, but protons *are* more massive than electrons, aren't they?"

9. ADVERBIAL MODIFICATION OF ASSERTION.

Some of the movements constituting an act of any kind may serve to conventionally modify *how* the speaker does what he means to do , that "modification" being achieved by his indicating conditions that are not conditions of success for that act. I call these "act-adverbial modifications". These modifications are "meaningful"; they are sometimes explicitly conveyed by such adverbs as "regretfully", but also by all manner of other expressions, as we shall see. Actions of most kinds are subject to act adverbial modification: one may just as "deferentially" open a door as ask a favor. (Frege disparaged this kind of meaning as "color" and "aroma"; the relegation was appropriate within the circle of his concerns, which were with truth and not with language.)

We are of course going to be mainly interested in the adverbial modification of assertion. The explicit disclaimers of certainty and of epistemic authority we have just discussed fall here, for they serve to nullify the normal implications of assertion without transfiguring the act in any essential way or altering "illocutionary force". One might, in answer to or to anticipate a challenge, also underscore the normal implication by affixing an "I am sure", "Well, I know" or "It *is* certain that". One would, in such ways, hesitantly or modestly, confidently or authoritatively assert what he does. One might, again, "conditionally", "concessively", "contrastingly", "as a conclusion", "as a matter of course", "regretfully", "apologetically", or "politely", assert, and to those ends employ such expressions as "I think", "if", "unless", "Well, yes...", "However", "but", "therefore", "and so we see", "of course", "I'm sorry to say", "It would seem", among many others. Such locutions are all sometime-employed to effect act-adverbial modification upon assertion.

Assertions may be "adverbially modified" in various ways by the assertor's employing expressions to indicate conditions that are not conditions of success for the assertion itself.

A much discussed instance of act adverbial modification occurs when a speaker annexes to his assertion of a fact F the little qualifier "I believe" Certainly, one may asseverate his opinions by insistently declaiming, "Well, I believe F". But that would be something other than an assertion of F. When one does qualify an assertion by saying "I believe F", then the resulting statement is seldom if ever to the effect that the speaker is in the state of mind of believing F. He asserts *what* he believes, not *that* he believes what he believes. This use of "I believe" adds nothing to the statement produced but serves only to disclaim the normal implication of certainty, or to abjure the implicit claim to knowledge, or to soften the appearance of dogmatism, or such like. Wittgenstein once put the point in a wontedly succinct way when he admonished us not to confuse a hesitant assertion with an assertion of hesitancy (*Philosophical Investigations*, Part II, x). These observations provide an easy solution to the problem sometimes called "Moore's Paradox". If "I believe that"^Sen were used to assert a statement about one's state of mind, as verbal appearances might suggest, then there should be nothing incoherent, as in fact there is, in the same speaker's concurrently saying "It's not the case that"^Sen. Why then does it seem so wrong for me to say "I think he'll be here, but he won't"? The answer is, simply, that when "I believe" occurs, as it normally does, in act-adverbial usage, "I believe that"^Sen would produce

The presence of "I believe" in the flow of assertional utterance is "act adverbial". That observation solves Moore's Paradox

the same statement as would be produced by simply saying Sen and that would stand in flat contradiction to what would be asserted by saying "It is not the case that"^Sen. This appeal to the adverbial status of "I believe" for the resolution of Moore's Paradox goes to a point different from that marked in our earlier conclusion that belief is an implication of assertion. It may be, however, that what *explains* the fact that "I believe" seldom adds anything new to the statement is that the flat assertion itself already carries that implication of belief. There is nothing of that kind left to be said by the use of "I believe", so it is used for something else instead.

Urmson's "Parenthetical Verbs". The general subject of adverbial modification, recently under investigation by a number of writers, was, to my knowledge, first seriously broached by J. O. Urmson in his classical paper "Parenthetical Verbs" (*Mind*, 1950). Urmson's contribution remains "best reading" on the subject at least for the issues of this chapter. Urmson justly observed a similarity in the usage of such first person, present tense, perfective aspect verbal forms as "I believe", with the usage of such grammatically designated adverbs and conjunctions as "admittedly", "however", "therefore", "if...then" and constructions of the "it is" ("probable", "certain"), and "if you" ("remember", "wish") forms. Urmson contrasted the possibly assertional usage of "He believes" ("thinks"), and "I believed" ("thought") with the common "parenthetical" usage of "I believe". He drew the important conclusion that, if one does, in the course of asserting a fact F, adverbially say "I believe", then the resulting statement is almost never to the effect that the speaker is in a state of mind of believing F.

While Urmson concentrated on the adverbial modifications prompted by the "evidential situation", he was quick to observe that one might also wish to adapt one's manner of assertion to the "logical" or "social situation".

10. ATTENUATIONS OF ASSERTION.

Adverbial modifications of constative utterance may by hybridized into or serve to nullify indications of conditions of success with resulting alterations of mood.

In our remarks of the previous section on the use of "I believe" we noticed that that phrase may sometimes be employed, not act-adverbially in the context of assertion, but rather to express a non-assertional constative mood or illocutionary force, e.g. to express an opinion or "performatively" to put oneself forward as a believer. This kind of adaptation and alteration in mood is sometimes achieved by the hybridization of an act-adverbial use, *viz* by the conversion of indicated conditions that normally bear on how a subject does what it means to do into conditions of success that affect the determination of what it means to do. In other cases, a modification of mood comes about by a kind of nullification of indicated conditions of success . In any event, the phenomenon is quite common within the realms of constative utterance, where adverbial uses readily thus transform into mood indications and also into other uses that may contribute to asserted statements. Thus (changing the case) "if" is employed not only for purposes of conditional assertion but also to propound conditionals and to assert conditional statements (see pp. 146-150, 383f.).

In this final section of Chapter 2, I'd like to draw attention to some cases of nullification of indicated conditions. Utterances which result from transformation of adverbial modifications into specific mood indications--of hesitant assertions, say, into expressions of opinion--commonly stand in for assertions that would have been enacted in other circumstances. The occasion is ripe for assertion, and that is clear to all. However, considerations of the "evidential, logical or social situation" (Urmson) which would normally induce caution or hesitation, may, in the speaker's estimate, call for greater adjustments than could be achieved by mere adverbial modification. These considerations take on illocutionary significance. A subject who doesn't think he knows something may, instead of asserting, simply express what he thinks. A mathematician, for example, who advises a student that such and such still unproven proposition, which he believes is true, ought to be investigated, is in a quite different position from the one he occupied the day before when he suggested that a certain result should follow from such and such proposition of whose recent proof he thinks he recalls reading in a late number of the "Abstracts". In both cases he is somewhat unsure of his ground, perhaps more unsure in the latter than in the former, but, while he asserts in the latter case, he hazards only an opinion in the former. In the latter case, the mathematician speaks as he does because he thinks he could be wrong about what he thinks he knows is so; in the former, he doubts that what he thinks can yet be known. Here then is one kind of case of non-assertional saying and meaning what one thinks is so There are others, e.g. *guessing* is saying what one thinks but knows he does not know, while allowing that others might.

"Attenuations of assertion" are utterances like assertions except for there being a finite number of conditions of success for assertion they do not indicate

Expressing an opinion and *guessing* are kinds of what I call "attenuations of assertion". In general characterization, an attenuation of assertion results when a speaker finds himself in circumstances where he would assert if he could, but yet thinks he cannot because of the known or supposed failure of some specific condition of success for assertion, e.g. that this sort of thing could be known. Our imagined mathematician by his own authority must be familiar with the limits of mathematical assertion, and therewith knows that he would overreach his authority were he to assert what he thinks; so he does not even try. But he must say what he thinks; so he proposes a substitute for the statement he would like to make. He lends his authority, not to truth flatly taken, but to a program of further investigation.

Attenuations of assertion are so-called because they borrow from assertion by specifically *not* indicating questionable conditions of success for assertion. The attenuation is commonly marked by the employment of an appropriate mood indicator such as "I "guess" ("suppose", "think"). Formally, an attenuation of assertion is an utterance of kind K, where

(i) Among K's conditions of success are the conditions of success for an assertional kind, A, save for a finite number of specifically excluded conditions

and

(ii) A condition (for doing) characteristic of the K-purpose is that, if the speaker had believed that the excluded conditions of success for A were satisfied, he would have done an A.

And for their having the additional condition for doing that subject would have asserted had he thought that the un indicated conditions were satisfied.

In this definition, I have deliberately left clause (i) open to allow for the possibility of annexing conditions as well as for excluding them. *Conjecturing* and *predicting* are such cases because they propoundingly indicate the non-observation of something. I am, however, strongly inclined to remove these more complicated deviations from assertion into a number of separated categories. Take utterances schematically formulated in English with sentences of the form "It may be that"^Sen, where Sen could be used to assert. Such "declarations of possibility", which I would gloss as cases of conventionally refraining from asserting *not*-p, do not indicate that a fact formulated in Sen could be known; but they also do indicate the additional second order condition that there is reason for not asserting *not*-p (see pp. 155). The latter annexation puts declarations of possibility at a distance from expressions of opinion and into closer proximity to conditionals, which are not necessarily attenuations at all.

On proving that a subject would have asserted if he could.

The second clause, that the speaker would have asserted if he could have, may seem hopelessly vague. I allow that it is "open", but I don't think that it is inadmissible. Our imagined mathematician, upon being informed that the hoped for result had already been published by Hilbertsky, would, having confidence in his confrere's sound reputation, immediately proceed to assert what he had previously only expressed an opinion about, thereby showing that his earlier attempt had indeed been an attenuation of assertion.

The natural history of attenuation is fairly evident though certainly diverse. In medical consultation, a doctor may *give his opinion* whether surgery is advisable; or a scientist may *propose* a plan for further research. Boldly to essay statements in such circumstances would reveal misunderstanding of the situation or of one's role in it. Again, one may withhold assertion because of certain ruling presumptions about the embedding activity, e.g. those that govern medical practice, the running of horse races or the conduct of mathematical research. Mathematics, regarded as a method, is devoted to explanation by demonstration and to a systematic, "linear", "deductive" organization of what is known.

What explains the existence of assertion-attenuations and of other stand-in's for assertion is the perlocutionary need to say something epistemically useful in recurrent circumstances which are epistemically deficient.

Undemonstrated statements would not serve the mathematical end of demonstrative explanation or underived ones find a place within the deductive organization of knowledge. A fact is not mathematically known if neither demonstrated nor derived. If the utterance product, by its very terms, is about mathematical goods, then demonstration or derivation becomes a condition for knowledge, and, in the supposed absence of either, one is not entitled to an assertion and may have to make do with a guess.

NOTES

[1]Or were. I've recently heard that the classification of guinea pigs is in dispute.

[2]With thanks to Tim Griffin

[3]Appendix B takes preliminary steps in that direction. For a rare anticipation, see W. E. Johnson 's *Logic*, Vol I, p.6, n2.

[4]Because assertions that the cat is on the mat and that the mat is under the cat are of the same fact despite having interchanged subjects.

[5]Putting aside that other sense in which one is "committed to" what he believes in.

[6]The question whether there is such an implication should be distinguished from Moore's Question, to be discussed in the next section, over the relation between assertions of F and assertions schematizable as "I believe F". Some have mistakenly supposed that the incoherence of "F but I do not believe F" owes to the fact that the assertion of F assertionally implies that the agent believes F.

APPENDIX A

FOUR OTHER THEORIES OF JUDGEMENT: RUSSELL, ARISTOTLE, WITTGENSTEIN AND FREGE.

I. ***Introduction***. Our account of assertion is a "theory of judgement" that descends from an ancestry reaching back to Plato, but which differs in assignable ways from those of its predecessors of which I have any knowledge. The main lines of my proposals for assertion may perhaps be better seen when set in implicit contrast with some alternatives. To that end, I shall, in this appendix, offer interpretative summaries, with bits of criticism occasionally thrown in, of theories of judgement from Russell, from Aristotle-cum-Plato, from Wittgenstein's *Tractatus* and from Frege, with some passing mention of Locke and Leibniz.

These several doctrines focus on truth and falsity. Now different kinds of things may be "true or false" in an understanding relevant to our inquiry--that understanding to exclude *true friends* and *true shots*--including thoughts, opinions, judgements and (equivocally) "things said". I believe (but cannot establish) that the conceptually "most basic" (but certainly not genetically most primitive) application of this understanding of "true-or-false" is to "things said", whatever they may be. To obviate inconclusive debate over priorities, I shall simply stipulate that we shall limit ourselves to the truth-or-falsity of *things said*, whatever they may be.

A judgement, for present purposes, is actual utterance, formulated in words or other "expressions".

I have urged that utterance is behaviorally distinctive, being, as I have held (pp. 12f, 47), "essentially conventional" in a special sense. I believe that most traditional writers, if questioned on this, would dissent from this opinion. Leibniz, for example, would probably have held that judgement differs from other "expressions of thought" only inessentially by its greater degree of articulateness and apperceptivity. It mattered to Russell that he should have been able to describe judgement in terms that required only incidental mention of language and no mention at all of "meaning"; for him any verbalization of judgement is the merest trappings. Locke, Frege and Wittgenstein would *perhaps* have favored my side; if so, that is an important opposition within the tradition. My opinion, (which I think differs from all other previous writers') is that those "things said" about whose truth or falsity judgements raise a question are not themselves those judgements; they do, however, depend for their existence upon assertoric language and are not to be thought of at all except in relation to a conception of language.

Theories of formulated judgement differ saliently over the nature of that "something" of whose truth or falsity a formulated judgement raises a question. Call that "something" a "truth-vehicle".

Now some theorists have indeed held that the judgement itself is the designated truth-vehicle. That was Russell's opinion. The same preference is featured in Wittgenstein's *Tractatus*, as I read that difficult work. Wittgenstein's *Saetze* are at once the designated vehicle (see *Tractatus*, 4.06) and verbalized thinkings or representations that such and such is the case (3.1, 4.). Wittgenstein's view differed from Russell's in demanding that such representations should also "show" or "present" (*zeigen, darstellen*) a possibility (*Sachlage*, 2.202, 4.031) or sense (*Sinn*, 2.221, 4.022, 4.031).

Others have opted for judgement-*type*s as their preferred vehicle for truth and falsity. That must have been the opinion of Plato and Aristotle who allowed that the same one affirmation or denial may be true in one instance and false in another, e.g. "The moon is new". The judgement-type, according to this view of the matter, is (presumably) to be identified from some string of words in which it could be formulated, hence in abstraction from actual cackle.

Those several doctrines that docket judgements or judgement-types as truth-vehicles invariably cast these judgements or types as being compositions of constituents and as being true or false according to how those constituents are combined. Such "combination-theories" of judgement may be austerely "one layered", as in Russell, "two-layered" in respect of instances and types, as in Aristotle or (very differently) "two-layered" in respect of Sense and *Satz*, as in Wittgenstein.

Other doctrines hold that the designated truth-vehicle is some such "second layer". That, I submit, is the case with Locke who averred that only "propositions" consisting of "ideas" in combination could be properly said to be true or false (*Essay*, II.xxxii). Frege, also, I believe, would have held that it is not the judgement itself ("*Urteil*") that is true or false but rather its "content"--a *Sinn* or *Gedanke*--that both determines and carries truth-value.

II. ***Russell's theory of judgement***, first published in 1906 ("The Nature of Truth", *Mind*, 1906, pp. 528-33, reprinted in *Philosophical Essays*), tied together all of his thinking about philosophical and logical matters through at least 1912, when it was meshed with his other doctrine of the distinction between knowledge by acquaintance and knowledge by description, a connection that had been anticipated in "On Denoting"(1905). The theory in pristine appearance--it was consequentially amended in the *Lectures on Logical Atomism*--is our literature's starkest, simplest and most striking version of an account of judgement as composition, and it differs from all the others by having no truck with representational elements. The theory certainly was advanced as an alternative to Frege's doctrine of *Sinn* and propositional content, which Russell had criticized in "On Denoting". Frege held that the "content" of a judgement is a complex "sense" sufficient to determine a truth-value, provided that the sense-constituents of the complex each determined a *Bedeutung*. Russell's unconvincing argument in "On Denoting" was calculated to show that there can be no proper determination of *Bedeutungen* by sense. Russell's competing account of "propositional attitudes" sternly eschewed propositions and appealed for constituency only to the fregean *Bedeutungen* themselves. Now Russell was not against dictionaries. He was properly indifferent to what he readily allowed, that the words we use in formulating judgements have meaning. He denied rather that these words attached to the objects of judgement by way of intervening ideas or meanings. Consider that the objects of judgement can be separately indicated by a sequence of "This"s all indistinguishable in meaning--as indeed is often nearly the case in everyday speech ("This was on this in this way."). The whole fact of the matter, as Russell saw it, is the judgement itself, directly constituted of the judger in relation to various items of its acquaintance. Frege held that judgement entailed a doxastically uncommitted "grasping" of sense. Russell allowed for as much and accordingly shifted his attention from judgement to *supposition.* But then supposition is just another true-or-false relationship between a subject and items of its acquaintance, and stands no more in need of propositions than does judgement. Russell's doctrine, to repeat, is compositional and designedly without "meanings". The theory, when turned to the analysis of formulated judgement or supposition, made a first "flight from intensions" (Quine's phrase) and is a classical expression of an "extensionalist" philosophy of language. The largest question about the theory is whether it succeeds in its extensionalist aspirations. I'll argue that it does not; but first the details.

Judgement in Russell's theory is a variably polyadic relationship between a subject and various items of its acquaintance; supposition is another such relationship. Russell said little by way of explicating these and other "propositional attitudes". He may have supposed that that was a task for psychologists.

The related objects of acquaintance must comprise at least one universal. There's no bar to them all being universals, as perhaps occurs when one judges that blue is a primary color.

Relationships of these kinds are simply facts consisting of the several relata bound together by the relation of judgement.

Relationships of these kinds may be true or false. They are true just in cases where there is another fact in which all the relata constituting the judgement save the subject and a designated universal are bound together by that designated universal. If the judgement exists without there being any such other fact, the judgement is false.

It is unclear whether the demand that the subject should have *knowledge by acquaintance* with the other relata implicates intensions. Fregeans may suspect that it does. (Frege said little more about his *Sinne* than that they were "Erkentnisswerte" see Chapter 3, p. 306.) Russell would have asseverated to the contrary, that knowledge by acquaintance is simply a dyadic relation that obtains between subjects and other things. There's much to argue about here, and the issue is still open (see pp. 188-191 of Appendix C). So let's put this objection aside and move to another.

Russell required that there be a universal standing ready among the relata to bind the other constituents into a truth-conferring fact. Some will suspect that these universals are Platonic Forms and "intensions", invoked (in Plato's phrase) "for the sake of all discourse". Russell asseverated to the contrary, that the universal is simply an object conceivable as a sometime constituent of concrete fact. Again, I am not ready to resolve this dispute, so let us leave it for another objection.

Now, by Russell's theory, in every formulated judgement exactly one of the constituent universals is designated as predicate or "relating relation" (Russell's term). Call that the "designated universal". Russell was prompt to allow that the very same universal could, in another formulated judgement, occur as an undesignated "unrelating" subject.

> This I think is the background to an objection Wittgenstein made but didn't explain when he declared that Russell's theory fails because it must be impossible to judge nonsense (*Tractatus*, 5.5422). The needed explanation is supplied in Wittgenstein's *Notebooks* when he argued that, while Socrates is human makes sense, Humanity is Socrates does not.[1] I didn't see the point of these remarks of Wittgenstein until I had already written out the objection soon to follow. But first take note that Wittgenstein, in his own theory of judgement, homogenized his style of

constituents (which are intensions) and denied that there is any real distinction between subject and predicate.

I think that Russell was right to insist upon a distinction between subject and predicate, although it is remarkable that he did so; I also think that the need for some such distinction results in the downfall of his "extensionalism". That is because *subject* and *predicate* are notions that have primary application to *language* and accordingly that Russell's requirement that exactly one of the constituent universals be "designated" implicates "intensions".

Consider: If we hold, with Russell, that predicate-universals are (in my term) "designated" and that subject-universals are not and also that (e.g.) the same *white* appears as predicate in *Snow is white* and as subject in *White is a neutral color*, how is that distinction to be made out? Since *white* (that object) is the same in both judgements, it does not occur (*a la* Frege) as an incomplete object when designated and complete when undesignated. The difference, as it seems to me, can only be marked either by the presence of a "copulative tie" (Strawson's term) attaching to designated universals or by some kind of type-restriction on the different argument places in the judgement, e.g. *neutral color* is a type of universal that occurs as predicable only with other universals of a certain type; *white* is a universal that occurs as predicable only with visible things. A copulative tie is not an object in a languageless world or a feature of language-independent commodities; it can only be an "intensional constituent". Type restrictions on arguments to a relation are equally intensional. Non-intensional constituents of judgement are given their places, as subject or predicate, by dint of the judger's familiarity with such "logical forms" as that of subject and predicate, and these "forms" are nothing if not intensional. There must be something of this ilk to relate a judger to the objects of his judgement. (There needn't be anything otherwordly about all this. In my own scheme, the conceptualizing agent secures the wanted relationship procedurally, and I see nothing to argue against procedures being behavioral and physical.)

Aftermath. Russell apparently took Wittgenstein's criticisms to heart. In material still unpublished when I was composing the above[2], Russell conceded that a judger must also have background knowledge by acquaintance with logical forms that do not figure as constituents of judgement. He also held that this kind of knowledge is available only to advanced thinkers. The "advancement", I suggest, must be to the level of conceptualization; the logical forms that a knowing subject needs in order to pattern objects into such logical molds as subject and predicate can be no less intensional than Frege's *Sinne*.

Historical reflection. This development in Russell's thinking recalls a crux in the classical debate over innate ideas. Russell's logical forms are included among the *syncategoremata* of classical metaphysics. These ideas figure indispensably in those necessary maxims, such as the law of identity, of which both Locke and Leibniz held we have intuitive knowledge. Locke argued, and Russell echoed him, that these ideas are among the last to come, hence not innate. Leibniz opposedly contended that without these ideas ready to hand a subject would be incapable of distinguishing other ideas. He described them as the sinews of thought. Russell's drift toward Locke gave Wittgenstein no satisfaction at all; he agreed with Leibniz that these forms are a-priori features of our representational scheme itself, implicated as "categories" in anything we might think or say about anything. He thence proceeded to the unleibnizian "fundamental thought" that these features can be neither represented nor representing things and that the "logical constants" that mark them stand for nothing.

III. ***Aristotle's*** scattered remarks on affirmation and denial must, I reckon, have been a reflection upon Plato's account of discourse[3]. Both philosophers allowed both thought and discourse as truth-vehicles. Aristotle, however, reversed Plato in giving pride of place to thought over discourse. To simplify our examination, I confine attention to expressed affirmations and denials or "discourse". I know no text attributed to either philosopher that speaks explicitly on the question of whether discourse is "distinctive" in the sense of pp. 46 et. seq. My gut-feeling is that both would have denied it[4].

Bits of discourse in the relevant sense are representations that raise questions of the truth or falsity of something. Such representations must be complex and unified. Aristotle (following Plato) accordingly held that items of discourse are complex representations put together from less complex representations. Note well that the constituency of discourse are *representations* and not (as in Russell) objects represented in discourse. Both concluded that discourse requires the existence of semantic imcomposites. (Plato; in the *Theaetetus*, found this "problematic"; Aristotle, in "Theta-10", took it as simply obvious.) Incomposite representations of themselves raise no question of truth or falsity. They are, however, in Aristotle, based upon a discoursing subject's knowledge-by-contact with a represented something. These theories do not seem to require that objects represented by incomposite representations should themselves be incomposite. Indeed, in Aristotle, the represented items are primarily if not always complex substances. Both authors held that discourse invariably incorporates constituents of at least two distinct kinds called "names" and "verbs". The texts suggest but don't actually say that verbs are for the sake of unifying the complex (as later in Frege and in Russell's *Principles of Mathematics*).

Aristotle distinguished both affirmations and denials into universals, partials and what we may call singulars. In singular discourse, the constituting name and verb are "subject" and "predicate". Aristotle seems to have denied that there was any "ontological" distinction between the individuals represented by the two complementary factors. He also says things that imply that only "universals" are predicated; from this it seems to follow that the distinction between particulars and universals, in the aristotelian development, is altogether a distinction "in meaning".

Both Plato and Aristotle thought that there are bits of discourse that are both sometimes true and sometimes false. They also seemed agreed that such variable judgements express only opinion and not knowledge. It seems from this that their truth-vehicles are not actual utterances but utterance-"types" fixed in some appropriate semantic manner. "The moon is new tonight" is such a bit of discourse, usually false but sometimes true.

For Aristotle, affirmations and denials, equally and alike, are combinations of terms. The difference is that affirmations represent combinations and denials represent separations. I suspect that Aristotle's explanation of denial as separation borrows from Plato's *Sophist* explanation of our conception of what is not so as a positive conception of distinctness in regard to existing things. Denials are representations that individual things are separate and distinct. Concentrating now on singular representations: Affirmations represent that the predicate-represented something is combined with the subject-represented something and denials that they are separate and distinct. There are three cases: (a) Substantial predication of substances (e.g. "That's a man or "That's not a man"), (b) Accidental predications of substances (e.g. "It sits" or "does not sit"), (c) Accidental predications of accidents (e.g. "The sitter weighs 200 lbs")[5].

Affirmational predications of accidents to substances represent that an individual is inherent in or combined into a substance. Affirmational predications of accidents to accidents represent that two inherent individuals are combined together in a single substance. Corresponding denials are trickier. I think that Aristotle held that denials represent that the predicated inherence is literally separate from an indicated substance. A problem here is that a denial, seemingly at least, may be false when there is no such inherence anywhere, and the theory accordingly loses one of the main advantages claimed by Plato in his account of non-being as distinctness; the same objection afflicts Aristotle's formula for the falsity of accidental affirmation.

It is not immediately obvious from the texts how Aristotle would have explained the primary case of singular substantial predication These predications say what a substance *is* or *is not*. Aristotle seems to have held that a determination of what a substance "is" implicates a determination of its identity. However, it is doubtful that he would have thought that the "is" of substantial predication is nothing other than the "is" of identity: Plato after all had already distinguished the "is" of predication from the "is" of equation; besides it is obvious that distinct substances may be said to be or not to be of the same substantial-species. I think that, for Aristotle, substantial predication says what something is *by name*. Aristotle seems to be saying almost just that in his remark that, of things which are essences and actualities, it is not possible to be in error only to know them or not to know them (*Metaphysics*, "Theta 10"). But surely he would have to allow that one could be wrong in saying what a substance is or is not, e.g. in my saying that a hyena is or that a chihuahua isn't a dog. It seems, moreover, that a discourser who outs with a substantial predication must already know that name and that such naming is accordingly more fundamental than predication. I reckon that Aristotle is assisted in this matter by some selective borrowing from Plato. Plato may be read to have held that affirmative predications are true just in case a referred-to object matches a named standard. Aristotle certainly allowed that, in cases of substantial predication, the predicate and the subject differ in their manner of *meaning*. The meaning of the subject term, surely, is to pick out or to refer to a substance. The meaning of the predicate could then reasonably be cast as an indication that another substance of "this name" (e.g. "dog") was to be taken as an object of comparison. An affirmative substantial predication could then be construed to say that the two indicated substances are "the same in name" and the denial to say that they are different in name. False substantial predication, whether affirmation or denial, must owe to the discourser's defective knowledge of what is in the name. Some such analysis may also be adapted to the case of singular accidental predication. Accidental predicates (e.g. "is white") also tell us what corresponding inherences *are* (e.g. *a white* , see p. 65, fn 2) and may, accordingly, be used as names for those individuals (e.g. the whiteness of this dog). The affirmation of a singular accidental predication says that there is an inherence in a substance which matches (or in respect of whose presence the substance matches) an exemplar named by the predicate.

Now a serviceable way of explaining what it is for something to *be* true or to be false is to say under what conditions it would be true to affirm that something is true or is false. (as Aristotle puts it, "We must consider what is meant by these terms", "Theta 10", 1051[b]). Both Plato and Aristotle cast truth as a kind of *Being* and falsity as a kind of *non-Being*. Aristotle was quite explicit that this *genre* of Being-with-non-Being are derivative and dependent. They pertain, not to things, but only to (complex) thought or discourse. The truth or falsity of affirmation or denial owes to how things are in fact connected and (fatalists take note) there is no determination in reverse. Aristotle's doctrine was a kind of "correspondence theory of truth". An accidental singular predicative affirmation or denial is true just when the represented individuals are combined or separated as they are represented to be and false if they are separated when affirmed to be combined or combined when denyingly represented as separated. Substantial predications would, in my reading of Aristotle, be true if a represented substance matched a named exemplar and false if it didn't, with the formula reversed for denial.

Such formulas, notoriously, produce an impression of futility, which I think may be overcome by more delicate and accurate formulations that make it obvious that affirmations and denials are assessed for truth value in relation to some particular substance.

(i) Substantial singular predication, affirmation: Of a substance, s, Af(S,P) is true (false) just in case S represents s and there is an s' (there is no s') such that P names s' and s matches s'.

(ii) Substantial singular predication, denial: reverse "true" and "false" in (i).

(iii) Accidental singular predication, true affirmation: Of a substance, s, Af(S,P) is true just in case S represents s and there is an individual, i, such that P names i and i inheres in s.

(iv) Accidental singular predication, false affirmation: Of a substance, s, Af(S,P) is false just in case S represents s and for every i and i' if P names i and i' is an inherence in s, then i is distinct from i'.

With appropriate interchanges of terms, (iii) is a formula for false denial and (iv) a formula for true denial. The formulas for accidental predications of accidents are more complicated but obvious.

This "aristotelian theory of judgement" is subtle and less vulnerable than both Russell's and Wittgenstein's "compositional" theories of judgement, but not altogether free of difficulty. I'm not clear how Aristotle would or could deal with vacuous predicates (e.g. "goat-stag") or how *distinctness*, *all-ness* and other *syncategoremata* fit in as constituents of judgement. Still, these problems do not yet amount to refutation; I continue to find the doctrine attractive, and second-best to my own.

IV.. ***Wittgenstein***, in the *Tractatus*[6], used "*Satz*" as a term for a judgement (*Bejajung*, *behaupten*, 4.064,4.21) that such and such is so (4.022); *Saetze*, in the *Tractatus*, are thoughts (*Gedanken*) expressed in signs (3.1,4.); thoughts are pictures (*Bilder*, 3.) and pictures representations of reality (...*bildet die Wirklichkeit ab*...2.201)

A *Satz* is a fact (3.14, 3.141, 5.542)--full-bodied and unattenuated, therefore not a propositional "content"--the fact, namely, of a metaphysical subject's (4.022, 5.641) thinking, picturing, representing or asserting what it thinks is so. A thinker cannot think (say etc.) that such and such is so without "showing" (*zeigen*) what it thinks to be so (4.022). It shows what it thinks to be so only by "presenting" (*darstellen*) a possibility (2.203) or a "sense" (*Sinn*, 2.221). One might put this, not altogether clearly or quite accurately, by saying that the judgement (thought, representation) represents a displayed possibility as fact. ("Show" in the *Tractatus* is of *sense* and resembles my "indication of conditions". pp. 66, fn 8) Consider, as an analogy, a situation in which a presented person is represented as the long-lost father to a once abandoned child. In the analogy, the presented object exists independently of the representation. *Satz*-presented possibilities or senses, by contrast, are only shown and shown only in the fact of assertoric judgement or thought. A picture or a thought, note well, is not a presented sense possibly *used* for purposes of assertion; a picture is a full-bodied fact and a representation that a presented such and such is so.

The fact of judgement consists of a subject putting together (4.031) and presenting a possibility as a representation of reality. A thought is a metaphysical subject's "modeling" (*Bild*, *Modell*, 2.12, 4.01) of reality as he thinks it to be.

In *Elementarsaetze* the fact of judgement consists entirely of names (3.202). Those names "reach out to and touch" (2.1511) simple objects; they may be thought of as formulated references to those objects. Wittgenstein's "picture theory" is another compositional theory of judgement. But here neither the judger nor the referred-to objects are comprised among the constituency of judgement. Wittgenstein's complexes have nothing in common with Russell's. Wittgenstein's names are like fregean senses. (This interpretation of *Namen* as senses is supported by the doctrine of *Ausdrueken* developed in section 3.3-3.318, esp. 3.314.) However, Wittgenstein's names (which are also objects), unlike fregean senses, would not exist if their *Bedeutungen* didn't. It is also to be carefully noticed that Wittgenstein's names figure directly as constituents of judgement and not as fregean constituents of judgeable content.

The possibility or sense presented in an *Elementarsatz* is a "*Sachverhalt*" (4.21). A *Sachverhalt* is a possibility of referred-to objects being in a relationship. These "possibilities" are not facts in the world. (If they were, they would be in "every possible world".) The whole fact of the matter of elementary judgement is the subject's representation of a presented *Sachverhalt* as being (if you wish) a fact. The fact of elementary judgement is the thought that a presented *Sachverhalt* actually obtains (4.21). If the referred-to objects do stand in the presented (shown) possible relationship, the representation (picture, thought, *Satz*, assertion) is true; otherwise, false. Here and continuingly in the extension of the doctrine to truth-functional judgements, it is always the full-bodied *Satz* that is true or false and not the shown or presented sense or possibility. Wittgenstein's tractarian metaphysic was "leibnizian" and his theory of judgement hearkens back to Leibniz' way of thinking about "expression". The "picture theory" has compelling abstract charm, but it lacks reality and was more vulnerable to critical probes than Aristotle's theory, and less true[7].

V. For ***Frege***[8], judgement (*Urteil*) is a kind of action. Every such act has a content" called a "thought" (*Gedanke*). Judgement is both perceptible and psychological; thoughts are neither, but inhabit a "third realm". The thought-content of a judgement is the sense (*Sinn*) expressed by the sentence used to formulate the judgement. That sense is a function of the senses of the expressions constituting the sentence. Senses are "supposed to" determine *Bedeutungen.* A thought that is at once the content of a judgement and the sense of the sentence used to formulate the judgement determines a *Bedeutung* just in cases where the senses of the expressions constituting the sentence all determine *Bedeutungen.* The *Bedeutungen* (if any) of the senses of sentences are truth-values[9]. The content-thought of a judgement is, by this, supposed to determine truth or falsity as a *Bedeutung*, and that thought is said to be true or false according to this determination. An act of judgement at once raises a question whether its thought-content is true or is false and takes a stand on the side of its being true. Frege described judgement as being a "recognition" (*Anerkennen*) of truth, and as taking a step from the sense to truth. We may speak here more familiarly, but no more accurately, of a commitment to truth. It falls out, then, that the designated "truth-vehicles" are thoughts. Here Frege's theory of judgement differs from all the others we have considered and from my own. The judgement itself is neither true nor false. However, since judgement does take a stand on the side of the truth of its thought-content, the judgement can be said to be "correct" or "incorrect" according as whether the thought-content is true or false. If any of the expressions constituting the sentence used to formulate a judgement have senses that determine no *Bedeutung*, then the thought-content of the judgement is neither true nor false. In that event, the judgement itself is not so much incorrect as inept. It raises an unanswerable question.

That's the theory in gist. Frege argued powerfully, on two grounds, that thoughts, not judgements, are true-or-false. First: truths and falsities may be brought to attention unjudged and unasserted, as arguments of asserted negative and conditional truth-functions. Frege was keen to establish a scientific basis for validating deductive inferences leading from premiss-judgements to conclusion-judgements. For any such inference to be thus scientifically warranted, the asserted truths and falsities must stand in objective relationships of logical entailment founded upon truth-functional dependencies. The truths or falsities asserted as premisses and conclusions in the course of deductive inference are accordingly determined by the unasserted truths and falsities that enter into negative and conditional truth-functions as arguments. The judgement which terminates deductive inference is a recognition of the truth of the conclusion.

Second: Frege observed that not only judgements but also questionings raise questions of whether something is true or false; he then argued that, since questionings are not said to be true or false in any relevant sense, judgements are not so either. Don't mistake the thrust of this argument. Frege was not opting for a general distinction between "illocutionary force" and "propositional content". (He inspired but did not father that doctrine of propositional content, which we shall examine in Part II of Appendix B.) Indeed, he expressly denied that those thoughts that figure equally as "contents" for questioning and judgement are contents for commands. Questioning and judgement are related but certainly different kinds of action: judgements provide answers for questionings, and questionings suspend judgements. The relation requires shared "contents". The decisive observation is that questionings are not true-or-false, with the intermediate conclusion that there's no reason to think that judgements are either.

I am convinced by Frege's arguments *contra* and subscribe to the conclusion that, while judgements may be right or wrong, these rights and wrongs are not the basic senses of "true" and "false". I shall now move on to consider how Frege did or might have responded to some other theories of judgement, to remark on some of the special features of his theory of judgement and on his views about truth in relation to judgement and finally I shall offer some criticism.

Frege would challenge the qualifications of judgemental-*types* to be primary truth-vehicles on grounds that such items may be true at one time and false at another. That criticism, taken bare without independent argument against variable truth-value, begs the question against the ancient preference. One such argument might be that such variation would ruin semantic determination by allowing single senses to lead us in opposed directions. To this one wishes to respond that there wouldn't be opposition if the movement or determination was at different times. Actually, the everyday "meanings" of "token reflexives" such as "this" and "now" (disparagingly dismissed by Frege) are of this sort. The judgement-kinds of classical philosophy actually seem to me to be qualified both as arguments for truth-functions and as answers to questions, and altogether to be rather like fregean thoughts.

Frege denied that judgement is a unified or unifying assembly of constituents. The truth-bearing thought can perhaps be taken as a "complex" of its determining senses; even the truth-value the thought determines as *Bedeutung* is variously representable as constituted of elements (see (vii), p. 307), but not the judgement itself.

Judgement, for Frege, was a single kind of action uniform across various "contents". Because judgement is uniform and judgements are distinguished only by their thought-contents, there is no distinction in judgement between (e.g.) affirmation and denial. Denial is simply the affirmation of negation, and negation is a difference in content. I don't think Frege ever said that judgement is simple. In fact he said very little about judgement, which he contemned as something merely psychological. His whole interest in *judgement* derived from his abiding concern with *truth* as the objective quarry of the scientific enterprise. (He never, in my recollection of his published writings, explained why we pursue the truth.) He credited those *Sinne* that affect the determination of truth-value and pretty much dismissed other "meanings". He just nodded toward judgement as the commonest way of registering and conveying truth. He thought of his own investigation as the objective scientific study of truth itself. His logic indeed is just that, a theory of the determination of truth-values by truth-values; it mattered, he thought, that these truth-values should also be determined by senses. The results of his inquiry or any other are registered in judgements that are recognitions of the truth of senses. He said almost nothing else about judgement. The "recognition" (*Anerkennen*) presumably is that the thought-content of the judgement determines truth. That of course is inept, since the thought-content could be false; the metaphor of a "step" doesn't help much either.

In order that a judger should make the step from thought to truth, he must (held Frege) "grasp" (*begreifen*) the thought. Now I think he's said too much! *Grasping* presumably is another act of mind, distinguished from judgement perhaps by not requiring a "step". But then, I suppose, graspings are distinguished from each other only by their thought-contents. Why then doesn't grasping require grasping as much as judgement does? Frege might have wanted to conclude from this that grasping is undefinable, never absent from cognitive action but almost never noticed. He conceded (perhaps under pressure from Wittgenstein) that the fact of judgement is also ordinarily unmarked. But here he could have argued that not every act of mind is judgement. The integrity of the scientific enterprise, which allows for hypothesis and retraction, depends upon that. So we must be able to mark judgement for what it is, e.g. by affixing an assertion sign. Still, it would seem, that judgement and its alternatives all equally implicate grasping, which itself must be forever fugitive and elusive.

From all this, I hazard to conclude that, for Frege, judgement is a two-phase mental act of grasping-a-thought-cum-recognition of truth. It is then perhaps composite. But it is not a composite of referents (as in Russell) or of referrings (as in Wittgenstein).

[Some of Frege's readers might suppose that for him judgement is itself a kind of thought, *viz* of the identity of the thought judged true with the truth. Frege would have none of that, and could argue that that equation would simply give us another thought to be judged true.]

What mainly gives interest to Frege's few unclear words on judgement is the accompanying discussion of *truth*--the main and perhaps sole object of logical inquiry--to which I now briefly turn (more to follow at #5 of Chapt. 3). We have seen that Frege held that *Sinne* are the primary vehicles of truth. Perceivable entities such as pictures, sentences, judgements and the sun may be said to be true in borrowed senses fixed in relation to the expression of thoughts. Truth and falsity, in Frege's opinion, while perfectly objective, are strictly undefinable. He argued (in a manner reminiscent of the early Moore and Russell), that any definition of truth must ultimately be circular, since any statement of the conditions for a thought's being true would warrant the judgement that a thought is true only if defining conditions *truly* obtain. (He added, as further criticisms of the so-called correspondence theory, that any correspondence-definition would misrepresent truth as a relation, would miscast it as something perceivable and, finally, since no correspondence is perfect, would destroy the sharp distinction between truth and falsity.) Truth cannot consist in the existence of facts, for (he held) we discover facts precisely by discovering that thoughts are true. (He inclined to the opinion that facts are true thoughts.) Frege also appealed to the (approximately true) observation that a thought and the thought that the first thought is true are equivalent, to support his thesis that truth is undefinable. I don't fathom his thinking here. Frege is the last person who could have been holding that truth is redundant (as the interpretation of truth as assertion would suggest) or that logically equivalent thoughts are identical. Truth is *something*, after all; we can say that thoughts are false. It would avail nothing to argue that the thought that another thought is false is equivalent to a negation, since (again), for Frege, certainly, negation is explained in terms of its truth values. Perhaps the bearing of the equivalence in question on the undefinability thesis is this: Any assertion of G would assert conditions for the truth of the thought that G is true. What better set of conditions could there be? But it is obvious that *that* does not provide a definition. Some may suppose that, for Frege, Truth *could* be defined as the *Bedeutung* of (e.g.) $0=0$. Against this, we must notice, first, that Frege didn't actually have a theory of the determination of *Bedeutung* by *Sinn*. Second, and perhaps more to the point: While it is unquestionably true that, for the author of the *Grundegetsetze*, the *Bedeutung* of $0=0$ is The True, that warrants no definition, since Frege had already explained the identity function ($\xi=\chi$) by its truth-values. (Frege's systematic practice in the *Grundgetsetze* was consistent with his later arguments about the undefinability of truth: truth and falsity are the only undefined objects he actually invoked for purposes of explaining his primitive functions.) Truth and falsity, for Frege, are ineluctable, fugitive realities.

Appreciation and criticism. Frege, in his writings on and around the topic of judgement, to me, seems right in his oppositions but unconvincing in his advocacies. I cannot go along either with his "platonism" or with his thesis that senses or meanings are the primary vehicle of truth and falsity.

I have no quarrel with meanings *per se*, and agree that the meaning of an assertion affects the objective determination of the truth value of something. I applaud his policy of restricting attention to meanings that do have that effect. Indeed, I have a theory of sense or meaning as indications of conditions of success that meshes with Frege's apparatus (Chapter 3, #4). My demurrer arises out of a point which Frege himself was perhaps the first to observe, which is that a thought may fail to determine a truth-value. That happens, in my scheme, when a full-bodied assertion with an unimpaired sense is unsuccessful and, on that account, fails to raise an answerable question about the truth or falsity of something. But now, if a thought may or may not determine a truth-value, then, if it does, something besides the thought itself--a fact perhaps--must contribute to that happy outcome. In our scheme that something else is precisely the satisfaction of the conditions of success whose indication is the meaning of the assertion.

Frege's aggressively propounded "third realm" "platonism" of thoughts and truths is hard to swallow, not because it is "third" but because it casts what we naturally think of as thought-dependencies as free-standing somethings. Frege's *thoughts* are "possibilities". Even Leibniz, who believed in possibilities if anyone ever did, freely allowed that possibilities exist, not in advance of conceptualization, but in the mind of God. Keeping now to our own human selves: It goes against the grain to think, for example, that the whole domain of transfinite ordinals was "already there" in the 4th century BC. in advance of any intellectual effort on the part of the likes of Cantor, Frege and Co. Frege thought that his "platonism" was required by the conception of the scientific enterprise as the pursuit of objective truth. Isn't it enough that these commodities should be there sometime as objects of speculative knowledge (a way of talking Frege occasionally fell in with)? Bank accounts, so far as we know, didn't exist in advance of human activity; but that is no bar to the scientific even mathematical pursuit of objective truth in regard to such matters, as in the keynsian theory of interest.

An argument against Frege's platonism of thought, and also in favor of my opinion that the primary truth-vehicles are products of human activity is as follows: Frege says that asserted thoughts must be "grasped". One must agree that the progress of science requires that thoughts be brought into our experience somehow. But then, if Frege is right, why can't we simply *name* them as we do other already-there things? We could then incorporate those names into our judgements without any need for grasping. Let's try it out. We do have names for some fregean thoughts, e.g. "The Riemann Hypothesis". Shouldn't we then be able to take the judgemental step from thought to truth by simply writing "|-The Riemann Hypothesis"? Well, I can imagine that a mathematician who proved the wanted theorem here just might register his triumph in that fregean way. That would, I'm sorry to say, convey nothing to me, for I have, alas, now quite forgotten what the Riemann Hypothesis is. One could not write down the judgement in the above "namely" manner unless one had "grasped" and retained the thought. That suggests that there would be no thought-object there to be named if it had never been grasped by anyone. That is my view: truth-vehicles are nothing if not "essentially graspable", and do not exist except in relation to conceptualizing activity.

NOTES

[1]In a letter dated June 11, 1913, on p.122. In an earlier, still unpublished letter, Wittgenstein had objected that the *variably* polyadic relations which Russell certainly needed to accomodate judgements of arbitrary complexity do not qualify, by Russell's own standards, as proper relations at all. Some may riposte that perhaps the relation takes *sequences* as arguments. I feel that that is an evasion and that Wittgenstein's objection was sound.

[2]I first knew of this from David Pears' article, "The Relation between Wittgenstein's Picture Theory of Propositions and Russell's Theories of Judgement", pp. 190-212 in *Wittgenstein Sources and Perspectives*, ed. C. Luckhardt, Ithaca, 1979. I can, on the penultimate occasion of formatting this text (January,1987), report that the material in question had at last been published as a book (*Theory of Knowledge*, edited by Elizabeth Ramsden Eames in collaboration with Kenneth Blackwell, George Allen & Unwin). A quick perusal of Chapter IX of Part I ("Logical Data") and of the early chapters of Part II ("Atomic Propositional Thought") reveals that Russell had at the time (1913) come to require acquaintance with (non-constituative) logical forms, apparently for the reason that something must give form to the relationship of the constituents. I think that this adaptation makes for a considerable improvement in the general doctrine Russell still stood by; Wittgenstein didn't applaud.

[3] For Aristotle I rely mainly but not exclusively on the *Interpretations* and Chapters "Gamma.4", "Epsilon.4" and "Theta.10" of the *Metaphysics*; for Plato, on the *Theaetetus* and *The Sophist*.

[4] Plato's preference for discourse over thought may seem to put him closer than was Aristotle to some such opinion; but then again, Aristotle's distinction between practical and speculative knowledge, of which only the latter is of *truths* formulated in affirmations and denials, may actually land him near to our distinctiveness thesis.

[5] I find no theory of "categories" in Plato, and on that count must disconnect him from several of the positions I now ascribe to Aristotle. The theories still remain fairly close.

[6]What follows is a synopsis of my interpretation of the *Tractatus*, which is controversially different from others. A full, documented but still imperfect statement may be found in my "Wittgenstein's 'Picture Theory' and Aristotle", pp. 171-204 of v.I, *The Philosophy ofWittgenstein*, edited by J. V. Canfield, New York, 1986.

[7]The critical probes touch on the unification of judgement, the manner of indicating a relation among referred-to objects, and the placement of "logical constants"; most important, Wittgenstein's theory was incapable of dealing with the problem of the exclusion of contraries, as he himself came to allow.

[8]I work mainly from Frege's later papers, "The Thought" and "Denial" with borrowings from the *Grundgesetze*. His related doctrine of "Sense and Referent" are expounded, examined and interpreted in #4 of Chapter 3 below.

[9]In the "early theory". There is no evident continuation or retraction of this in the later papers.

APPENDIX B

CONSTATIVES, PROPOSITIONS AND EXPLANATION

I. SOME CONSTATIVES, DESCRIBED.

We shall name, exemplify and describe a selection of constative moods, with the aim of marking contrasts and co-ordinations with assertion in their combined support of the advancement of human knowledge.

I have contrasted *assertion* with other "constative" moods. These various congeners of our intellectual enterprise find common ground in the human concern for knowledge. A first objective of this appendix is to identify and describe a selection of these moods. The effort is a preliminary to something better, which I rather suspect will never actually be achieved. My inquiry here will be wide ranging but shallow; it will also be dull, as "linguistic philosophy often is, put in as illustration, with an incidental hope that it may shelter my backside from the lashings of familiar "anti-verificationalist" objections. Little is argued. Some readers may complain that what follows is but a string of stipulations. That is not so. My identifications and descriptions are responsible to the facts of the matter, adequately exemplified, I hope, in specimens. It is a wearisome business of whittling away at cases; I have been humbled by exceptions, and lack confidence in my proposals; indeed, even in preparing final revisions, I have been distressed to uncover serious deficiencies, especially in the matter of conditionals; still, I feel it is a task to be broached, if only for illustration's sake. My descriptions and definitional formulas are mostly framed in the expected terminology of *knowledge* and *thought*. I mark contrasts and trace liaisons, especially with *assertion*, taking notice of the varying appropriateness of the ideas of evidence, proof, truth, etc to considered items and of how these several utterance kinds fit into the acquisition, conveyance, preservation and organization of our knowledge of truths, in commonsense, science and mathematics. I stress that the relationships between these different constative kinds and the presumed central case of assertion are various and sometimes distinctive, and on this point our approach differs from the familiar blanketing appeal to an alleged common propositional content, which we return to in part II.

In Chapter 2 we considered *attenuation* and the mechanisms by which adverbial modifications of assertion are transformed into distinctive indications of mood. Other constative kinds require that the speaker should actually conceive a specific kind of assertion. Thus a condition (of doing) for the *dissembling* of an assertion of what one does not think is so, is that the speaker should believe that what he seems to assert should *not* be true, and a condition of success is that the imposter actually be false. A pretender must know what it pretends to be. Again, it is plausible to assume that propoundings of conditionals and declarations of possibility require that the speaker be able to make actual (second-order) references to assertion-kinds associated with the embedded clauses: one says something about the statements that might have been produced in those terms. Metalinguistic interpretations of conditionalization and modality are not accidental curiosities but informed responses to a specific kind of dependence. Other constatives do not depend upon assertion, but simply resemble and differ from it in specific ways. Conventionalized, reportorial expressions of pain, intention, etc--sometimes called "avowals"--are cases in point.

The broad-ranging classifications to follow should, except when brought under common genera, be exclusive. I fear that they may sometimes fail to be so[1]. I worry most here about "illocutions" that may, in various ways, be general or universal and not be generalizations, for I have previously wanted to docket some of them as generalizations.

We begin with four kinds of a subject's saying what he thinks but thinks he does not know, namely *expressing an opinion, guessing, conjecturing* and *predicting*.

Expressing an opinion.

1. *Expressing an opinion*, of which schematic examples are,

(i) "Our best opinion is that the rate of inflation will fall to between six and eight percent for the upcoming year."

(ii) "In my opinion, chemotherapy offers the best chance of sustained recovery."

(iii) "Well, I *think* John is already on his way."

(iv) "*I* think that was a heron."

(v) "Well, in my judgement, he's an excellent prospect for the job."

(vi) "I estimate that the costs will be less than $6 per unit."

is simply a matter of one's saying what one thinks is so.

Remarks. The specimens illustrate that expressions of opinion are characteristically neither propoundings nor reportings; the opinions expressed may be true or false and evidence and arguments may be brought to bear on the issue at question. We allow, however, that each of the consulted parties is "entitled to his own opinion" and expressions of opposed opinion are not reckoned as contradictions. Speakers can express only their own opinions or, as spokesmen, the opinions of a particular group. Expressing an opinion is a kind of "performative", and that feature of this mood supports the hypothesis that this kind of expression arose as an adaptation of a familiar style of adverbial modification upon other constative moods (pp. 103).

Guessing.

2. *Guessing*, of which schematic examples are

(i) (In a TV "contest") "My guess is that the goods are worth $5500."

(ii) (At the races) "My money's on Dobbin."

(iii) (Playing bridge) "I just have a hunch it's a singleton queen."

is conventionally selecting from a specific (though not necessarily discrete) range of alternatives.

Remarks. A condition of success for guessing is that the specific range be populated and the guesser's selection have a place on the range: if the race is cancelled or the selection scratched, the bet is off. The guess is *right* or *wrong*, perhaps even *true* or *false*. It is a condition for guessing that the speaker should not have known the outcome; he may have even thought that his selection was positively unlikely on the evidence. In guessing, a speaker certainly allows that he may be wrong; but of course he doesn't guess that he's wrong about the matter before him. Guessing, as it seems to me, is a kind of propounding.

3. *Conjecturing*, of which actual examples are

(i) Propoundings of the celebrated Fermat Conjecture or of its denial

(ii) Pre-lunar-landing conjecturings about the age of the moon

(iii) Conjecturings about Alexander the Great's complicity in the murder of his father or about the disease he succumbed to in Babylon,

is propounding what one thinks may be so and could be known but thinks that no one actually knows is so.[2]

Remarks. Conjectures may be proven out or otherwise established as right or wrong. Conjectures are often highly innovative proposals, and are never mere selections from specific ranges of alternatives. The "may be so" and "could be known" in my formula for *conjecturing* requires that speaker can "entertain possibilities". Conjectures, in contrast with expressions of opinion, find their natural place within the well-codified disciplines of mathematics, science, history and the like, and therefore look forward to some kind of regularized resolution. However, conjectures (unlike some predictions) cannot be simply calculated or otherwise "generated" by the apparatus of the discipline in question.

Predicting

4. *Predicting*, of which actual examples are found in

(i) Predictions of currency devaluations

(ii) Astronomers predictions of eclipses

(iii) Percival's prediction of the existence of trans-neptunian planets

(iv) The prediction of trans-Uranium elements

(v) Predictions that the Minoan-B script derived from one source or another

(vi) Predictions that the Russians attempted to send a man into space before Gugarin made it [example from 1980].

(vii) Predictions that further information would soon be gained about the genetics of cancer,

is propounding of what one thinks will be known but is not yet known among some relevant community of interest.

Remarks.
"community of interest":

a) With reference to (vi): The Soviet government already knew (1980), if it is so, even though we (still, 1991) do not. I may predict what others already know, e.g., that they already know certain things.

b) Different canons of knowledge govern the activities of different communities. A newscaster who has private information may think he knows there will be a devaluation and assert as much to his wife over breakfast, although, in his place as newscaster, he may be forbidden to hazard anything more audacious than a prediction. Again, astronomers may only predict future eclipses even though I might assert same on the basis of their announced predictions.

"will be known":

c) The futurity of predictions is of knowledge not of predicted fact, as is clearly shown in examples (iii)-(vi). Prediction is not prophecy or always forecast, and not all forecasts are predictions.

d) *What* is predicted, however, is the fact and not the knowledge. If I predict that something will happen within the month, and it turns out two years later that it didn't happen, that refutes my prediction. (The fact of knowledge, past present or future, can also be predicted, as in example (vii), but that is only one kind of case among others.)

e) Similarly, prediction is not assertion about future knowledge. I can assert (not predict) on Monday that we will know on Wednesday what Tuesday's weather will have been.

f) People sometimes talk as if they thought that all things said in the future tense are predictions; the above example shows that it is not so. Assertions in the future tense say what one thinks one now knows will be so.

g) In predicting, I say what I think will be known. To the objection that someone might predict the end of the earth or the onset of an ice age which we know no one could witness, I reply that we might gain advance knowledge of such facts.

h) Predictions can be wrong or right and more or less accurate; I don't think that they are literally "true" or "false".

i) Predictions may be confirmed, borne out and fulfilled, but are doubtfully said to be verified.

Verification is a process applied by an intelligent agent, whereas fulfillment comes about of its own.

j) Predictions typically result from the application of theories. I
cannot make the predictions of the Astronomer Royal because I am incompetent in celestial mechanics. I base my assertion on his predictions. This curious circumstance, that I may be able to assert what the astronomer can only predict, illustrates the difference between those canons of knowledge the astronomer in his professional office employs when he undertakes to project known data and canons available to all readers of the almanac. It also illustrates prediction in its most typical phase, as resulting from the application of theory.

k) The typical dependence of prediction upon theory lends plausibility to the literally false thesis, refuted by some of our examples, that predictions are always implicitly general.

A *resume*, with attention to the conditions of success for assertion.

a) "There are authorities": *Prediction* and *conjecturing* require it and *guessing* leaves it open; while we may solicit expressions of opinion from a panel of experts, who are authorities, we then, so to speak, put them all on a par, and in that way deny them special authority.

b) "A sort of thing that could be known": *Predicting* and *conjecturing* require it, *guessing* and *expressing an opinion* leave it open.

c) "Sometime applicable procedures are available". All four moods are generically noncommittal on this most important condition of success for assertion.

I proceed to some other moods.

5. *Hypothesizing*, of which examples are found in formulations of *Hypothesizing*

(i) The Caloric Hypothesis for heat

(ii) The Undulatory Hypothesis for light

(iii) A gardener's hypothesis about the cause of a blight on his daffodils

(iv) A hostesses' hypothesis about the non-appearance of a guest,

is propounding what one thinks may be confirmed by facts hitherto unobserved by propounders of the hypothesis.

Remarks.

a) Hypotheses are put forward to be "worked with", for the prime perlocutionary purpose of guiding the progress of investigation. Isolated from such connections, they stand in their own right as provisional theories. Scientific hypotheses may come to be confirmed as laws or relegated to museums for the history of thought or oblivion.

b) An hypothesis is confirmed by displaying facts that would verify statemental consequences of the hypothesis.

c) Hypotheses in themselves are neither verifiable nor falsifiable by application of procedures.

d) A condition of success for *hypothesizing* is that the hypothesis itself has not yet been either established by confirmation or disestablished.

The hypothesis itself, which can no longer be made because established or disestablished, does not cease to exist on that account, and we may continue to speak of it as an hypothesis, as we nowadays do of the Caloric Hypothesis. "Big hypotheses" when established transfigure into Laws of Nature[3].

Describing

6. *Describing*, produces descriptions, of which examples are

(i) Chemists' descriptions of stuffs

(ii) Witnesses' descriptions of faces

(iii) Geologists' and travelers' descriptions of places

(iv) Reporters' descriptions of battles.

Describing is sequentially reporting the observed features of something--a stuff or a place, an event or a face. Call that object the "descriptum".

Remarks.

a) "something". A condition of success for describing is that a *descriptum* should exist. *Descripta* can be almost anything[4].

b) "observed features". A second condition of success for describing is that the describer should have actually observed the *descriptum*. If he was looking at something else, his describing fails.

Describing should be contrasted with what we might call "repeating a description", as might be done, for example, by a diligent student who, in taking a chemistry examination, sets out a description which he has learned by rote. *Repeating a description* presupposes the existence of a description yielded as a product of *describing*.

c) "reporting". Not propounding.

d) "sequentially". Describing is not the assertion of a conjunctive predicable to the *descriptum*; it is a sequential report. Describing always stops at some point, but there is no predetermined point at which it must stop. The investigating officer requests from witness a description of "indefinite length".

e) It remains that describing often follows familiar schedules of description: One describes a face by saying how it is with the hair, the nose, the eyes, the skin, the mouth, and perhaps a touch of psychological salt is added with the glint of the eye and the turn of the lip. One describes a chemical element by giving its atomic weight, valences, density, color, spectra, melting and boiling points under standard pressure, temperature, etc.

f) If the describing was successful and a description is produced, the describer must have been there. Though we may question his abilities as an observer or a reporter, we cannot ask him how he knows that the thing was as he says. Questions about evidence and confirmation are also out of place.

g) Descriptions imply statements because of their "indefinite length", but are not themselves implied by anything. One couldn't tell from any list of statements where an otherwise given description ends or how long it is.

h) The existence of authorities--ones capable of producing descriptions--is a condition of success for describing (see b and f above). Another is that the *descriptum* not merely could be but is known.

i) Produced descriptions are subject to various characterizations and assessments, e.g. as *accurate, faithful, detailed, rough, careless, involved, routine, vivid, perspicuous, revealing;* also, descriptions can be *verified* by subsequent observations, but not by application of available procedures.

j) Descriptions are not literally "true" or "false", but their statemental consequences are.

k) Every false consequence is a mark against a description. It remains that a description might be accurate and useful in the bargain while implying a falsehood or two.

Avowals

7. *Avowals*, including expressions of pain, resolution, intention, emotion, sentiment, and preference, of which these are schematic examples,

(i) "That hurts."

(ii) "I *won't* eat any peanuts this evening."

(iii) "I intend to be there before the meeting starts up."

(iv) "I'd like roast beef, rare."

(v) "I love you."

are conventional expressions and self-displays of states of the subject.

Remarks.

a) "conventional". As Wittgenstein was wont to observe, avowals may replace more primitive non-conventional expressions, e.g. wincing. Some but only some avowal-displayed states may presuppose capability for conventional action, e.g. pain certainly doesn't, but *wishing* may[5].

b) "expression". To exclude demonstrations of abilities, wounds and other such "external" things. Expressions are "of the mind". They may or may not be "conventional". My chancy proposal for what distinguishes *expression* from other self-displays is that the "mental phenomena" they are of may be dissembled (by pretended expression).

c) "self". Speaker might conventionally or non-conventionally display *other* things, e.g. a ring or a case in point.

d) Avowals are typically the instruments by which speakers clarify their states, for the perlocutionary end of showing others how they "feel" or for their own better comprehension. What is avowed is the sort of thing that could (but needn't) be known.

e) A condition of doing for "genuine" (= candid) avowals is that speaker be in the displayed state.

f) Because they are responses to self, avowals are *reports*. They may stand as premisses from which conclusions can be drawn. At the very least, the fact of avowal stands as evidence that subject is in the displayed state and as a formulated reason for believing that he is.

g) A speaker cannot be "wrong" about being in the state he candidly avows, and interlocutors cannot request evidence or credentials from the speaker. The speaker may be wrong about the "character" of the displayed state, e.g. in saying that it's a "burning" pain, and one may ask the speaker for clarification, e.g. whether what he feels is "really" what he says he feels. ("Do you really feel that bad about it?")

h) One may also doubt and confirm the candor of the speaker and the genuineness of his avowal.

Generalization

8. *Generalization* is my next topic. First some specimens:

(i) "All birds are warm blooded."

(ii) "All elements whose atomic number is greater than 120 are unstable."

(iii) "No Japanese is discourteous."

(iv) "Women generally live longer than men."

(v) "Every year over 10% of federal income tax returns are fraudulent."

Not all "universals" are generalizations. Here are some contrasting specimens:

(i') "All men are mortal." (an axiom),

(ii') "No polynomial function has π for a root."

(iii') "All terrestrial mountains higher than 8000 meters are in Asia." (universal statement).

My thought here is that one comes to the formulation of a generalization as a result of generalizing upon cases. My formula for this species of utterance, formulating a generalization, is that it is an act of one's propounding what he thinks will be found to be the case on the basis of reports already in.

Remarks:

a) Generalizations are combinedly reportorial and propounding. They look ahead to future falsification, but may also be defeated by reported, but previously disregarded, cases.

b) "will be found". What is propounded is not the finding but the cases-to-be found.

c) Our examples illustrate that generalizations, like statements, are of many kinds--affirmative, negative, universal, and statistical--and they pertain to different interests and fields of inquiry.

d) Generalizations differ from statements--universal statements included--chiefly in respect of verification. In understanding the formulation of a generalization, we also understand that no verifying procedure could have been applied. Typically, generalizations are falsified by finding contrary instances.

e) Generalizations have entailment liaisons with "instantiated" statements and with other constative products. So, for example, a generalization *All S is P* will entail an indefinite number of conditional predictions to the effect that if an S is found, predict it will be P.

f) Generalizations abound in both science and in everyday life. In science, they are interesting thoughts until rationalized as laws or by demonstration. In everyday life, they are utilities, available for the use of parents, priests, employers, advertisers, social workers, physicians, travelers, as guides for making decisions or recommendations. (Take caution, however, against assimilating generalizations themselves to this way of using them.)

g) Generalizations are true or false, and can be known to be either. Somewhat differently, generalizations may be said to "hold" or not to hold ("Your generalization about life expectancy was true, but it no longer holds since men have begun to take more exercise.") Generalizations can be believed, doubted, supported, and confirmed.

h) I find the *truth* of generalizations a particularly hard nut to crack. The metalinguistic rule that a generalization is true just in case all statemental partial generalizations are does not suffice as the wanted explanation. Generalizations, no less than statements, may or may not be "true to the facts". But of course there may be many distinct "truths" answering to the same one fact. So *truths* are distinguished by something other than their facts, and (if I am right) most evidently so in the matter of generalizations. I long ago urged that there is nothing in the facts corresponding to a generalization that is inaccessible to statement (pp.

2f). Now true generalizations, and true universal statements too, always entail the further *truth* that *These are all there are.* Suppose we got a complete inventory of the books on my office shelves, ticked off in a longish batch of singular statements: we would then have exhausted the facts of the matter of the books on these shelves. But there would still be nothing in that batch of statements to entail that *these are all there are.* Our earlier remark that a generalization is true just in case *all* statemental partial generalizations are, would already, in its evident circularity, have produced dissatisfaction about "general truth". Pretty clearly, we aren't going to do much to explain that notion solely by appeal to "correspondence with the facts", and the matter of the truth of generalizations is still pretty much up in the air. Something but not enough will be added to bring it closer to the ground when, in the next chapter, we try to spot a basis for our concept of generality within the ambit of our investigative repertory (pp. 266ff.) and, later on (in Chapter 12 of Vol. II, see pp. 410f.), we consider the verification and falsification of universal statements.

I now move on to consider formulations of two sorts of "principles" or first truths, namely axioms and natural laws.

Axioms

9. *Axioms* are certitudes which follow from "what things are". Examples are

(i) The axiom (not so dignified in Euclid) that two points determine a unique straight, which follows from the euclidean explanation of a straight as a line that lies everywhere even with itself

(ii) Frege's axioms for propositional logic

(iii) The "Group" axioms for rotations or the "Field" axioms for rational numbers under addition and multiplication.

(iv) The axiom that elected representatives are responsible for the interests of their constituents

(v) *That all men are mortal*, a "first truth" that permeates the whole of casuistry.

In formulating an axiom one reports that a truth is of this kind, and a proof is derived from a consideration of what things are.

Remarks.

a) The formulation is at once reportorial and "second order" in respect of the truth reported on. A condition (for doing) for such a formulation is that the subject know how to assert or otherwise formulate the truth in question. The report that is the axiom is accordingly to be distinguished from the possibly assertional report that it is an axiom. (The latter might be verified even by one with little grasp of what the axiom is all about by looking it up in a book.)

b) The truth of the reported-on formulation is a condition of success for the formulation of the axiom.

c) Axioms, though they be in respect of "first truths", are still expoundable in consideration of what things are, and may require defense and explanation: Is this really axiomatic? So Frege shows by semantic analysis that his axioms were of statements that followed from his exposition of the connectives. Again, the unique determination of a straight line by two points emerges from the eudoxian definition of a straight as a line that lies everywhere even with itself. Children are quick to understand that organisms are bound to perish.

d) *Postulates*, in the classical sense of Euclid, may be distinguished from axioms in that they are not truths or of truths but propounded stipulations or demands. Most typically, what is postulated is that something can be *done*, but sometimes also that something should *exist*, e.g. a coordinating function. Postulates are made to play the role they do in the deductive enterprise by way of the *assumption*, that the indicated stipulations are met or satisfied. If, for one reason or another, one cannot extend a drawn line segment or draw a parallel, then one cannot automatically assume the applicability of Euclidean Geometry.

Laws of Nature

10. Examples of *Laws of Nature*[6] are

(i) Newton's Laws

(ii) Galileo's Law of Free Falling Bodies

(iii) Kepler's Laws

(iv) The Law of Supply and Demand

(v) Mendel's Laws

(vi) Balmer's Law of Spectral Distribution for Hydrogen

(vii) The Principle of Least Action

(viii) The Principle of Natural Selection.

This illustrates a variety that leaves it almost impossible to say anything about natural laws in general which isn't merely callow. Desperate for any kind of formula, I leap to the perlocutionary use of laws: In using a natural law one brings a problematic case within an established principle for a domain.

Remarks.

a) Laws are "first truths" for a domain. At the time Balmer's Law came out, there was little else of a general kind to be said about spectra.

b) "for a domain". We apply the Law of Supply and Demand in Economics not Genetics, and Newton's Laws in Mechanics not Economics. Laws are more naturally said to "hold" within a domain than to be "true".

c) Although I, for example, can be said to know what Newton's Laws are, one is said to "know a law" only if he knows how to apply it. One knows the laws of mechanics, for instance, when he knows how to handle those second-order differential equations which the laws prescribe.

d) Natural laws themselves provide a mechanism for describing cases, even deviant or exceptional ones. The Law of Supply and Demand is useful for describing lapses therefrom, e.g., in circumstances where a basic foodstuff rises in price so much that an impoverished community can no longer afford to eat much else. Geneticists continue to calculate with Mendel's Laws, all the time knowing that their conclusions may be frustrated by particles from outer space, unpredictable cross-overs or transposable genes[7]. The theories of relativity have not abolished the use of vector addition for velocities or of Newton's Law of Gravitation for calculating orbits.

e) One may "discover" laws "empirically" (as was the case with Balmer's Law) or by surveying a body of data (reportedly, the manner of Newton and Maxwell) or simply by excogitating the matter (Einstein's "Elevator Experiment"). Scientists, however, seldom feel comfortable with a law unless it can be rationalized either directly (the Law of Supply and Demand) or within a theory (Balmer's Law and Mendel's). So, while laws are often discovered and established "a-posteriori", in typical cases we eventually come to understand that this is how these things must be. Einstein took us back to Galileo and Bohr made us understand Balmer.

f) Laws are "established" by their predictions, which show that test cases are as the laws would have them be. This "methodology" confirms that the theorists reasonings have not overlooked important considerations.

We now take a perfunctory look at the recently heavily controverted "big case" of conditionals and then conclude with a longer examination of the comparatively neglected matter of "May be" utterances.

Conditionals.

11. Examples of conditionals are[8],

(i) "If you pass your exams you should (may, might, can, could) get a scholarship."

(ii) "Had Kennedy lived, we would have gotten out of VietNam earlier."

(iii) "If Bizet and Verdi were of the same nationality, they both would be Italian."

(iv) "If Bizet and Verdi were of the same nationality, they both would be French."

(v) "If my wife were unfaithful, I'd never know it."

(vi) "If Shakespeare didn't write *Hamlet*, someone else did."

(vii) "If Shakespeare hadn't written *Hamlet*, someone else would have."

Now I wish to contrast conditionals with "if-then"s generally taken, and with conditionalized utterances, with conditional statements and with disjunctions.

I believe that "if p then q " generally taken is used to convey that a fact that would be formulated in the words of the protasis clause, p, is or would be in one way or another pertinent to "constating" or otherwise "saying" (ordering, betting, etc) what would be identified in the words of the apodasis clause, q.[9] Usually but not always, the pertinency introduced by the protasis clause is the sort of thing that would stand as a *reason for* saying the kind of thing indicated by the apodasis. [(ii) above seems to be an example counter to my previous opinion that the pertinency is always that of being a reason for saying.] Of course, reasons for asserting and other pertinencies are various: *sufficient for the truth* of is the reason most considered in books of logic; *your interest in hearing this* is another. Our little formula for "if" is meant to cover all these, including such cases as that of "There are some biscuits on the sideboard if you want some" and adverbial modification ("If my telescope is working, a comet just entered Orion"), as well as "material conditionals" and the assertion of bona fide conditional statements (e.g. *"If Dobbin wins, I'll collect $66.00."),* which we shall consider more particularly in Appendix D, #17. I certainly don't reckon that all of these qualify as *conditionals.*

In the matter of *conditionals*, I submit (flinching from previous failures) that what the protasis clause marks as pertinent to "constating" is *also* given either (i) as a *reason for believing* what would be formulated in the apodasis clause or (ii) as marking a certain fact as being *an explanation for* something to be picked up from the apodasis, where, under both formulas, it is understood that the truth of what would be formulated in the apodasis clause does not require the truth of what would be formulated in the protasis clause. My formulas for conditionals, obviously yield no general rules of calculation such as that for the "material conditional".

Material conditionals do fall under my explanation of "if...then" but are not *conditionals* by the above formula, for it is no part of their "truth conditions" that the *p* should be *a reason for believing* or stand as an *explanation* of anything. They are simply material disjunctions[10] Let me say, however, that I do not share the widespread dissatisfaction with the broad use of "*not-p or q*". The material conditional is an instrumentality of logic. Logical inference at its most basic, I believe, is a matter of eliminating alternatives, and most naturally proceeds by the excision of disjuncts. *Modus Ponens* is precisely a rule for excising negative disjuncts from material conditionals. No wonder then that material conditionals are so prominent in the expositions of logic and the logical expositions of theories. Yet, it seems, everyone must sometime come forward with his own variant form of conditional statement, and I shall eventually add one of my own. *Not-p or q* alone holds a place in all our kits, simply because we are all now and then beholden to logical exposition.

Conditionals are various. Some "license inferences"; others, esp. those formulated with cognates of "can" are "schematic". The latter do not readily yield to model-theoretic analyses according to which (roughly) the conditional is true if the proposition expressed in the apodasis clause is true in all possibilities for which the proposition expressed by the protasis clause is true. That, for me, is but a special case. My two little formulas for conditionals are meant to be only provisional, for I would like to find another to "unify" them, and are not apt to satisfy any of that numerous body of authorities on conditionals. I hope the remarks now to follow will give some sense of where I would come down on some disputed issues.

Remarks.

a) The formulation of a conditional, *if p, then q*, under both the *reason for believing* and *explanation* formulas is "second" or "meta-level" with respect to both the *p* and the *q*[11].

b) Example (ii) is a case that does not seem to fall under the *reasons for believing* formula but which does seem to fall under the *explanation* formula. The fact indicated by the apodasis is that we stayed in VietNam as long as we did; Kennedy's assassination is the protasis-indicated explanation. It is worthy of notice here that no one who believed the conditional in question would ever think to counterpose it. That supports my claim that it falls under the *explanation* rule, for the formulation of an *explicandum* seldom if ever indicates an explanation of the *explicans*. Now explanations are also reasons *why*; the difference of this case from the others cited, suffices to show, I believe, that not all reasons-why are also reasons for believing. This observation, if it holds up, stands as criticism to another theory of conditionals that has had some currency in recent literature. The view is that *all* conditionals are equivalent to statements of causation. An additional point of evidence against that thought is that causes may themselves be mentioned in either clause of the conditional, e.g. when a lawyer argues that if deceased had not expired by cause of cardiac arrest, he would have been killed by the crash.

c) Our "reasons for believing" formula is illustrated by the Bizet-Verdi examples. *Being a reason for believing* is not closed over conjunction. That Bizet and Verdi were both of the same nationality would be a reason for thinking that they were both Italian and equally a reason for thinking they were both French, though hardly a reason for thinking that the two of them were both both French and Italian.

d) "Would be a reason for" is "objective". It usually isn't anyone's actual reason and, per example (v), may even be inaccessible to the speaker by the terms of the conditional.

e) Conditionals can be true or false (see vii), known, supported with evidence (my wife is a master at cover-up), and perhaps sometimes even verified, by deriving the statement *q* from the statement *p*. More commonly, the consideration identified from the protasis would be shown to be a reason for believing or an explanation of what is identified in the apodasis only in relation to other accepted facts, e.g. that there is this play in the corpus of english literature.

f) One may also support conditionals by showing that the sort of fact indicated by the protasis clause is also a fact of another sort (Bizet's nationality was French), or by bringing it under one or another formula. (The department has a policy of awarding scholarships to first year students who pass their exams; this policy makes your passing the exams a reason for believing you will get a scholarship.) The notorious inconclusiveness of conditionals may owe to the availability of plural explanations and reasons for believing and therewith of competing redescriptions and subsumptions.

g) The fact that both explanations and reasons for believing, hence conditionals under both formulas, may be *supported* by appeal to general considerations by no means implies that those conditionals themselves must be implicitly general. A conditional is no more a formulation of its supporting evidence than a statement is a formulation of its criteria.

Formulas for "may" and "can". We are mainly interested in the force of "It may be so" utterances or "declarations of possibility".

12. *Declarations of Possibility.* Much of the literature on modality which I have chanced to read is fogged by the tacit assumption that "may" and "can" afford speakers of English with alternative expressions of *possibility*. But these two little words behave very differently indeed, and, if both are used to convey a sense of possibility, they do so differently. In anticipation of what follows, my provisional conclusions about them are as follows: constative may-be-*p*-utterances are non-assertional propoundings in which the subject conventionally refrains from asserting not-p; H-can-happen-utterances, by way of contrast, may be assertional in their own right, in which event the asserted fact, C, would normally be a condition for the obtaining of an indicated other fact, H (*what* can happen). I shall below return to the question of how closely these formulas fit what philosophers have dished up as explanations for the assertion of possibilities. My main interest for

now is to explore the (English) sense of "It may be so" in actual utterances, which I shall want to call "declarations of possibility".

Specimens:

(i) "The cornice may fall."

(ii) "He may do the job."

(iii) "He may not do the job."

(iv) "General war may not come within the next fifty years."

(v) "The first Berlin airlift may have averted general war."

(vi) "He may take it or leave it" (= "He may take it or he may leave it.")

(vii) "You may go now." (uttered as a "volitive"; you're not required to stay).

Two "might" specimens will also be useful for purposes of comparison

(viii) "He might do the job."

(ix) "General war might have come in 1948."

I believe that all of the "may" specimens save the implicitly "molecular" (vi) are paraphrases as "Maybe"^Sen, e.g. "Maybe it will fall"; otherwise, "Maybe: p". "Might" differs minimally from "may" by being sometimes "contrary to fact". "Might", which philologists tell us once was a past-tense for "may" and (connectedly, for English) a subjunctive, now functions in the present tense as an autonomous "auxiliary", often with a sense little different from that of "may" (as in viii) and sometimes, echoing its subjunctive past, to convey an implication contrary to fact. Nowadays, "may have" is the only past tense "may" has.

Contrasting with "can".

Listen to the difference in the corresponding "can" specimens.

(i')	"The cornice can fall."
(ii')	"He can do the job."
(iii')	"He can't do the job."
(iv')	"General war cannot come within the next fifty years."
(v')	"Israel can (or could) have built an atomic bomb several years ago."
(vi')	"He can take it or leave it." (= "He can take it and he can leave it")
and	
(vii')	(coincident in sense with (vii). "You can go now." (said as a "volitive").

Despite some overlap, best seen at the (vii) examples, there is an undeniable difference in the meaning of "may" and "can". Some languages play it down (in my american ears, the French "pouvoir" seems to, although the contrast may be found in a differing usage of "se peut" with "peut-etre" and "il peut"). Others throw it into high relief [notably Japanese, in which the idiom translated as "may be p" is revealingly literally rendered as "It is not known whether p" ("p-u ka mo sire-masen") and the idiom for "He can H" as "He effects H-ing" ("H-u koto ga dekimasu")]. I have found the lexicographers and grammarians of English next to useless on these two verbs. They are both called "modal auxiliaries", I suppose because they take verb-phrase complements. This classification covers over an important difference. "He may C" paraphrases into "It may be that he C's", whereas "He can C" paraphrases into "He is able to C". That suggests that the personal reference is subordinate to the "may" but is the subject of the "can". (I read the conspicuous difference in the corresponding Japanese idioms to yield the same conclusion.) I submit that "can" is a verb usable to assert something of a subject, e.g. that a cornice is loose or that a man will have a later opportunity; the subject term is *not* buried in a complement clause to which "can" is tied; "can-H" is a compound predicate of the unsubordinated subject. "May", by way of contrast, is used to take a certain posture toward possible circumstances indicated in a complement clause. This emboldens me to represent "may" cases uniformly as "It may be: p", and to represent "can" cases uniformly as "A can-H, where "can-H" is a schematic predicate and "A" a schematic subject term. The difference also comes out with negation. "It may not: p" goes into "It may be: not-p". On the other hand, "It cannot do" always paraphrases, *not* as "It is able not to do" (= "It needn't do"), but as "It is not the case that it is able to do. The negating term, like the subject, should be represented as falling inside the complement for "may" but outside the "can".

My thoughts about "can"-sayings appear undefended at pp. 39f. and 139f. I have explored the matter at considerable length in another place ("Meaning and Saying", *Foundations of Language*, 1972, pp. 66-97, esp. pp. 70-76), where I arrived at this kind of formula.

Constative A can-H utterances are assertions of specific but unspecified facts of a sort that would normally be conditions for the occurrence of happenings of kind H. A is asserted to be in that condition. So, for example, I may assert that a cornice (A) can fall (H), meaning that the plaster is rotten or that there are tremors nearby; having time, a vehicle, strong legs, permission to leave work, etc may be asserted of the expected late arrival. The obtaining of any of these conditions can be denied by saying A cannot H.

Remarks.

a) "normally". A specific condition asserted in A can-H is only normal, not necessary for H-ing. That H happens in respect of A does not imply that A can H. I borrow Austin's example of a poor golfer who closes his eyes and just happens to hole a 90' putt. Again, a poor shooter may sink a last second desperate shot from the opposite end of the court perhaps because the ball was deflected into the ring. The cornice might be blasted off. It goes even further than that.

b) "specific but unspecified". This comes out in negations. I say "We can be late", meaning time is getting short; you say "We can't be late", meaning that that privilege is not available for this event; there's no contradiction. Consider the two understandings of "You can't go too far with me". "Can"-assertions are neither ambiguous nor necessarily vague with respect to the asserted but unspecified conditions. A can H is not paraphrastically equivalent to "a condition for A's H-ing is satisfied", in the sense of at least one such condition. Nor does A can-

H assert a disjunction of conditions for H. A can-H is also not paraphrastically equivalent to "All conditions for H-ing are satisfied. Finally, although A can-H asserts a condition for A's H-ing, it does not assert the conditional that, if that condition or some other is satisfied, A successfully H's. Assertive A can-H's are always, simply, inexplicit assertions of specific conditions for H-ing. These conditions may have to do with strength, training, opportunity, or almost anything. The words of an assertion that A can-H do not mean any one of those conditions. The meaning of the word "can" could be explained by saying that it is assertionally employed to assert specific but unspecified conditions.

c) This explains the apparent equivalence of the volitive "can" with the volitive "may". "May", which is constatively employed to refrain from asserting can also be used to refrain from commanding. Nanny was right: "may" is a natural and an excellent word to use for granting permission or giving leave to. We have also several times noticed that A can-H is often used to assert that A has been granted permission to do H. Now if we transform "You can-H" into a volitive by converting the reference *you* into an address[12] the most conspicuous condition for A's doing-H which the speaker would seem to be able to effect by volitive speech, would be to cause his respondent to have permission to do-H; hence, by a very natural devolvement, one could give permission by saying "You can-H". But one might, in the same manner, also impart resolution, confidence, or any other agent centered condition for doing-H amenable to the influence of conversational exchange. Thus, "You can do it (jump in the water), just try".

d) The schematic meaning of "can" creates an illusion of "assertoric modality". "He can-E-or-F" may seem to say that the possibilities of his E-ing and of his F-ing are both open. Well, if the "can"-saying is true, then those two possibilities are indeed open; but that is not what was said. Rather what we have here is the assertion of a condition for both his E-ing and his F-ing, confirmed in specimen (vi').

e) There just could be a language without our use of "can", one that employed only expressions equivalent to our "opportunity", "skill", "permission", "luck", "composure", "determination etc. So far as I know, there is in fact no such language. Its vocabulary would have to anticipate too many "can"-assertible condition sorts.

f) This argues that there cannot be what Austin once called the "all in" sense of "can". (See "Ifs and Cans", *Philosophical Papers*, p. 177). We don't need it, anyway: we simply say A H-ed.

g) What explains the widespread availability of the equivalent of "can" is that it makes it possible for us automatically to report upon the satisfaction of hitherto unremarked conditions for the occurrence of happenings. In this respect "can", like demonstratives, enables us to mean unsaid things in saying familiar ones. It is one literal alternative to figurative or stipulative innovation.

Now, returning to our topic, I submit that "Maybe: *p*" utterances are acts of conventionally refraining from asserting the negation of *p*[13].

"Maybe: p" utterances are acts of conventionally refraining from asserting the negation of p.

Remarks.

a) We could ask speaker why he thinks it may be so. He could defend his utterance by giving almost any reason for leaving the question open and not asserting *not-p*--perhaps by citing reasons for believing *p*, or for not believing *not-p*, or perhaps by general allusions to the current state of knowledge. He opts for leaving things open with a conventional refrainment from asserting *not-p*. These "refrainments" are not assertions; nor are they expressions of resolution not to assert. The speaker may very well intend to assert *not-p*; he just doesn't do so. The declaration of possibility is a constative mood on its own; a device for "leaving things open", for indicating what we do not know, a "negative" sort of thing perhaps, but still an identifiable form of action which is a species of restraint.

b) This formula nicely fits an earlier cited datum: "He may be coming, but my guess is that he won't" makes sense while "He may be coming but he won't" doesn't. *Guessing* isn't assertion, so there's no evident conflict in both refraining from asserting and guessing; saying that he won't does, however, convey an assertion, and that does palpably conflict with refraining from that assertion. (Those several other kinds of "saying what one thinks" that we contrasted with *assertion* will work as well as *guessing* , e.g. "He may come, but I don't think he will" makes good sense.)

c) How can refrainment be positive action? I doubt that one who is troubled by this will achieve satisfaction without going rather deep into the general theory of behavior. One way of refraining from doing something (e.g. taking a piece of cake) is simply not doing it, or anything else; one may refrain from asserting *not-p* by saying nothing at all. But it is well understood that not doing anything is often a very difficult thing to do. Restraint and forbearance are hard virtues, common enough perhaps among the downtrodden, but in our society too often the victim of indulgence. It is no surprise that forms of behavior have evolved to stand in as substitutes for sheer inactivity. Even dogs will learn to *refuse* food by walking away; and the anti-patriot will *not* sing by sitting down when the anthem is played, or not go to war by burning his draft card. Refraining may be a kind of active non-doing, an act of *braking* action. Nowhere are active substitutes for inactive doing more needed than where the inactivity is conventional. Here there is no external hook on which subject can hang inaction. Although some people are rather good at giving and understanding silent answers--the English have more conventions for this than do Americans--the phenomenon is still as derivative as it is rude. Small wonder then that we have devices available for showing that we are not asserting (commanding, etc) and what we are not asserting (commanding, etc).

d) That this is the correct analysis of *Maybe: p* utterances is confirmed by the sense of volitive "You may V" utterances, e.g. "You may go". In thus *granting permission* to his attendant, the officer refrains from making further demands.

e) Declarations of possibility incorporate two distinct kinds of "negations": (i) the *non-assertion* of (ii) a *negative* statement. One who reports that another has made a declaration of possibility implicitly says that the other did *not* assert *not-p*. But, of course, in enacting such a declaration, the speaker does not report or otherwise assert that he does not assert-- he actively refrains from assertion.

f) "May" has no proper "contrary". There is nothing answering to the negation of a declaration of possibility which is itself a declaration of possibility or impossibility. (A "contrary" perhaps is assertion!) This is amply confirmed by the unique sense of "It may not p" which can only mean *Maybe: not p*.

g) *Objection.* Sentences for *Maybe*: *p* can be embedded as "that" clauses, as in "I know that John may be there already". But surely one cannot sensibly know that a refrainment from assertion. If we gloss the problematical sentence to mean "I know that it's alright to refrain from asserting that John is not there", then, it would seem, we should have *originally* paraphrased "John may be there" as an *assertion* that it is alright to refrain from asserting that John is not there.

Reply. The way round this difficulty is to point out that it is not unique. One may vote for John by saying "I vote for John"; in a moment of indecision one may also procrastinate by saying, "I think that I vote for John", meaning, "I think I have reached a decision to vote for John", without endangering the "performative" force of "I vote". Similarly, an assertion that it's alright to refrain cannot obliterate refraining. Deletion systematically stands companion to subordination in all these cases.

h) In perlocutionary use, *Maybe:p* utterances commonly subserve assumption: "Since he may not come, assume that he won't, and take it from there."

"Could" (in English) is more serviceable than are either "may" or "can" for the "philosophical assertion of possibility".

On the formulation of possibilities. "May" and "can" have different qualifications and liabilities for asserting the existence of possibilities in the "philosophical sense". Among "may"s qualifications are, first, that it (with "might") does serve to "leave things open", which is what putative assertions of possibility are supposed to do. Though I think the lawn is wet because watered by the gardener, still, weather being what it is, I say "It may have rained", leaving that possibility open. Second, both assertions of possibility (if there are any) and declarations of possibility involve a "second-level" review of "propositions". The marks against "may"s candidacy for being the "philosophical modal" are, first, that it is not assertional and second, it is not obvious how one could ever negate a declaration of possibility into something corresponding to *necessity* with the sense of "It is not possible that". It is just the reverse for "can". "Can"-sayings close things down a bit by *asserting* conditions for the occurrence of happenings. "Can"-sayings can be externally negated yielding the reverse of "Must-not", which corresponds to *necessity.* The idea of necessity taken in connection with that of *possibility* would find a better footing in the field enclosed by "can" and "must" than in the field of "may be"s. "Can"s disqualifications are mitigated by the consideration that the *possibility* of A's H-ing is *one* kind of condition that can be asserted by saying "A can-H".

We are missing the best candidate. The old word "could" sometimes serves as a past-tense for "can", as in "The USSR could deliver an atomic bomb in 1952". But sometimes it appears in its own person with its own present or timeless sense: "We could land on Mars by 2000"; "There could be a solution to this equation"; "could" borrows an external negation from "can": "He couldn't be there yet" says *It's not the case that he could be there yet.* Furthermore, "A could-H" in the present tense seems to differ from "A can-H" by drawing attention away from the subject A to the "possibility" of A's H-ing. If that is so, then, in "could" we may have a vernacular instrument almost ready made for the assertion of possibilities[14].

Assuming as a kind of behavior is a kind of positing as a premiss. Assuming so explained is "perlocutionary" and not purely conventional.

13. *Assuming, a non-case.*

Assuming in the sense of

(i) The "Let us assume the contrary" which typically initiates *reductio* demonstrations in mathematics

(ii) The "We must assume the worst" declarations of police investigators

(iii) The "Anyway, assume"s of advocates, who provisionally wish no greater concession than that it could be so

and

(iv) The "We may assume"s of reasonable likelihood,

is certainly a form of action, conventional in nature, for which a fair general formula is *positing as a premiss*[15].

I originally took it ("assumed") that assuming was alright as a constative: The produced *assumptions* are true or false, aren't they? I then reflected that "premiss" is a strange word to find in the characterization, since premisses are identified as such by their place in arguments or discussions, of which, in general, perlocutionary characterizations are inevitable. I was, furthermore, unable to characterize *assuming* by its conditions of success in relation to *assertion*, since *assuming* seemed to vacate *all* the conditions of success for assertion--an assumer indicates neither authorities, knowledge, procedures, nor their denials. The next stage was to see that, if assertion had its assumptions, so too did generalization, prediction, declaration of possibility and every one of the whole tribe of constatives capable of producing true products. Trouble followed troubles: If, as seemed reasonable, every statement, generalization and so on had its corresponding assumption, the opposite was not true, since there are assumptions for which no statement etc exists: We say, "Assume that 5/0=n" or "That stones were mortal". I applaud those who take formal logic as the study of the consequences of assumptions. No wonder it works widely! It occurred to me that that German tradition that calls our *propositions* "*Ausnahme*" was well directed. But what *are* assumptions? Earlier we argued that *questioning* could not be a mood, since almost every mood has its questions (pp. 61ff.). So too, most every constative mood has its assumptions. *Assuming* could only be a hybrid of language into something else, essentially conventional no doubt, but just as much essentially perlocutionary. We have found it natural, along the way, to say that postulates and declarations of possibility are "used" by being assumed. Here "used" can mean only used for the further purpose of drawing consequences. Altogether then, and with apologies for autobiography, *assuming* is a kind of hybrid of constative language into investigating consequences. It is, as we put it, *positing as a premiss*. This at once explains why assumptions may be true or false and why we can find no special place for them in the field of constatives.

II. PROPOSITIONAL CONTENT: AN ALTERNATIVE TO STATEMENTS.

I wish in this section to consider a rather widely held opinion to the effect that different "constatives" uniformly involve (different) "attitudes" toward common "propositional contents". Such a doctrine if true would certainly simplify matters and, in the bargain, would be immensely more appealing than my own

The doctrine of propositions represents utterance as the expression of a "propositional attitude" toward an indicated "propositional content".

doctrine of "products". This modern doctrine of "propositions", most generally taken, holds that utterance consists of an expression of a distinctive "propositional attitude" toward an indicated "propositional content". So, for example, one may **assert that**: *John will come* or *Mary will*, or **wonder whether** either thing, or **order that**: *John should come* or *he shouldn't*. The expression of attitude is an indication of mood; the content comprises everything else that is distinctively linguistic in the utterance, including the realization of referential, predicative and denotative uses. Following terminology of R. M. Hare, we style as "neustic" that phrase which, in fully articulated utterance, indicates attitude; what indicates content we call the "phrastic": "I assert that", "I predict that", "I assume that", "My opinion is that", "I conjecture that", "I propose that (to)" and "I order that" are neustics; their complements are phrastics.

The doctrine strictly taken holds that all logical relationships are among "contents".

The strict doctrine holds as a second principle that logical relationships obtain primarily among propositional contents, whence they are transmitted to utterances. If we symbolize mood indications as M_1, M_2,... and propositional contents as P_1, P_2,... and a possible combination as M_i-P_j, the doctrine, at its strictest, maintains that logical relationships of entailment among utterances obtain entirely by virtue of relations between the component propositions, in accordance with the following general rule: Infer M_i-P_j from M_i-P_k in case P_j is logically entailed by P_k. So: If "M_1" represents assertion (**I-**) "M_2 prediction ("I predict that") and "M_3" command (!), and P_1 is a proposition schematized as p, P_2 a proposition schematized as q and P_3 a proposition schematized as "p v q", then, since the proposition p entails the proposition p v q, the following holds: From M_1-P_1 infer M_1-P_3 (**I-**p ∴ I–: p v q); from M_2-P_1 infer M_2-P_3 (I predict that-p ∴ I predict: p v q) from M_3-P_1 infer M_3-P_3 (! p ∴ !:p v q).

"Contents" are "senses".

The full doctrine holds, as a third principle, that the propositional content conveyed by a phrastic is the sense or meaning of that phrastic.[16].

The principles may be relaxed to allow for inferences from conditional utterances.

These principles may, reasonably, be relaxed at several points. Some authors ask, first, that we permit the indication of "contents" unbonded to any neustic indication of force or mood, simply as utterance conditions e.g. to allow for conditional commands such as "If it's raining, take your coat" (If P_1, then !P_2); such "unbondedness" is needed to provide sanction for inference to full utterance from the assertion of the "condition", e.g. one infers that he is to take his coat ($!P_2$) from the assertion that it's raining ($\mathbf{I}$-P_1) Generalizing, a new rule appears: Infer M_2-P_2 from if P_1 then M_2-P_2 and $\mathbf{I}$-P_1.

Further rules may be added to govern inferences from neustics.

The proliferation of operator logics, by which formal methods have been extended to the analysis of modality, tense, obligation, belief, knowledge, interrogation and command, have loosened the doctrine further to allow that propositional attitudes may have their implications too; with that, however, we shall need "closure" rules by which attitudes are incorporated into "richer" contents. Moreover, if every neustic were logically distinctive, the complementary propositional content would lose its appeal.

We agree with those who have argued that the strict rule of entailment does not apply to commands.

A "non-constative" instance of this has been debated in print. B. A. O. Williams observed that one could not infer a command to mail the letter or not to mail it from the command to mail the letter. Hare's answer to the objection was that the entailment holds, but that it would be inappropriate to draw the inference because that would run counter to gricean "conversational implicatures"[17]; it is understood that no one without very special reasons would command a person to mail a letter or not to mail it; so of course we do not infer that command from the command to mail the letter. The riposte is unconvincing. Surely it makes no sense to command a person to do as he wishes; here the specific force of the "neustic" forbids the inference. Williams was himself inclined to explain the difference along these familiar lines: Inferences are sanctioned by rules having to do with the truth-values of things we say. Things said can be relevantly true or false only if we aim to adapt what we say to what we find in the world; what is said is then responsible to the state of the world, and its truth value is determined by an observation of how things are. That fits assertion, prediction, conjecture, and other "constatives"; it does not fit commands, where the relation between language and the world is reversed; here the respondent would normally aim to adapt what he does to what is said by the speaker, and respondent's action is accordingly deemed to be correct or incorrect. Rules of inference which hold for assertions are, for that reason, only precariously carried over to the case of commands.

Like counter-examples can be found among constatives.

I agree with Williams' conclusion but not with his explanation. First, it all but banishes actual cases of imperative inference (from the command "Come here!" I infer the command "Don't go away!"). Second (and this matters more), we can apply William's arguments to non-assertional constative moods in which what we say is unquestionably responsible to the world. My prediction that it will freeze tomorrow does not imply the prediction that if the temperature falls below 10°F tomorrow, it will freeze tomorrow, notwithstanding that the first alleged proposition (it will freeze) is supposed to entail the second (if it goes below 10° it will freeze). Perhaps my prediction that it will freeze does imply the prediction that it will go below 10° or the prediction that it will freeze--an inference not sanctioned by the rule of propositional content. It does not entail the prediction that if it goes to 10° it will freeze; that is no more the kind of thing that can be predicted than that if today is Tuesday tomorrow will be Wednesday.

Utterance wants a less simple-minded theory of "valence".

The doctrine of propositions superficially resembles the theory of valence in inorganic chemistry: Just as a negative valenced radical must be bonded to a positive, so attitudes to contents and neustics to phrastics. So far, however, it has only pretended to be a theory. No one in my knowledge has told us how to use it to determine whether propositions exist here or there and which propositions these are; we are still not in position to decide whether (e.g.) the supposed proposition involved in my promising to be home by seven is the proposition that I am home by seven, or the proposition that I have promised to be home by seven, or the proposition that I have kept my promise to be home by seven, or.... My own prejudgment of the matter, which I hazard in advance of the detail and elaboration I frankly doubt will be supplied, is that the idea is too simple to work. We have already found reason to think that "neustics" too have "content", which materially affects the potential entailments of utterance. We may perhaps go along with the analogy to the chemical bond, but only if we allow that the components of compounds may be variously related. The hydrogen atoms of water might bond together to give an acid ($CO_2 + H_2O \rightarrow H_2CO_3$) or be separated to yield a base ($NH_3 + H_2O \Rightarrow NH_4OH$). I have already suggested that what typically happens in the case of "volitives" is that referring and tense indicating elements are converted into addresses and indications of "aspect" as in "John, you will come!". What were referential and time indications are now conjointly appropriated to the mood indication itself, though still bonded to the uses which indicate what kind of action is required; only now the major break comes at a different place. This is a very natural transition because it is in the nature of a command, generally taken, that the required deed is to be done by someone. The supposed propositional radical is partially dismembered and its elements redistributed.

This picture of the situation obviates the need to invoke duplicated elements in the "deep structure" parsing of the utterance. According to the duplication thesis, a command "You shall go!" has the form "I order you: you go" and an expression of intention "I will go" the form "I order me: I go". I speculate instead that the connection between the mood indication and the identification of the intended act is naturally mediated by the tense indicator. Particular acts are done at particular times somewhere indicated in the utterance; but it is also part of the sense of command that what is required should be done subsequent to the occasion of speech. A time indicating future tense thus becomes a "aspectual" in the mood indicator or "neustic" while retaining a connection with what indicates the required action sort in the "phrastic".

My "products" are better based than propositions are.

My own doctrine of "products" allows that these various statements, predictions, conjectures, promises, etc have various properties and relationships that owe as much to the realization of mood indications in their producing acts as to the realization of those referential, predicative and denotative uses that are licensed to play into the expression of propositions.

Our doctrine of products is at least as well adapted as is the doctrine of propositions to explain the extended applicability of formal logic.

Some will contest our faith in "products" on the ground that the blanket undetailed assumption of propositions has been magnificently sustained by the impressive and continuing achievements of formal logic, which can, with only an occasional hitch, be applied pretty directly to every kind of constative and sometimes even to other kinds of utterance. The extended applicability of formal logic, surely a source of satisfaction, is an easy mark for the proposition doctrine but a sticky problem for us. An accommodation is suggested by the question we touched on above over determining what proposition is promised. The extended applicability of logic confirms the centrality of assertion. With every non-assertional kind of utterance are associated a nest of different but related statements in which are formulated reports to the effect that the performance was attempted, or succeeded, or was "fulfilled", in a variety of ways, and the inferential potentials of these statements, very likely formulated in the same words as would be employed in the reported-on non-assertional utterance, may, if caution is exercised, be imposed as patterns to codify the inferential potentials of the resulting non-statemental products. If we know that a statement that a person does A is inconsistent with a statement that he does B, then we may suppose that he is sanctioned to infer the command that he is not to do B from the command that he is to do A[18]. A statement that a prediction of E is fulfilled implies the statement that E occurs, and perhaps also another statement that another event E' does not occur. We may use this to show that the original prediction implies the different prediction that E' does not occur as well as the statement to the effect that the predictor did not or (differently) should not have predicted E'. It's all there. Why lose any of it? But what does all that have to do with the application of formal logic? Only this: the application may be justified and perhaps it may have to be. So, for example, the logical relations which obtain between universal and conditional statements may, with due caution, be employed to standardize the parallel relations between generalizations and conditionals (which can only be non-assertionally propounded) and other constative products. If a generalization is true only if any corresponding but circumscribed universal statement is (pp. 141f.), then the inferential rule of specification will also apply to the generalization. Thus you might infer from "All swans are white" that my pet swan is white; however, it is less clear that you could infer that, if my pet raven were a swan, it would be white. Again, we might, for the purposes at hand, simply replace a conditional with a formally analogous conditional statement--a useful and safe pretense so long as we know where we are. But if we lose our direction here, we may be stopped in our tracks by the kind of argument Lewis Carroll employed to show that conditionals are not everywhere replaceable by their statement proxies ("Achilles and the Tortoise", *Mind*, 1895, pp. 278 ff). Actually, as it now seems to me, the doctrine of statements, because it reaches to exceptions and limitations, has the advantage over the doctrine of propositions in this matter of the applicability of formal logic.

Two last gasps for propositions: assumptions and parentheticals. It may be countered that the role of proposition is in fact played by *assumptions*. We anticipated this thought in the last section, where we *argued* that assuming is a perlocutionary category of action. The conclusion, if right, blocks this attempt to revive propositions as locutionary contents.

Some may now, finally, urge that unasserted propositions are obviously adverbially posited as conditions for assertion, prediction, command or whatever. Are we to take it, then, that propositions are fixed in such parenthetical clauses? The proposal is desperate, but the challenge interesting. Let's look at a case. My wife and I have been arranging a bridge party. I hang up the phone and report, "Well, now we've got our four tables" (*q*). I hesitate briefly and then add, "Well, we do if everyone shows" (*p*). I wouldn't have said the last if I were not talking to some further purpose--to inform my wife or to guard our plans. That we now have four tables is something I think I know, and I say so: Conditions of success for assertion are indicated: people can know this sort of thing; there are authorities (namely, myself) and the report can be proven out either way by applicable procedures. Now I wouldn't think I knew we had our four tables if I thought there *would* be a no-show. The "if" clause, for perlocutionary purposes, guards the assertions by indicating that there is no reason to think we don't know something for *this reason: p*. Here the "content" of *p* could be thought of as identified by reference to the *statement* that everyone shows. However, I don't assert *that* statement, perhaps because I don't think I know that everyone will show. The "if" indicates that I am *not* indicating as a condition that this is the sort of thing that could now be known; the "content" is simply that of an utterance-kind analyzable as an attenuation of assertion (see pp. 103f.) This is *one* case. "*q*" might have represented something other than assertion; generalization perhaps. Similarly, the *p*-content might have been fixed by reference to some other utterance-kind (e.g. "If all birds *are* warm-blooded"). The point is that my description of the case (which I hope seemed perfectly natural) did not "posit" a proposition for *p* to be, but *fixed* the "content" with reference to full-bodied utterance.

III. ON EXPLANATION.

Thesis: Explanation is a kind of action that could have resulted as a hybrid of "constatives" into making-comprehensible.

1. Constative language appears usually in the service of the acquisition, dissemination, control and organization of knowledge and for the enhancement of understanding. Most constative kinds, taken pure, result by abstraction from such "richer" deeds as that of informing others of the facts. *Informing* is a more "natural" kind of action than *assertion*; it would therefore be wrong to say that informing *is* a hybrid of asserting into getting others to know. It seems to me, nonetheless, that we could explicate *informing* as a kind of action that *could have* resulted as such a hybrid, in the sense that the conditions of success for informing are a union of the conditions of success for asserting and for getting another to know[19]. My first and main thesis in this section is that Explainings-why or how-come[20] can similarly be explicated as actions that "could have" resulted as hybrids of constative utterance into something having to do with the enhancement of understanding. After arguing for that thesis, I go on to consider putative assertions both that one thing explains another and that one thing causes another.

There can be no explanation unless something in particular stands in need of explanation; the explanation is in relation to what is known.

2. I begin with some observations on the "nature of explanation", which I hope will seem obvious to everyone.

First, we could not succeed in explaining anything unless something specific challenged the understanding and needed explaining. The specificity of the *explicandum* should be stressed. We explain the ringing of a bell perhaps by observing that a photoelectric circuit has been rigged, not by citing Einstein's theory of photoelectric emission. On the other hand, Einstein's theory does provide an explanation of photoelectric emission. The fact that a family is eating its dinner at dinner time needs no explanation-why. On the other hand, one may reasonably wish for an explanation of why Americans dine earlier than Spaniards.

Second, explanation is relative to what is known. There is no explanation without a "need"; and then adducible factors are limited by what we know. On both counts, conditions of success for an act of explaining are fixed only in relation to what is already known, either generally to the world at large or specifically to interested parties. If we all know that a photo-electric circuit is rigged, then that fact does not explain the ringing of a bell.

Explanations are various.

Third, explanations are of many kinds. A body falls, and we explain that occurrence by saying that the thing broke loose from a bracket. That is one thing to explain, and one kind of explanation. The equivalence of gravitational and inertial mass is quite another order of fact wanting quite another type of explanation. It is unlikely that we shall ever achieve a complete taxonomy of explanatory "factors".

Fourth, the same one fact may be variously explained. The bell rang, we explain, because the circuit was rigged, or because the power was on (we didn't expect it to be), or because someone interrupted a controlling circuit. ("There must be someone there.")

Any explanation may be "sufficient"; none is "total".

Fifth, any such particular explanation could be "enough" in itself; on the other hand, there can be no "total" explanation of anything, any completed list of necessary and sufficient conditions for the ringing of the bell, except that fact itself.

There is no explanation without constative utterance, but understanding can be achieved with nothing said. Utterance by itself is never enough to secure explanation.

3. *Defending the thesis.* These observations do not encourage the hope for achieving much in the way of a single general characterization for acts of explaining. I believe, however, that what I have proposed and shall now argue for, that explaining is an action kind that could have resulted as a hybrid, can be generally defended and has significant consequences. First, defense. We show that *explaining* could be a hybrid.

Coming to understand is not necessarily the result of utterance. Facts can be displayed without a word being said along the way, and one might thereby come to see the situation correctly for what it is; we can come to understand why things happened as they did, and even see the explanation, but in none of these cases has an explanation actually been provided.

No explanation is given until something is actually said. There is no obvious limitation on what kind of constative utterance might have been harnessed to the task. You explain his manner of action to me by pointing out that he is left-handed, or by remarking on his limited education, or by offering some generalization about Turks.

Some kind of constative utterance always contributes to the giving of an explanation.

So there is no explanation without constative utterance. But utterance alone is not enough. What I am told may for good reason fail to quell my curiosity or ease my understanding. A near proof of this is the following: Explaining at least requires the use of language. Suppose that it be assertion, for that sometimes figures and no other appears a better general candidate. Now the assertion may succeed and yet produce a false statement. But in that case the act of explaining would have failed. Hence explaining cannot be asserting, and presumably no other use of language either.

4. We have now argued that neither utterance nor *making comprehensible* are alone sufficient for explanation, but that taken together they are. That's the sense of my claim that explaining-why could be a hybrid of constative utterance into providing understanding. That utterance by itself is not enough has consequences for philosophers.

Various constative moods can contribute to explanation and no one of them can be resolved entirely into its explanatory uses.

First, no constative mood of language can be identified with or explicated solely in terms of explaining. This distinguishes *explanation* from *generalization, description* and *prediction*: these are purely and simply language; explaining is not.

Second, if explaining is not a constative mood, then there is no advance limit on which constative moods contribute to explanation. In fact, many do. That is a hard charge for those who have simplicistically held, for instance, that explanation is logical derivation from generalizations, or subsumption under laws of nature or the testing of hypotheses. These theories of explanation are at fault not so much for their reliance upon derivation, subsumption and testing, for these types of performance are no less perlocutionary than is explanation; nor is it their tendency to override distinctions between different sorts of explanatory factors which is to be criticized at this point; their error lies simply in neglecting the variety of constatives which sometimes figure in giving explanations.

The achievement of understanding is a result of making familiar.

5. If it is correct to hold that explaining could have resulted by the hybridization of constative utterance into something else, we must next consider what that something else could be. I am not sure that there is an answer to this question, but I think there may be, if only because successful explaining must have something to

do with providing understanding. How is that understanding to be secured? I have no better candidate-formula for this than the old if vague idea of *making familiar*. *Familiarity* is as "relative" as the need for explanation. The familiar fact of perception, for example, becomes "problematic" when thought of as a relationship between an experience and a *perceptum*. Some explanatory theories of perception try to make somesuch relationship seem familiar and compelling; others appeal to photo-electricity. Again, no explanation is normally needed for a ball falling to the ground when dropped, or for its not falling if it's caught in a chair. We are all familiar with that sort of thing. If, contrastingly, in the course of theorizing, we wonder why things don't fly off in random directions or simply persist where they are, we may look for and find an explanation of that general circumstance, and then use it to explain the motions of the planets as a kind of falling, at which point different orders of phenomena begin to look alike. (Hence the somewhat confusing appeal to "generalization" in theories of explanation.)

Making familiar may be secured by fitting a phenomenon into a pattern, or by fitting a pattern onto a phenomenon.

6. If making familiar is not, as I have argued, simply derivation from general rule, in what other way are we to explicate it? I believe that we should think of it as establishing a correspondence between the *explicandum* and a familiar pattern of occurrences, states and circumstances.

That is commonly done by finding some feature of the *explicandum* which answers to one strand of the pattern, e.g. the circuit is energized, or the cell is installed, etc. With this, we can better comprehend why the explaining may succeed by mention of a single such factor. The others are so to speak given in place as soon as we see the *explicandum* as an instance of the kind of thing which might be fitted into the familiar pattern.

Differently, one may impose the pattern as a whole on the phenomenon and then, to test for success, look for factors in the phenomenon corresponding to strands in the pattern (e.g. "Regard a planet as a body falling into the sun").

In summary, then, my proposal is that explaining may be analyzed as an action kind which could have resulted from hybridizing constative language into the purpose of establishing a correspondence between a problematic phenomenon and a familiar pattern of types of (explanatory) factors.

My proposal is compatible with Mill's remarks (*Logic*, chapter 5) about commonsense explanation, though not with his notion of philosophical or scientific causation explicated in terms of sufficient conditions.

Supposed assertions that "say" that a C explains an E produce statements that are verified by establishing a 'correspondence' between E and some feature C of an explanatory pattern. The obtaining of E is a condition of success for such an assertion, and the obtaining of C a condition of its truth.

7 *Explaining* a fact E by reference to a fact C is one thing, and *saying* that C is the or an explanation of E is another. Performances of the latter kind may stand as utterances pure and simple. It is unlikely that such utterances ever qualify as assertions strictly taken; however, I shall now simplifyingly assume that they may.

My proposal in the matter of such supposed assertions is that, when successful, they produce statements that would be verified by exhibiting a correspondence between the E and an explanatory pattern either representing C or incorporating a factor-type corresponding to the factor C. Let us confine our attention to the latter sort of case in which both C and E are facts.

A condition of success for such an assertion is that E should actually obtain: If E were not so, the supposed assertion must fail. If, contrastingly, C does not obtain, it would seem that the produced statement would be false. (It could be false for other reasons too.) The obtaining of C is, accordingly, a condition of success, not for the utterance, but rather for the verification act.

"Explains"--contexts are "opaque".

The contexts "C explains...", "---explains E" and "---explains..." are all "opaque", in the sense of Quine. "Why is he here?", I ask; you reply, "Because he is the building inspector." If that is correct, then it would be true to say "The fact of his being the building inspector explains the fact of his being here". But if he is the building inspector, which is a condition of truth for the statement, then the fact of his being here is the same as the fact of the building inspector's being here; but it is clear that this statement, *The fact of his being the building inspector explains the fact of the building inspector's being here*, is not true in the same circumstances as would be the original one. Also, if the building inspector is the man who just arrived, then the fact of his being the building inspector is no different from the fact of his being the man who just arrived; but it is not true that his being the man who just arrived explains his being here. More abstractly, if it is true to say "C is the explanation of E", it would normally not be true to say "The explanation of E is the explanation of E"; that utterance, far from being a tautology, probably says nothing at all, for it undercuts the possibility of verifying a statement of explanation by establishing a correspondence between pattern and explicandum. "...explains ---", though grammatically a predicate, should not be thought of as used to ascribe a relation to pairs of things; in this respect, it is like such locutions as "---is to the left of---" and "...knows that---".

Is Causation a relation?

8. Is "causation" a relation? An asserted fact like a is to the left of b might also be differently asserted by saying "a is north of b"; that assertion, which produces a different statement of the same fact, ascribes a genuine relation to the pair <a,b>. So, too, if *C explains E* is a fact, perhaps that same fact could be otherwise asserted by ascribing a relation to <C,E>. A candidate statement of this kind would be that *C caused E to happen.* We have, in this transition, moved away from "explained" to a specification of explanatory factor. Generally, we hypothesize that any fact of explanation would also be a relationship between a specific explanatory factor and the *explicandum.* Of course, the reverse conclusion would not hold, since (e.g.), although C caused E to happen, C may not explain E, perhaps because something else does or because E needs no explanation. Explanatory factors may be present and operative when not appropriate as explanations. An argument for this is that, while to say "The explanation of E (which may be the cause of E) explains E" scarcely makes any sense at all, "The cause of E caused E" seems unexceptionally true. Also this: if a = b, then, if E causes a, E causes b; but if E is the explanation of a, E may not be the explanation of b; e.g. if the dissolution of the sugar = the sugar's dissolving at such and such a rate, dropping sugar into the water may cause that event under either specification; but surely dropping the sugar in the water does not explain its dissolving at such and such a rate.

Causes may be identified as such without having to be invoked for purposes of explanation.

It might be objected that explanatory factors such as causes and beliefs are selected and classified as such because of their explanatory roles, and that therefore the concept of (e.g) a cause is inapplicable except in relation to explanations. I doubt the "therefore": The conception of gravitation was first introduced in connection with the explanation of the trajectories of neighboring bodies; but surely bodies have the gravitational potentials they do without that necessarily explaining anything at all. The situation resembles that of *sortals*: While the introduction of specific sortal characterizations is a reflection of human interests, still an animal may be said to be of the sort it is, without the speaker having any particular interest in that sort of thing at all; he identifies a robin for what it is even though he doesn't care a tweet about robins.

Are causes "phenomena"? A traditional crux in philosophy.

9. Berkeley, for one, would have denied the supposition that *...causes---* is a relation among observables and that causation is a perceivable phenomenon (=idea); Hume followed. Their argument was a request to notice. So watch me as I strike a match. Allow that rubbing the match on the surface causes the flame. But what do you see? Well, the movement, the flame and perhaps other smaller things, but nothing that qualifies as the *causation.* Saying that the rub caused the flame may be correct as an expression of inference or well-grounded "transition of belief"; but that inference was no part of the phenomenon you observed. The Berkeley-Hume "consideration" is persuasive; but the conclusion unargued. We have already seen another side to the story.

The "opacity" of "...causes---" argues that causation is not a phenomenal relationship.

10. A relationship, however dignified, as "cause" or otherwise, would obtain if and only if the *relata* are thus related *however* those *relata* be referred to. Dagfinn Føllesdal has argued that we cannot freely "quantify into" a context controlled by "causally necessitates" taken as an operator[21]. That almost establishes that the *predicate* "...causes---" is opaque, and accordingly, that *causation* is not a phenomenal relationship. We could illustrate this where a cause and its effect are identical but asymmetrically so-called. Classical considerations of *causation* give passage to that conclusion.

"Classical maxims" about the cause reaching to the effect imply that causation is not a phenomenal relationship.

11. *Thesis*: In every case of efficient causation, the "proximate cause" and the "proximate effect" are the same. I do not hold that all causes are identical with their effects, only that whenever there is a cause that has an effect there is a cause that is identical with one of its effects, which effect cannot be truly said to cause its cause. Take the exemplar of two colliding billiard balls: the movement of the impacting ball causes the impacted ball to move; those are certainly distinct movements; but there is causation here only if the one ball impacts upon the other; the "proximate cause" of the movement of the second ball is the impact of the first upon it; that will happen only if the second ball is impacted upon at the time of impact; that episode of being impacted upon is the proximate effect of the movement of the first ball. My argument illustrates ancient maxims that a cause must reach to its effect and that there is no action at a distance; I hope it will be allowed me that the proximate cause of the second movement is also the proximate effect of the first movement; the proximate cause and the proximate effect occur together at the place and time of impact; such impacts are "boundary" events and are unique to the movements they separate; the impacting upon and the being impacted upon are one and the same boundary event, at once proximate cause and proximate effect. The first ball impacting upon the second causes the second ball to be impacted upon by the first, not the reverse; but the "two things" are the same. That seems to me to establish that we cannot always substitute identicals into the context "...causes---" and, accordingly, that *causation* is not a phenomenal relationship[22].

Causation occurs within a "nexus" where connections are always made at some point. The representation of such a nexus is an explanatory pattern.

12. I have now defended Hume's thesis that causation is not a phenomenal relationship by arguing that the predicate "...causes---" is "opaque". I am not ready to endorse Hume's further opinion that the "necessary connection" we sense to obtain between happenings respectively denominated as "cause" and "effect" is grounded only on a "transition of belief". The movements of the two billiard balls are, upon impact, connected into a single happening. The movements in question are of bodies and our description of the phenomenon depends upon our being able to identify and separate those bodies. It may be that true statements of identity and distinctness are inescapably necessary. (See Chap. 11, #3, p. 40) So it may be that the connection of the two happenings, the one called the cause and the other called the effect,

cannot be conceived or observed except as a necessary connection. But now this "nexus of events" is in itself a single happening that may be observed without any observation of or calling of causes or effects. However (and this summarizes my conclusion), if we do call the nexus "causal", then we see it not only for the connection it is but also as being represented by a causal explanatory pattern[23].

NOTES

[1]Before ultimate revision, my text included several paragraphs on *judgement* and *estimate*, expunged upon Jesus Ilundain's and Nancy Kendrick's observations that these were, all of them, cases of expressing an opinion.

[2]"Could be known" may be wrong. A reader worries about conjecturing that something will never be known. My guess is that such expressions would better be called guesses than conjectures.

[3]So-called hypotheses in mathematics, e.g. The Riemann Hypothesis, are better called "conjectures".

[4]But not everything. A reader wonders over the description of the geographical relations between Italy and Greece: I wouldn't cast a geographer's cartographic delineations as descriptions. I'm not clear under what separate heading they should be domiciled.

[5]See Wittgenstein's *Philosophical Investigations*, Part II, #i. His (poorly selected) case was that of *hoping*.

[6]Not to include such mathematical truths as The Law of Large Numbers.

[7]Written in the late 1970's; the scene had expanded a lot by 1991!

[8](iii) and (iv) are adapted from Rescher, with a credit to Quine; (v), attributed to R. Thomason, was suggested to me as a counterexample by P. Maher. [I was late in coming to appreciate that the "if" in (v) may be understood in two quite different ways, first, to emplace some words incidental to a remark on my general innocense or, second, to say that my wife would see to it that I would never pick up on her infidelities; I originally took it in the second way.) (vi) and (vii) are late additions, stolen from J. Bennett's "the Phlogiston Theory of Conditionals", *Mind*, 1988, pp. 509-28, an interesting paper I mostly disagree with; Bennett's own thesis actually has little bearing on what now follows.

[9]Since composing this note, I have profited from reading a paper of Bede Rundle simply titled "If" (*Proc. Arist. Soc.*, 1984, pp. 15-30)--a fine-textured example of "good old fashioned linguistic philosophy"--which I reckon "best reading" on the general sense of "if". I concur with Rundel's chief if provisional conclusion that "if" is not always *conditional*. I think that he wants to hold that "if", taken most generally, is used to present a consideration, possibility or supposition. I prefer "pertinencies", chiefly having to do with reasons for saying, partly because we have lots of non-*iffy* locutions to do the jobs Rundel mentions and partly because my formula permits cases (of which *he* suggests examples, e.g. "To put it briefly, if crudely, it stinks") which do not involve the idea of a suspension of assertoric force, an idea to which Rundle appeals.

[10]Melnick reminds me that some disjunctions may be "non-material".

[11]That conditionals are "second level" is already found in the writings of Reichenbach, (*op. cit.* #62), Quine (*Math. Logic*, p. 29), and Nagel (*Structure of Science*, p. 71), and sounds like what Ryle was getting at when, in considering Lewis Carroll's puzzle of "Achilles and The Tortoise", he proposed that conditionals are "inference tickets". ("'If', 'So', and 'Because'", in *Collected Papers*, Vol. 2.)

[12]"**I**-You smile" => "You, smile!" and so "**I**-You can smile" => "You, can smile!"

[13]One may, I believe, also non-conventionally refrain. In "Maybe" utterances the speaker indicates considerations of success for refrainment.

[14]I cannot forebear from an etymological speculation: We noticed that "could" serves, not only as a "modal auxiliary" on its own,. but also as the proper past tense of "can". I speculate that this originates from the familiar germanic transmogrification of pasts into subjunctives, where subjunctives may convey the force of an unactualized possibility. (Cf. earlier remarks, on p. 155, *re* the relation of "might" to "may".) A subjunctive of "can", which is what I speculate "could" once was, would have the grammar of "can" and very nearly the meaning of "may".

[15]*Supposing* may occur more generally on the side of an argument or discussion, and is not restricted to premissing. I would characterize it as a "positing as if known".

[16]Frege, who fathered the modern doctrine of propositions, did indeed subscribe to all three principles, but carefully restricted the application to assertion and interrogation. See pp. 120f. above.

[17]Williams, "Imperative Inference I", *Anal. Supp.* 1963, pp. 30-36; Hare, "Some Alleged Differences Between Imperatives and Indicatives," *Mind*, 1967, pp. 309-26.

[18]But this may have things backwards. A rule of exclusion for command-copliances may be a "deeper fact" than the exclusion of contraries or the Law of Contradiction. See p. 262, where we shall review Melnick's reading of Kant on this point.

[19]It dawned on me, too late, in the course of preparing index, that I could have more clearly put this by treating the language as an attenuation of the "could-be-hybrid".

[20]In contrast to explanations-what, how-to, who, where, when etc.

[21]"Quantification Into Causal Contexts", *Boston Studies in the Philosophy of Science*, Vol. Two, pp. 263-274.
My own argument to follow achieves a fuller presentation in "Hume Was Right, Almost; and where He Wasn't Kant Was", pp. 435-49 in *Midwest Studies in Philosophy* 14, (Minneapolis, 1984).

[22]"Phenomenal" matters. The conclusion is compatible with a doctrine, one to which my colleague Melnick subscribes, that *causation* is a relation between *kinds* of phenomena which (in my presentation) may figure as entries in those causal explanantory patterns to which I appeal in the main text.

[23]Unless we docket every *dynamism* as a case of causation. That way of talking, which is (perhaps) continuous with Kant, lines up pretty well with contemporary physicists' accustomed use of "causal law"; Melnick also favors this "non-explantory" notion of *causation*.

APPENDIX C

KNOWLEDGE, INFORMATION, ACCESS, CERTAINTY AND INQUIRY: PRELIMINARIES TO A RATIONAL EPISTEMOLOGY

Our explanations of assertion at several points presuppose an understanding of "know".

1. "Assertion" is a word for meaning and saying what one thinks one knows to be so, and we have, within our theory of action, provided a formula for *assertion* in terms of conditions of success having to do with there being subjects who could know what could be known. Assertion as a mood stands in evident perlocutionary connection with the life of knowledge--a matter of everyday human concern that overlays and mingles with such other "ways of life" as getting others and ourselves under control, cultivating spiritual integrity, making ourselves attractive and letting off steam. Our analyses of *assertion* and its congeners disconnects them from these larger matters. Still, to fix a central position for assertion within the uncertain boundaries of constative utterance, we take our bearings upon the surrounding ranges of knowledge.

Knowledge is "problematic" and, for systematic purposes, stands in need of explication. This appendix is a preliminary to such an investigation.

I once felt unconstrained to enter upon the bailiwicks of "epistemology". *Knowledge*, after all, is a "common notion" without which we could not speak of or otherwise deal with creatures in our accustomed ways. It seemed to me that, by thus fastening onto knowledge, we were securing our explanations to rock bottom biological fact.

That is not good enough. "Common" though it is, *knowledge* is notoriously "problematic" and likely to trip us up in our theoretical endeavors. I would have been gratified to fall back upon some established "rational epistemology" for assistance. It doesn't exist, and I must head off on my own to scout those forbidding territories. I shall be satisfied to draw a number of distinctions, inexact and impressionistic, but obvious enough to keep us out of pitfalls and also (I hope) serviceable as a kind of large scale map onto which the general terrains of knowledge can be fitted. The presentation, though perhaps too long for most readers, is only a "preliminary to epistemology", burdened with distinctions of language and laced with talk about "know", its cognates, stand-ins and complements; it is destined to be as inconclusive as "linguistic philosophy" always is. It's here to protect the formulations of Chapter 2 and Appendix B; I pray that it also may assist the cumulation of results in the theory of knowledge itself. Readers innocent of epistemology and confident of their grasp of "saying what one thinks one knows", "S could know F" and "F could be known" are invited to forego this appendix as something unnecessary for their understanding of the main text. Other readers, less happily disposed, are advised to consult the marginal summaries, dipping into the text only as they feel a need for amplifications.

Knowledge is a fact, and may be better or worse.

2. I have no doubt that human knowledge exists as a matter of fact. I am unable, however, to prove that we do have this knowledge, not, anyway, knowledge of *truths*, and the systematic developments of this treatise are meant to be consistent with philosophical skepticism.

> The "philosophical" or "cartesian" skeptic doesn't worry over the "practical knowledge" of dogs and cats, but does express doubt that we humans ever have knowledge of truth and falsity. I have argued that any such "expression" implicates "know how" and I shall urge that know-how is authentically knowledge and always companion to knowledge of fact. It might seem from that that we shall have achieved a "transcendental refutation" of philosophical skepticism. I would be glad to have it so. But, alas! The factual knowledge implicated in skill or "know-how" may be "merely practical", e.g. Fido's knowledge of his mistress or of the way home. Philosophical skepticism, which exclusively concerns our speculative knowledge of truth and falsity, so far makes unrefuted good sense.

Still I am sure that philosophical skepticism is false and that knowledge is an ungainsayable fact of everyday life. Some doctors know and others don't that certain drugs are counter-indicated for patients with a history of renal deficiency. Is it only prejudice for me to prefer a doctor who has access to such information and who reads, comprehends and remembers over one who has lost touch? Knowledge and ignorance about drugs and kidneys really matters. We sometimes seek a "best opinion" on medical urgencies. But knowledge whether a certain drug is counter-indicated is not just a best opinion founded on the best evidence. How, you may wonder, can knowledge be better than a best opinion? Well, knowing may itself be "better or worse": "She knows him (or "that") better than I do"; "I know that as well as I know my own name". While there may be no "best" grade of knowing, there are lots of gradations here, all contrastable with and some less worthy than a "best opinion". (A urologist's best opinion costs more than a pharmacist's knowledge.) Knowing something may perhaps be confident opinion based on evidence--we'll see more about that later on; still, a subject's knowledge is not graduated solely on a scale of confidence or with reference to a balance of evidence; whether and with what qualifications one knows a fact is at least partly a matter of subject's access to established information; the certainty of that information is, then, at least in part, a matter of how that fact fits in with other known things.

3. *Remarks on the meaning of "can-know"*. Among the conditions of assertion are that something can or could be known and that there are authorities who can or could know. Similar formulations were featured in my attempts to draw lines across the field of constatives. Such "can-know" sayings couple two of the peskiest of English verbs, whose refracting semantics creates linguistic illusions that would baffle our enterprise if left unnoticed. I shall accordingly lay over to argue for a dull but important point, namely, that both of these verbs have what I call "schematic meaning".

"Can-H" has a meaning that is "schematic" for asserting the obtaining of conditions thought normally required for a subject's H-ing.

The "amphiboly" of negative "can"-sayings is familiar and systematic. Consider "The load can't be too heavy". Should the speaker be taken to be saying that there are no limits on what the axle will bear or something else that implies that there are such limits? Well, he may mean either thing with those words. It's part of the very meaning of "can" to enable "can"-saying speakers to mean different things and, accordingly, different things may be denied by saying "can't", e.g. that there are limitations on load tolerance or that there is an ample margin between present load and load limits. These different conditions, in any given circumstance, have a common connection with *what* it is then said can or cannot happen. So, in explanation of the *meaning* of "can", I submit that one who says "A can-H" means to assert the obtaining of a condition thought to be required for the occurrence of A's H-ing. Note well that, by this formula, while the meaning of "can" is certainly "dispositional" with respect to H-ing, the meant or asserted fact is simply one or another condition on A. The "can"-saying speaker brings our attention to a condition in connection with and as disposing to happenings of the H kind; still the condition is itself no more or less a disposition than is any other fact; it is a state, circumstance or transaction separate from whatever sequels it may have.

The "schematic meaning" of "can", which makes it so very useful a word in everyday conversation, is an open invitation to inquire what a "can"-saying speaker *meant.* So you may wish to know Shwayder's meaning when he says that a speaker can or could know, or that such and such can or could be known. Well, so far, I don't mean anything in particular! My "can know" rules are only *schematic* stipulations contrived to cover a variety of conditions.

More about "can" can be dug out of the side-passage to our examination of "It may be: *p*", on pp. 151-155. I now proceed to some similar observations about the meaning of "know", which is etymologically cousin to "can".

4. In saying that a person knows the trail back to camp, speaker may mean that subject has been there, has studied the map, remembers what he saw or was told, or that he once learned and will remember when he spots it, or that he once learned although he doesn't remember, or that he once learned and could remember if he tried, or that he could figure it out; again, a subject may (or may not) *notice*, *realize* or "*see*" what he otherwise may not (or may) know. Certainly one may learn, even with a sense of discovery, what he *already* knows.

I suspect that Aristotle touched the same point in both the first and the ultimate chapters of his *Posterior Analytics*; again, Plato, in *The Meno* could be read to be arguing that the slaveboy must already have

> otherwise known what he now comes to realize in the matter of constructing a doubled square, and the aviary passage of *The Theaetetus* seems to me to have like implications. Familiarly, Aristotle used such distinctions in knowledge to advance his discussion of *akrasia* in Book VII of *The Nichomachaen Ethics.*

Briefly said, one may know in one coverage and concurrently in another coverage know or not know the same thing.

"Know" has "schematic" sense.

What kind of "coverage" is this? There are, I think, only two plausible options: either "know" is *generic* for such things as *notice, realize, be acquainted with, have learned, remember* etc or (in the manner of "can") it is *schematic* for these. Now the use of a generic name such as "mammal" does not convey an identification of any one or another of its species; the use of a schematic term, by contrast, is used to mean one or another specific item of its domain. I ask the reader to check his use of "know" against that test, and hope he will decide with me that it's schematic for such things as remember, realize, notice, understand. For example, if I notice a person's departure, I know she has left. There are two further considerations (not proofs) against a generic understanding, and nothing I know of against the schematic understanding. The *first* consideration starts with the observation that the different species or subgenera falling within a genus are supposed to be exclusive, e.g. no equine is both a horse and a zebra; but certainly a subject may well realize and understand what he remembers, so these are not proper species of a single genus of knowledge. *Second,* I don't see where we are prepared to classify and survey these or other appropriate "species" of knowledge, in the style of an entomologist; we lack and seem destined forever to lack the subclassifications needed to make sense of "generic knowledge" in present understanding. In contrast, the use of a schematic "know" may be openly and explicitly available for the helter-skelter assertion of knowledge-conditions including ones that have never yet been noticed.

The great advantage of having (unitary) schematic meanings for such words as "can" is that they enable speakers (unambiguously) to "mean" different kinds of things. I say that he can hit the basket from there; you say he can't; we both use "can" in the same sense, but we each unambiguously mean to assert quite different things: I assert that he has the skill and you deny that he is open. That is what we have found with "know": you may deny he knows because he has forgotten that fact which I assert he knows because he learned it once.

For asserting of subjects the obtaining in them of conditions for the utilization of information.

5. What are the conditions we variously assert by saying "know"? What binds them together as knowledge?

They are all of them, as it seems, *states* of a (knowing) subject. Now, I submit, such a state is called "knowledge" only if it can also be seen, by the observer, to be a kind of *readiness state for action.* The meanings of "emotion words" (e.g. "anger", "pity") fall under a like formula, as signifying readiness states for action. We therefore need further explanations for "know". I achieve that by adding that the action in question should be a *utilization of information* If I am right in all this, then knowledge is always so-called in relation to information. Note well, however, that knowledge, by these formulas, remains what we said it is, a *state* of the knowing subject; it is *not* a *relationship* between that subject and that information in relations to which it is identified[1]. So, "knowledge", by this, covers a variety of states of a subject, that are conditions for that subject utilizing information; e.g. that he has acquired, retained, comprehended that information. In saying that a subject knows something, we mean to assert the obtaining of *some* such condition. A subject knows something only if it is in *some* such condition for the utilization of information; and it doesn't know if there is *any* such condition it fails to be in. "He knows" and "He doesn't know" may be compatible, as "He has met her" and "He hasn't met her" are not. *If* it made sense to contemplate a "totality" of such conditions, then a subject would undeniably know something if and only if he was in *all* those conditions. That "all in" state of knowledge, however, is an unrealizable figment, because only schematically definable.

My formula for knowledge--that it is one or another condition of a subject marked as being for that subject's utilization of information--calls for commentary. I shall try to be brief[2].

My first comment is a big one and touches on issues that are bound to be controversial. A state of a subject, we said, is truly called "knowledge" *only if* it is also seen as a state of readiness for the utilization of information. States of a subject that are truly called true beliefs also meet that condition, which observation effectively defeats the prospects of converting our formula into a *sufficient* condition for knowledge. Now beliefs may *always* be qualified as being "of" something. That is also true of certain but perhaps not all knowledge states. I am myself inclined to let in as qualified states of knowledge certain of those conditions of subjects that do meet our condition but which are not properly said to be "of" anything.

Knowledge "for" the utilization of information may or may not be "of" that information.

"Perceptual Knowledge" always implicates an "innate" state of knowledge that is not "of" its information.

A particularly interesting putative kind of knowledge which I doubt can properly be said to be "of" its information arises in relation to so-called "perceptual knowledge". Looking at the beast and seeing it there, I come to know that there's a horse on the road. Seeing the horse, puts me into a condition for utilizing the information that there's a horse there: I stop or drive around it. But sight sometimes produces no such knowledge or, again, may do so in a curious manner. I was riding along in a bus, looking out the window; later, a police officer stops the bus and inquires whether any of the passengers happened to see the assault that occurred on a corner we had passed. Well, I didn't notice it at the time and had then no other kind of perceptual knowledge of the matter; but now something clicks, and I report that I did see the assault; so now I know. The earlier perception did not then produce knowledge, but now it does. The first consequence of this little story is that perception of itself is not yet qualified as knowledge. Perception, to be sure, can produce perceptual knowledge; but then only if the subject is in condition to "convert" the perception. That "converting condition", I submit, is a state thought required for the utilization of information, but not to be characterized in terms of that information. *Knowing-how-to*, as we shall presently notice, is another kind of state that so far qualifies as knowledge by our formula, but is not "of" "anything". Perceptions themselves, note well, do not qualify as knowledge under our formula, for perceptions are not states at all, but rather happenings. By way of contrast, *vision* may indeed qualify, though I rather hope it would be ultimately thrown out by appeal to the sense of "readiness state for the utilization of information". The "converting condition" I have been promoting as a kind of knowledge is, some of it anyway, "innate". Leibniz would have likened it to instinct.

Leibniz argued against Locke that we could not have "ideas of sensation", e.g. colors, unless we had a capacity to distinguish colors;

> this, he held, showed that we had to have an innate idea of distinctness, which he also likened to an instinct for inference. (That likeness recalls Lewis Carroll's famous argument that every deduction evidences knowledge of a "principle of inference" that does not figure explicitly as a premiss to the deduction.) Leibniz also seemed to want to hold that this capacity to distinguish--this "idea of distinctness"--, which is for me a form of knowledge is also knowledge "of" distinctness. I disagree with him on that point, and agree with Locke that proper knowledge "of" distinctness is a late comer. (I do, however, concede that the question remains open, for the Lewis Carroll example does suggest that some forms of "instinctual knowledge" may be *of* "abstract truths".)

In what follows we shall usually (not quite always) be concerned with those knowledge states of readiness for the utilization of information that are also said to be "of" that information.

Second Comment, (recalling pp. 38ff. of Chapter 1, and p. 180 just above): While the *meaning* of "know" is "dispositional" in respect of *utilization*, the *fact* of knowledge is always and simply a state of the subject.

Third, these states are variously classifiable. I have named a few of these classifications. e g. noticing and understanding. I have no general idea of what they all are except to say that they are conceived as knowledge by a conceptualizing observer in explanatory connection with what they are conditions for, *viz* the utilization of information. Following a fashionable analogy from computing machinery, these conditions may be likened to the installed presence of a chip, the loaded presence of a disk or a tape, power-on etc. "Knowledge" may be the satisfaction of any one or another such condition. So (e.g.) a computer with a chip installed but with its power off may be said "to know" the multiplication table; again, its power being on may be enough for it to "know" it does or does not have a disk in place.

> Strictly speaking, my computer can no more be said to "know" anything than could an abacus or a transit. The same arrangement of "gates" could "code" for quite different things. Does the computer then know all these different perhaps incompatible things? Here it is apposite to notice that it is very unlikely that the arrangement of closed

"gates" that produces "2" on my screen is represented by the binary numeral for 2 ("10"); 2's "ASCI Number", by the way, is 50. The condition of my neurological circuitry that *may* be my knowledge (e.g.) that "10" is the binary representation of 2 cannot of course be said to "code" for anything.

Fourth comment: Although we named several kinds of knowledge-conditions in the familiar terms of "learning", "memory", "recognition", "noticing", "understanding" and the like, I see no *a-priori* reason why they should not also be actual states of our neurological circuitry, as the computer analogy suggests.

Fifth: What I would presume to call the "empirical theory of knowledge"--nowadays fashionably called "cognitive science"-- is concerned to establish the nature of states ascribed to subjects as knowledge. Physiological and psychological inquiry into this matter of knowing, research guided perhaps by theories of computation, finite automata and the like, seeks to identify and to describe various cognitional mechanisms; because those mechanisms are cognitional, the examined data must check-out against our everyday understanding of "know"; but such considerations of usage matter only to establish boundaries for an enterprise that, notoriously, is disconnected from what the ascribed knowledge-conditions are said to be "of". Conceptions of *truth, evidence, access* and *certainty* do not play into these investigations. Rational epistemology works the other side of the street. It presumes that there are these knowing-conditions for which it has some names but of which it knows little; its inquiry is dominated by the consideration that knowing-conditions are often (not always) said to be "of" something else. What we have so far adduced says too little about the meaning of "know". As against "too little", we have said *nothing* about what knowledge *is*. For that we await results from cognitive science. We rational epistemologists have hard questions enough, questions which the cognitive scientist is right to shunt. Let me continue to concentrate on the meaning of "know". Everyone will agree that the meaning of "know" differs from the meaning of "think" and, accordingly, that ascriptions of knowledge to subjects differ in meaning from what I shall call "ascriptions of belief". "Condition for the utilization of information" does not yet distinguish the condition of knowledge from the condition of true belief, no more for us than for the physiologist. We shall, later on, in our pursuit of rational epistemology, annex conditions having to do with *access, evidence* and *certainty*, all in relation to what a subject's knowledge is said to be "of"; these further elaborations will facilitate the explanation of the distinction between our conception of knowledge and our conception of belief.

Information is "fact" and knowledge "of" is of facts. An ambiguity on "knowledge".

6. A subject's state truly said to be knowledge is a condition for his utilizing information, and the ascribed state of knowledge also may (or may not) be said to be "of" that information. Information is always a fact, and knowledge is of facts. Now, in everyday usage, known facts are themselves sometimes called knowledge, e.g. it is common knowledge that Alexander the Great perished in Babylon, even though many of my students don't seem to know that fact. We must then be cautious of an ambiguity in our use of "knowledge", to mean both (as hitherto) a condition of the subject and also known facts or information. When appropriate, I shall distinguish bits of information as "items of knowledge". Knowledge ascribed to a subject as a condition for the utilization of information is not the same fact as is that information or item of knowledge. The two kinds of knowledge are related of course, and neither would exist as knowledge in a world without animate subjects.

Facts are called "information" only in reference to a formulation and only if "registered in a fund" under that formulation.

7. Not all facts are information of course: only "known facts" qualify. More to be remarked, known-facts qualify as information only in relation to a *formulation*. It may or may not be a fact that a son of Philip of Macedonia who was complicitous in Philip's murder perished in Babylon; if it was, it may or may not be the same fact as that Alexander the Great perished in Babylon; however, if those facts are the same, the fact it is, is qualified as information only under the second and not under the first formulation. We just don't know what Alexander's complicity in the murder of his father was. Proceeding: I said that facts qualify as information only if known, without necessarily being known to anyone in particular. I now put that by saying that information must be *registered in a fund* and (reverting to the first qualification) under "an appropriate formulation".

"Registered to a fund under an appropriate formulation" or the equivalent is an inescapable but "problematic" qualification for *information*. Here's the problem, which I cannot solve: If the observer ascribes the knowledge by saying that the subject "knows that such and such", then he does say what that knowledge is "of" and he certifies its registration under what he fancies to be an appropriate formulation. If he is wrong, as he may be, then his ascription of knowledge to the subject is faulted. Now a conceptualizing observer who ascribes knowledge "of" to a subject may not himself have knowledge of that fact. He might then say that subject knows "when", "where", "whether", or something like that. It may even be that *only* that subject knows the matter in question; I see no objection to the idea that a *conceptualizing* subject may himself certify new registrations to the fund of information. The difficulty, which I can only acknowledge and not really resolve, arises when the subject of whom the observer says "It knows wh.. ..." is not a conceptualizer. The known fact or information is identified in relation to the observer's formulation; but there may be a question of its certification. A bird spotter says that the condor knows where its nest is; but the spotter himself doesn't know, and maybe no other conceptualizer does either; it is not obvious how the allegedly condor-known information is to be registered. The best I can do with this is to say that observer's ascription anyway creates a presumption that the fact the subject is said to know is registered in the fund as an item of knowledge. The presumption may perhaps be validated by saying, e.g. "Its nest is where it rests tonight; so follow it there."

8. Ascribed knowledge which is also "of" something comprises several kinds. I begin by summing over two of these.

Knowledge-"that" with Knowledge-REL = Knowledge of Fact.

We noticed that an observer who ascribes knowledge "of" to a subject may or may not identify what that knowledge is "of". The information-identifying case is commonly conveyed by the "know-that" construction and the non-identifying case with a relative pronoun in place of "that". (Knowledge-REL) Let us call ascriptions of either kind, "ascriptions of knowledge of fact".

Knowledge of Fact with Knowledge-PERS = Factual Knowledge.

9. "Know" may take a direct object, as when we say "Jones knows Smith", "Hume's *Treatise*", "England", "The Theory of Relativity". Call ascriptions of such knowledge ascriptions of *Knowledge-PERS*. Some writers, thinking perhaps of what I shall presently dub *Knowledge-"that which"*, have held that the distinction between knowledge of fact and knowledge-PERS is deep and important. I'm sure that it isn't. Anyone who knows that S is P must have Knowledge-PERS of S; on the other side, someone who

knows Smith or Hume's *Treatise* must also know something yet unspecified about Smith or that celebrated work. Ascription of knowledge-PERS is statistically more often ascription of familiarity or acquaintance than is the ascription of knowledge of fact, but this is only a matter of more or less. ("Were you familiar with the way he ate with his feet?" asks about knowledge of fact.) Ascriptions of Knowledge-PERS are simply vague in respect of what knowledge is of. Knowledge-REL leaves one or another piece of the known fact unspecified; knowledge-PERS leaves everything unspecified except one piece. Ascriptions of knowledge of fact and of knowledge-PERS together comprise what I shall call "ascriptions of factual knowledge".

Knowledge "that which" is factual knowledge whose ascriptions are "transparent" in relation to the identification of known fact.

10. Aristotle and Russell, as we saw in Appendix A, both made appeal to a non-judgemental kind of knowledge. Aristotle called this "contact" ("*tigein*"), where Russell spoke of "knowledge by acquaintance". This quantity is not simply knowledge-PERS, as Russell once suggested. Ordinary ascriptions of factual knowledge identify what the knowledge is "of" in relation to formulations and are, for that reason, normally "opaque" in respect of determinations of the elements of the known fact. I may truly be said to know that man but not to know Smith, even though that man is Smith; again, I may be truly said to know who drank the pernod but not to know who pinched the silver lighter, even though they are the same; again, I may know that that man drank the pernod but not that Smith did. Aristotle's "contact" and Russell's "knowledge by acquaintance" are, in their respective theories, special and fundamental either because they bring the intended object itself into the judgement or because that object serves to determine a "basic meaning.[3]" Are there such basic meanings? They would have to be "transparent" within those formulations that identify what an ascribed bit of knowledge is of. The claim that there are such kinds of knowledge would be suspect if there were no such formulations. Those formulations, I believe, are clumsily available in an English idiom we may generally schematize as 'knowledge-"that which"', e.g. knowing "the book which" or "the person who"[4].

Examples (all of them fictional): Oedipus knew Jocasta but he didn't know his mother even though Jocasta was his mother; yet he did indeed know the person who was his mother, too late and too well.

Fido knows the person who is the weekday postman; he does not yet know and now menaces the person who is the regular postman's Saturday substitute. The regular postman enters the house one night disguised and wrapped in a sheepskin coat, and is unknown to a snarling Fido who still hasn't lost his knowledge "that which" of the man who is the regular postman and who is the burglar.

Cyril knew the person who left and he knew the place she went to, but he didn't know that Jennifer went to the bedroom (otherwise he would have followed).

The pretended examples are meant to illustrate the possibilities of knowledge-"that which" being either "speculative" or merely practical and of being either PERS or of fact. It is, in every case, "factual".

The chief fact about knowledge-"that which" is that, although it is of an object that is identified in relation to a formulation of information, the terms of the ascription secure a "transparent reference" to that fact. Fido's knowledge-"that which" is the same no matter how the postman is identified, just as Oedipus's knowledge-"that which" of his mother is independent of any reference to Jocasta. In this respect, a subject's knowledge-"that which" resembles a subject's perception of the object. (More on this below.)

Knowledge-"that which" is knowledge alright--a state that conditions a subject's utilization of information, identified, in an appropriate formulation and registered in a fund--but it is special.

Commentary is now in order. The *first* comment is to answer an objection by a onetime colleague that knowledge-"that which" isn't really all that special but only one instance of the schematic use of "know"; Fido simply *recognized* the regular postman; he *noticed* but didn't recognize the Saturday substitute. I reply that Fido still had unrecognizing knowledge-"that which" of the disguised regular postman when he came to burgle.

The *second* comment has to do with the difference between knowledge-"that which" and registered perceptual exposures. Both may afford a basis for transparent reference. A difference is that knowledge-"that which" typically gets better with experience. Thus Fido was certainly aware of the substitute postman on the occasion of their first encounter; as the months roll by, Fido comes to have improving knowledge-"that which" of this man. A subject's registered perceptual exposure to something may be necessary but is certainly not yet sufficient for subject's acquisition of knowledge-"that which" of the object in question. Knowledge-"that which" is not in a subject until an observer could truly report its more-or-less presence as a condition that would affect the subject's utilization of information.

My *final two comments* draw out consequences of the "transparency" of knowledge-"that which". First, a subject may have knowledge-"that which" of an object and not know the identity of that object or otherwise know anything "about" that object. It might be objected to this that Fido must have known that the postman was a human being. I don't think so. Consider: While Fido may know that smell which emanates from grilling steak, he needn't, even in the practical sense of "know", know that the smell is a smell.

The last comment is difficult to formulate and more speculative: Knowledge-"that which" must take its object "simply". Any "articulated" knowledge ascribed to a subject would in its formulation be knowledge "about" something or other. Since a subject may have one such item of knowledge about a particular something and lack another, any such "information" stands to create a case of referential opacity. The "simplicity" of knowledge-"that which" rationalizes Aristotle's and Russell's theories of "contact" or "acquaintance". Plato likened a thought to a word and the elements of the thought to the letters of the word: we learn to spell words by assembling letters; but then, if knowledge is likened to spelling, knowledge-"that which" is like spelling letters, which is a rather strange thing to do. But now caution! An object of which a subject is said to have knowledge-"that which" may itself be as complex as a person or any other fact. It is even doubtful whether the ascription of knowledge-"that which" would be true unless subject were able to put the object into connection with other things. We claim only that an ascription of knowledge-"that which" neutralizes all implications about what those connections are. In our imagined case of Cyril's knowledge-"that which" of Jennifer's departure for the bedroom (p.

189), we had to break the formulation up into knowledge-"that which" of a person and knowledge-"that which" of a place. It looks from all this that transparent factual knowledge cannot but be "of simples". That, if I am right, is a "conclusion", not a point of departure.

Knowing-how-to is knowledge too.

11. We earlier allowed for knowledge conditioning the utilization of information, but not being of that information. We cited mechanisms mediating perception and perceptual knowledge, and also mentioned the case of knowing-how to or *skill*. I wish now to say something more about this matter of skill or knowing-how-to. Skill is nothing if it's not an "ability-factor" in behavior, a kind of condition, assertible by saying "can", thought to be sometimes required for the success of action. Thus we say "He can play the piano", "read" or "juggle", meaning that he knows how to. The absence of this factor may variously result in undone or in faulted action: Some measure of know-how is required if one is even so much as attempt to juggle, read or play the piano; that is not so for biting. Now (to echo a familiar theme), while the meaning of "know how to A" is "dispositional" with respect to A-ing, still, facts ascribable in this idiom--the skill itself--are simply conditions of subjects, contrastable with such other ability-factors as vision, strength, opportunity and permission. Knowledge-"that" also schematizes such ability factors, but ones which contemporary philosophers (following Ryle perhaps) are wont to distinguish from knowledge-"how". I myself believe that Plato's contrary-tending equations of virtue with both skill and knowledge was closer to truth than his detractors have fancied. The decision, for me, would come down to a question over whether skill is always a condition commonly thought required for the utilization of information. It would be evidence that skill or know-how is indeed so regarded if, whenever an action were faulted by cause of lack of know-how, it was also faulted by lack of factual knowledge. I believe that that is indeed the case.

The concurrence of lack of skill with lack of factual knowledge is obvious for such activities as reading Greek and integrating differential equations. What of piano-playing or juggling? It seems that faulty technique in piano playing is always traceable to some kind of failure to secure a correct match-up between keyboard and score: a beginner who can't do "3's against 2's" doesn't know when to put his fingers down. Similarly but less obviously for juggling: I would have less trouble in lofting one of these three balls while catching another if I had better concurrent knowledge of where my hands and the balls were ("co-ordination"). Furthermore, common explanations for lack of skill, such as physical impairment, lack of training and practice, sleepiness and distraction, also explain a subject's failure to take in relevant information. I have urged (p. 42) that those kinds of "proficient action", such as *reading* and *solving equations*, which "logically" require some degree of skill, are precisely ones for which subject must be able to seek out relevant information of "conditions of success" by making an investigation.

Now, of course, there may be facts and information in regard to a subject's execution of a skill whose utilization by that subject is not conditioned by the subject's having that skill. Highly skilled tennis and basketball players may not have or ever utilize expert physiological information in regard to swings and jumps. However, such knowledge may enter into their skills and most especially so if their physiology is in decline.

Now, coming back to where we started: Certainly, ascriptions of knowledge-"how" differ from ascriptions of factual knowledge in that they do not characterize the knowledge as being "of" some item of information. It is also to be remarked that knowledge-"how" is conceived as habit and thus as a mechanism that is set and attuned--better or worse, it should be said--in the course of experience; but so too may certain mechanisms of factual memory. I have so far found nothing to defeat my proposal that skill is an epistemic mechanism more or less fit for the utilization of information, and therefore so far qualified as a kind of knowledge that may be ascribed to subjects.

The message matters a lot for our enterprise. Investigative skills, I shall hold, are a kind of knowledge indispensable for conceptualizing behavior and, accordingly, "most important" for the developments that follow in this treatise.

All knowledge is "practical". "Theoretical knowledge" is (practical) knowledge of the truth of formulations.

12. A subject's knowledge comes out and is usually noticed in his actual utilization of information. That jibes with our account of the meaning of "know". Knowledge, by this, is prospectively evident in the information-utilization behavior of the subject and is, accordingly, always "practical".

Different knowing subjects command different behavioral repertories. What subjects can be said to know varies with what they can be said to do, and the scope of their knowledge varies with their behavioral sophistication. Fido hears a sound and knows that his master is at the door; again, Fido may know his way home rather better than his master does. It would be odd, however, to say that Fido knows (as master does) that Race St. is west of Vine; for, while Fido, from his perambulations, may know those streets better than does his master, still not by name.

The example, I believe, illustrates the distinction, first drawn by Aristotle, between knowledge which is "theoretical" and knowledge which is "merely practical". But, we noticed, all knowledge is practical: So what is the distinction?

Plato, I think, would have maintained that theoretical knowledge is anyway more useful than practical "true opinion". That is likely so for molding a certain quantity of gold; but Fido, though he be innocent of geometry and map reading, will commonly get home more readily than I do.

Aristotle sometimes wrote as if he thought that practical knowledge is always "singular". But surely dogs may have general practical knowledge of the behavior of other dogs.

Recent writers have suggested that practical knowledge is a matter of fitting deed to word and theoretical knowledge a matter of fitting word to deed or other fact[5]. That too narrowly requires that a knowing subject be endowed with language.

Knowledge may or may not be "reflective" in the sense that an ascription of reflective knowledge entails that a subject knows that it knows what it knows. It emerges from the formula I shall give that theoretical knowledge is reflective. The converse is doubtful. Certain kinds of investigative behavior in beasts seems to evidence their practical knowledge of their (practical) ignorance of where something is.

Some items of knowledge or bits of information can be known only theoretically, e.g. my knowledge of streets, the Pythagorean Theorem and that the sun is approximately 93,000,000 miles away. Are there items of information of which subjects may have both kinds of knowledge? Well, Fido and I may both know the shortest path across the room. Suppose I had come to know this by use of the theorem that one side of a triangle is less than the other two together: Wouldn't my knowledge of this fact then be theoretical? Yes, because bound up in my knowledge of the theorem (unfamiliar to Fido), since theorems are "of truths"; so, if I got my knowledge from the theorem, then I must also know that it's *true* that such and such is the shortest path. While Fido may indeed know the fact formulated in such a truth, he doesn't know such truths. (This, by the way, makes reply to those ancient cynics who ridiculed the theorem, which actually takes some proving.) There seems, then, to be a difference in what our knowledge is of.

With that observation, I can now enter a formula for theoretical knowledge: Theoretical knowledge of a fact is (practical) knowledge of the truth of some formulation of that fact. The consequences of this proposal are welcome and confirming.

Since the truths in question are of formulations, a subject who knew such a fact--a truth--would also have to know how to give locutionary expression to that truth and to know that it had done so when it did. But then it would also know that it knew this truth (see pp. 52ff.).

Such knowledge is a condition for the utilization of truths, characteristically in theoretical enterprise but also in "vouching", agreeing, opposing and other locutionary activities, and also in "non-locutionary" activity such as molding gold.

I notice in passing that the distinction between theoretical and practical factual knowledge is not the same as the distinction between factual knowledge and skill. That is so, not just because theoretical knowledge may enhance skill in all activities, but more directly because theoretical activities, like others, have their skills.

There certainly are difficulties in the distinction I have now defined between the "merely practical" and the theoretical knowledge of a fact. The one I now find most trying is to reconcile what appear to be opposed orders of dependence. While theoretical knowledge is presumably genetically dependent upon practical knowledge, practical knowledge is itself always identified in relation to a formulation, and is presumably therefore "conceptually dependent" upon a theoretical knowledge of the known fact by some articulate observer of the knowing subject. To reconcile these apparently opposed dependencies, we may rather desperately suppose that mere practical knowledge is "brought down" as a kind of debased "attenuation" from "real" theoretical knowledge of fact (see my *Stratification*, esp. pp. 210-215). That formula is well adapted to the explanation of "belief" (see below); it doesn't do for knowledge, because it leaves us still to explain how theoretical knowledge is "brought up". So, I think, a better reconciliation will be found there, in the idea of bringing subject up to the truth-formulating use of language; the resolution of the difficulty of explaining the difference between gross mere practical and exalted theoretical knowledge will then require nothing less than the explanation of this use of language as a kind of behavior. Chapters 1 and 2 of this treatise provide that explanation, not for an empirical cognitive scientist to be sure, but for "rational epistemology".

The distinction between knowledge claims and ascriptions will not affect our examination of knowledge.

13. We shall in the ensuing pages be mainly, though by no means exclusively, concerned with *theoretical knowledge of*. There are many distinctions within that confined *situs*. One we have met with at #7 of Chapter 2 but which we shall not have to worry over henceforward is the "austinian"[6] distinction between "first-person", "quasi-performative" "claims to know" and ascriptions of theoretical knowledge.

> The main question over "I know *p*" is how saying that differs from the straightforward unmodified assertion of *p*. "I know *p*" *inter alia* conveys an assertion of *p;* on the other side, one does not assert what one does not think one knows. Speaker might have some reservation; if he doesn't, then perhaps he would do well to say, "I know". But normally speaker's assertion of *p* gives us to think that he knows (the fact) *p,* and would say so. According to that, "I know" parenthetically conveys an intensification of the normal implication of assertion. That surely happens; but sometimes something else is wrapped up in "I

> know". I follow Austin on this matter: In asserting *p* speaker S gives his interlocutor, A, reason to think that *p* is true; speaker's assertion of *p* is evidence of *p*, for other parties. S is responsible to A for the truth of *p*. But S's commitment may stop there. If A goes on to assert *p*, then, though he may have been misled by S, A is still liable. If, however, S had said "I know *p*" he would not merely have given A reason for thinking that *p* was true, but would also have taken the further step of *authorizing* A to assert *p*. Now if A goes ahead, S must have a share of the responsibility for A's misleading his audience in case *p* should turn out false. ("I got it from a reliable source"). Normally, S's assertion of *p* would imply that S would enter a claim to know *p* (pp. 99f.). When A hears S assert *p*, A may wish some assurance before he takes his chances with the message, and he may ask S how he knows. S's affirmative answer confirms his authority, and, if it is a good answer, should serve to reassure A. The difference remains: while my assertion of *p* provides a ground for your reassertion of *p*, my saying "I know *p*" expressly authorizes assertion of *p*. It is, in my jargon, a wholly conventional "hybrid" of asserting into authorizing.

The cross-cutting distinction which Austin noticed between answers to questions of *how* a subject knows and *why* it believes something to be so--a distinction equally applicable to claims and ascriptions--will matter a lot.

"Ascriptions of belief" are "attenuations" of ascriptions of factual knowledge; the subject is reported to act "as if it knows".

14. *On the Conception of Belief.* An observer may notice and report that a cat or a person has thoughts, ideas or notions of or about something rustling in the shrubbery or that said cat or person thought that there was a bird in the bush. Such thoughts, ideas and notions are, in the jargon of contemporary anglo-saxon epistemology, commonly dubbed "beliefs"; I follow, and call the imagined reports "ascriptions of belief". Such ascriptions are various. They may but needn't imply that the subject was aware of its belief; they needn't imply that the thought even so much as crossed the subject's mind--a listener believes that his blaring radio is connected to power without that thought having to "occur to him" at all. Ascriptions of belief always imply that the belief in question could be further characterized in respect of what it is "of". (There are no beliefs corresponding to knowledge-how or to the innate mechanisms underlying perceptual knowledge.) The ascription may or may not actually secure that further characterization: those with "that" clauses do; those with "of" or

"about" phrases commonly don't. While ascriptions of belief may imply that the ascribed belief is true-or-false, they neither imply that the belief is true nor imply that it is false. Ascriptions of belief, in contrast with ascriptions of factual knowledge, do not imply that the belief is "of" actual information.

Our inquiry has so far been and will continue to be concerned with the sense of ascriptions of belief and with the meaning of "believe"--that word succeeding such terms as "thought", "idea"" and "notion". The fact of belief is, like knowledge, a condition of a subject; that condition, whatever it may be, is no more and no less a disposition than any other condition; it remains that the *meaning* of "believe" is characteristically "dispositional" in respect to the behavior of the subject. Now the further "of" characterization of an ascribed belief is always in terms that would also be appropriate to the characterization of factual knowledge in regard to the information which that factual knowledge would be "of". That raises the question of the relationship between an ascription of belief and a counterpart cognate ascription of factual knowledge. Since (as we have just noticed) the belief need not be "of" actual information, it is open to doubt whether the manifesting behavior is properly to be characterized as utilizing information. What then *is* the relationship between the senses of "He believes that *p*" and "He knows that *p*"?

The consensus among contemporary epistemologists is that "knows that *p*" is, somehow, to be explained as "truly believes" **plus**. I demur[7], with provisos. I believe that the proper route is to explain the sense of "believe that *p*" as "knows" **minus**. One may protest that if belief is knowledge **minus** then surely knowledge must be belief **plus**. Perhaps; but then beware of leaping to the conclusion that the knowledge and belief in question are always "cognate", *viz* invariably to be characterized as being "of" the same thing. Some knowledge may have no counterpart belief, and our talk of **plus** and **minus** should not be taken to imply that a knowledge-condition in a subject is always identical with some belief-condition in that subject. The existence of knowledge without counterpart belief would diminish the prospects of explaining "knows that *p*" as "truly believes that *p*" **plus**, without yet having to depress our hope to explain "believes that *p*" by subtraction from *some* "knows that *q*". I follow that lead, and shall treat ascriptions of belief as "attenuations" of ascriptions of factual knowledge.

Since it's not within reach of rational epistemology to say much about those conditions that *are* knowledge or belief, we instead concentrate, in our examination of the meanings of "know" and "believe", on the manifesting behavior. The sense of "attenuation" must be that a conceptualizing observer sees a believing subject as one who "acts as if it knows"; subject, while perhaps not actually utilizing information, is reported to act as if it were.

Objection: A believing subject could know he didn't know. Answer: We must take care to get the right identification of the "content" of the belief.

I know from experience that there will be resistance to this formula for belief. Here is a *first objection*:

A subject could believe that *p* and *know* that he didn't know that *p* and show all of that in his behavior; so it is wrong, in these instances, to say that the subject both believed *p* and acted as if it knew *p*. "As if it knew" is not entailed by belief.

Three cases: betting on large likelihoods, expressing opinions and making conjectures[8].

Discussion of the first alleged counterexample. I select a ball from an urn I know to contain ninety-nine reds and one white. Although I won't know in these circumstances that the selected ball is red, I would certainly seem to believe that it is, and if offered even money on the outcome, I'd better bet on red. Yes, but if offered a hundred to one against the selected ball being white, then I'd better bet on white. Knowing all this, I now must doubt that I do believe that the selected ball is red; I believe, rather, that it is likely to be so. I also know that, and certainly act as if I do. Canons of rational betting behavior argue that my belief was in regard to *likelihoods*.

"Baysians" will of course resist this way of putting the point. They would deny that I believe it highly likely that the selected ball is red but would say rather that I have a high degree of belief that the ball is red. F. P. Ramsey's objection[9] against my way of putting the matter was that there is no unique, objective selection of units for determining possibilities. To be sure. But there is no belief to be assigned, with or without degrees, absent some kind of "understanding"; the understanding assumed in the example was that the probability of selecting a red ball is 99/100.

Looking now to *expressing an opinion*, which is the second alleged counter-example. Here subject simply says what he thinks is so. So he certainly thinks it is so. But he also by implication tells us that he doesn't think he knows it to be so; so (one may argue) he cannot be acting as if he did. I reply that saying what one thinks is actually one way of acting as if one did know and, indeed, the subject may actually know. But, of course, in this kind of case, the subject doesn't think *it knows*. So here the "content" of subject's belief cannot be that it knows, but is rather *what* it knows, if perchance it does

We handle *conjecturing*, the third alleged counter-example, along the same lines. Our formula for conjecturing was propounding what one thinks *may* be so and could be known to be so but thinks that no one actually knows is so (pp. 131f.). According to that, the conjecturing subject believes, not that it is so, but that it may be; certainly, in making his conjecture, he acts as if he did know of that possibility.

The alleged counter examples do not yet wreck my proposal that a believing subject acts as if it knew. They do council circumspection in saying what an attributed belief is "of".

Second objection: acting as if one knew cannot be sufficient for belief, since a one who pretends to know acts as if he did, but then doesn't believe what he pretends to know.

A second, more fruitful objection argues that the "acting as if it knows" characterization is appropriate to cases that aren't cases of belief, e.g. when a subject pretends to know what it doesn't believe. The first objection argues that "as if it knew" behavior is not necessary for or entailed by ascription of belief; the second "pretending" objection effectively argues that the behavior is not sufficient for the ascription. That seems right.

If we are to retain my formula for *belief*, we obviously must tighten the terms of "as if it knew". My guiding thought here is that belief is like knowledge in the *immediacy* with which it induces action[10]: one who knows p is set to use that information without further investigation; similarly, a subject who believes p, unlike one who merely thinks it may be so, enters upon "appropriate" action without further investigation. That is "vague", and requires a good bit of eking out. Ground already prepared in this appendix gives footing for the effort.

We must say more particularly in what way a believing subject acts as if it knows.

Now, *pretence* is an action available only to rather sophisticated subjects. A cinched-up rule of "as if" should squeeze them from consideration, since we want a notion of *belief* no less ascribible to cats and condors than to ourselves. The appropriate "as if it knew" behavior should be something already evident in the actuation of brutes and persistent in our own. Here are two examples, one merely "practical" and the other "theoretical".

First, "practical" example: Ewes are "programmed" to accept only their own lambs, and an ewe's belief that a lamb is or isn't her own will be evident in her behavior. But an ewe can be induced to mother an alien lamb by the ranchhand's covering the orphan in the skin of the ewe's own dead offspring. Here the ewe is brought to think that the alien lamb is hers.

Second, "theoretical" example: An examinee may be brought to an erroneous *opinion* that Capetown is south of Buenos Aires from its knowledge that Capetown lies near onto the southern tip of Africa.

The ewe is presumably aware of and knows that odor which is of her own lamb; that knowledge-"that which" induces her belief that this is her lamb.

Similarly, the ignorant examinee was brought to his belief that the proposition that Capetown is south of Buenos Aires is true from its knowledge of the truth of the proposition which is an answer to the question "Where in Africa is Capetown situated?".

Belief is a knowledge-like kind of condition induced in a subject by its knowledge "that which" of something.

The counter-examples and the acceptable cases taken together suggest the following expanded formula for belief (please take a breath): An ascription of a condition, B, to a subject is of a condition of belief just in cases where the ascription implies the obtaining of a condition (*K*) that the subject has knowledge-"that which" of a fact and there is an explanatory pattern within which the obtaining in subjects of a *K*-kind of condition explains the obtaining in subjects of a *B*-kind of condition, where *B*-kinds of conditions are commonly asserted as knowledge, and the *K* in question does explain *B*.

Our stipulation restricts the "as if it knew" behavior to some information-utilizing *kind*. That was the sense we were seeking. The adapted formula fits the ewe and examinee examples and seems to disqualify the pretenses and dissimulations. Now our talk of "kind" leaves matters pretty loose. Still, the stipulation that the belief be induced by some actual item of knowledge is fairly restrictive. And here, note well, the knowledge-"that which" in question is neither cognate with the ascribed belief nor is it of itself a kind of belief.

Call my formula, if you wish to, a causal theory of belief. An ascribed belief, by this account, is simply a condition of the subject, but one that the observer casts as being of a kind that commonly explains subject's utilization of information. The behavioral sequel is therefore going to be for the subject to "act as if it knew"--a connection which must certainly be "selected in" for its survival potentials. It matters no less that the cause involved in the implied statement of explanation should allude to some actual item of knowledge-"that which". A subject's belief, by our formula, is always based on knowledge, and does not obviate the use of cognitional-characterizations.

I know of no counter-examples to the formula. It applies equally to belief-ascriptions that do and do not explicitly say what the belief is "of". The formula admittedly remains insufficient for the characterization of ascriptions of belief if only because of "kind". A more serious objection is that it cannot distinguish true belief from actual knowledge.

A true belief is a condition for the utilization of information.

On the sense of "true belief". Rational epistemologists certainly must be concerned for what is meant in saying that a belief is true. The idea applies to the beliefs of brute creatures and plainly does not require that true beliefs are always "of truths", e.g. propositions, statements, reports and the like, or even beliefs. A subject's belief is true surely only if that subject is in a condition that is "right" for the facts, in the sense of being in a right set for action. True beliefs are "of reasons for action". What does that mean? Well, our formula for ascriptions of beliefs makes beliefs be like knowledge-conditions, dispositionally characterized as being for the utilization of information. A subject who is in such a condition is in a right set for action. I leap to the conclusion that when a subject's belief is said to be true it is then implied that the belief-condition is a condition apt for the utilization of information. Of course the belief inducing knowledge-"that which" is also for the utilization of information; but the inducing knowledge and the induced belief are different conditions and the associated informations are seldom if ever the same.

Consequence: Conditions of factual knowledge that are induced by conditions of knowledge-"that which" are true beliefs.

We need something "more" about factual knowledge to distinguish it from true belief.

Objection: Our formula allows that all those conditions that are true beliefs are also conditions of factual knowledge, because they are conditions for the utilization of information. Empirical epistemology can endure this consequence. Rational epistemology will not tolerate it and especially not in regard to theoretical knowledge of truths and counterpart opinions. We regularly claim and credit opinions as true even in default of knowledge. We must now finally look to see what further conditions must be imposed upon our conception of knowledge. The further implications we have to draw on have particularly to do with the possibility of giving good answers to the question of *how* a subject knows what it does and *why* it should think it to be so. A knowing subject must have *access* to the fund of information and have *evidence.*

15. *Of Why? and How?* Austin, in a paper we have cited (pp. 195f.), fastened attention on the difference between answers provided to the questions "How does S know?" and "Why does or should S think (or otherwise "believe")?". "How does S think?" is obviously different, and "Why does or should S know?" is seldom asked. What then is the significance of the difference between "How?" and "Why?"? Less than you might suppose.

"Why should S think such and such?" is a request for S's evidence that such and such is so.

A lead to the answer is found in the observation that we do *sometimes* ask "Why does he know that?" and would expect a reply that would also provide answer to the "How?" question: S knows about the deliberations of the Executive Committee because Blabbermouth told him. In asking "Why?" instead of "How?", I intimate that S's knowledge of this delicate matter is surprizing if not actually illicit. "Why?" questions suggest that the *explicandum* is, if not quite what it should be or actually in doubt, anyway unexpected: Why does he think, feel, do...that? Why is it boiling; it shouldn't be?; again, if I ask why the solstices fall on these dates, I may suggest that maybe they don't; but there is no such suggestion if I ask how they come to fall on these dates. "Why?" (unlike "How?") suggests that something in the way of a "justification" or (anyway) a *reason* is called for. Returning now to "Why does or should S believe F?": I take it that one's beliefs ought to be true; evidence is evidence for truth; so when we want to know subject's evidence, we naturally ask why he thinks it's so; if that evidence is strong then we may tentatively allow that subject thinks as he ought. We have now rationalized Austin's conclusion that "Why should or does S think?" is a request for evidence.

Answers to "How does S Know F?" questions commonly say something about S's access to information F.

Why is the question "Why does S know?" restricted to cases like those of privileged information? In those rare cases, perhaps S ought not to have known and the leak needs accounting for; the whole question then has to do with how S came to know and it has no proximate bearing on the matter of the evidence for his belief. But, normally, *knowing* is not the kind of thing of which it makes sense to say that one ought or ought not. "But surely 'oughts' are in offing somewhere here, for inquiry is something that can go wrong!" I think that the answer is that "know" itself *already* conveys an *ought*. "Ought to what?" then? Not "to know", surely; but only and I think always "ought to think" (or otherwise "believe"). It's the thought or "belief" and not the knowledge that may go wrong for lack of evidence or truth. Knowledge is something for which justification is uncommonly needed, and "How does S know?" is a request for something other than justifying evidence. We observed that in the exception (e.g. our subject's illicit knowledge of the committee's action), answers to the "Why?" and "How?" coincide and supply an explanation of how S came to know. The "How does S know?" question presupposes that S thinks as he ought; an answer, while it may incidentally convey an indication of evidence, commonly says something of S's *access* to the information.

True answers to a "How does S know F?" question certify S's access to a fund of information containing F.

16. *On access.* An "How does S know?" question is always appropriate whenever an "S knows F" is said. That holds uniformly for ascriptions of all orders of factual knowledge--"that" and PERS, theoretical and merely practical, "that which" and otherwise. Instance: How does Fido know that there's steak cooking in the patio? From the smell, I say. Here he also has knowledge-"that which" of the smell. My answer to the question of how Fido knows that smell is by having smelled it. Instance: I

tell you how this unphilosophical and oblivious acquaintance of ours, despite his seeming ignorance, knows that work which Hume called *A Treatise of Human Nature*, by recalling that we used the book in a class we once took together. Instance: I claim to know that Charles Martel lived before 800 AD; asked how I know, I may say that I *learned* it in school or that I *inferred* it from my school knowledge that Martel's grandson Charlemagne was crowned in 800 AD. Classroom exposure also gave me lasting knowledge-"that which" of the answer to the question of when Charlemagne was crowned. The known F in these cases is information our knowing subject is conditioned to utilize, e.g. by heading for the barbecue, by saying, "I think I've seen this book" when a copy is thrust into his hands or by showing off in Trivial Pursuit. The information in question pertains to a fund of knowledge, and answers to the "how" question certify that the subject has access to or drawing rights upon that fund, established by general experience, education, hearsay or perhaps by original deposit.

Access has various "position" and "tellingness" components. Tellingness components are items of knowledge-"that which".

Access has two components, one having to do with subject's "position" and the other with the "tellingness" (Austin's term) of what it has position on. Factors of either kind may be cited in answer to the question of how the subject knows. How do I know that there's a whippoorwill in the garden (so early in the year, too)? From the sound ("Tellingness") that I hear ("position").

Positions are various, to include such things as observation, established expertise, education, study, general experience and hearsay, and may be better or worse: Tarski had better position on The Deduction Theorem and Jespersen may or may not have had better position on a point of american usage than I do.

Tellingness components are facts of which the subject has knowledge-"that which".

Tellingness components are also various, to include the smell of steak and the sound of a whippoorwill, answers to questions such as that over when Charlemagne was crowned and indeed Charlemagne himself (for Leo II certainly knew that man whom he crowned).

We remarked that both the position and tellingness components of *access* are various. So, for example, one may gain access to a fund of information by virtue of what one has perceived or done, by virtue of special gifts, or by one's position as an expert, or one's general reputation, or one's acquired competencies, association, education, background, exposure, experience, and general "feel". A person might substantiate her claim to know by saying "I saw it", "I proved it", "I have absolute pitch", "I was Einstein's assistant", "Don't you know who I am?", "I know how to titrate", "I know what bitterns are", "She's my daughter, after all", "I have a Ph.D. in History", "I lived in France when I was a girl", or simply "I just know". Again, tellingness components may be traces, anticipations, circumstances, or the fact itself.

A rational epistemology should provide a code and a canon of access--a schedule of position and tellingness factors by reference to which a subject's knowledge may be qualified and assessed. It appears that "best grade" knowledge of truth by this canon is knowledge "that which" of a fact obtained by the verification of a truth formulating that fact.

These several factors may be "better or worse" for knowledge. So here we may find a basis for the assessment of knowledge according to a "canon of access". Rational epistemology should be in the business of developing such a code. As an example of what might emerge from that inquiry, I propose that a subject's knowledge of truth is "best grade" or "perfected" according to the canon of access if said subject comes to have knowledge-"that which" of the fact formulated in that truth, not by passing observation, but by deliberate and controlled test or "verification". (I stress that this proposal is only for our knowledge of truths. I have no proposal for assessing my dog's knowledge of the presence of a stranger.)

17. It's commonly though not always appropriate to ask a knowledge claimant *why* he believes what he does. So we may ask a gardener who claims to know what is wrong with his oak or who claims to know the state of his oak why he thinks it's rotting in the center. I doubt that the question would be appropriate in relation to the gardener's claim to know that state which his oak was in. I jump from these examples to a general thesis that we can always challenge a claimant to factual knowledge that is not also knowledge "that-which" to say why he believes that what he claims to know is so. With an eye on "austinian" distinctions, I submit that a "why-believe" question when appropriately addressed to a knowledge claimant asks for a "justification". That submission certainly fails when we ask the "why-believe" question in relation to an *ascription* of knowledge to a subject, for then the answer might be that subject is superstitious or deranged.

An answer to a question of why a subject, S, said to know F should believe that F obtains argues for the truth of S's belief by providing evidence that F does obtain

We can easily make the "why-believe" question asked of the knowledge-claimant fit for ascriptions of knowledge by rewriting it as "why *should* S believe---?", and I shall so take it in what now follows. The question addressed to a reporter asks that reporter to "justify" the subject in believing what the reporter said the subject knows, where the ascribed knowledge is factual but not "that-which". You may here wonder why the ascription of such knowledge should entail the presence of a "cognate belief". We've prepared the answer: The subject is presumed to have position on some tellingness factor, and has been brought into his ascribed state of knowledge by his knowledge "that-which" of the tellingness factor. But then he must also have been brought into a state of belief, presumably no different from his knowledge (p. 202).

Now, I submit, a belief *is* justified only if it is true. But surely the reporter's answer to the "why-believe" question can never properly consist in his saying "because it's true". What his answer, if it's a good one, will provide, rather, is a reason or reasons for thinking that the belief is true. Such reasons are, I believe, what we usually mean by *evidence*.

Now evidences are various. Smoke is evidence for fire, so is a charred beam or the testimonies of history; a "mark" or the presence of Miss Smith may be my dog's evidence for the nearbyness of Miss Smith's dog; the gardener supports his belief that the tree is rotten by citing the evidence of cankers on the trunk; I support my belief that it's so by citing his authority.

Is there anything common to all of this? Well, evidences, which are reasons for believing, are facts. Those facts can be cited to justify a belief only if they are somehow "connected" with that fact that "makes the belief true". The only proposal I know of as to what the nature of this connection might be is that the believed facts and the cited evidences are instances of entries in an "explanatory pattern" known to the reporter or claimant. Good answers to the "why-believe" question--ones that do give reasons for thinking that the subject's belief is true--are so far "objective" in that they needn't be known to the subject himself, and indeed the subject need have no belief in regard to the cited evidence. It remains that the reporter must be familiar with the evidence and the pattern, and that these patterns are themselves, not observed "phenomena", but creatures of intellection (see pp. 169f.).

A subject's belief is "well-founded" if it arises from its knowledge "that-which" of a fact that is also evidence for the fact it believes.

Now another point: A subject's belief may be, not only well supported by evidence even true, but also "well-founded" in the subject's own experience. My dog's belief that this other dog has been by is pretty well founded if it arises from my dog's sniffing and taking in the "mark" which the other dog left on this tree. My belief that there's a fire over in The Downs is pretty well-founded if it arose from my sighting of smoke which is ascending from that place. Here my knowledge "that-which" is of a fact that is also evidence for the fact I believe.

We hypothesize that belief is "best grade" according to be canon of evidence when it arises from the subject's knowledge "that-which" of believed fact.

How well-founded may well-founded be? Is there some "best grade" belief in the mix, to be determined by appeal to some notion of evidence? I think so. A subject's belief may always be "fixed" in a formulation of what it is "of", and is "opaque" over "extensionally equivalent" substitutions into that formulation. By way of contrast, the formulation of the knowledge "that-which" that supports the belief is "transparent". Now the very same fact may be twice registered within an explanatory pattern under different formulations of that fact. My proposal is that belief is best grade when just that is the case, *viz* where the subject has knowledge "that-which" of the fact he believes, where the known fact is also evidence for the belief and where, finally, it is additionally required that the fact of which the subject has this knowledge "that-which" makes it the "object" of the subject's "position". So, for example: My belief that there's a fire burning is "best grade" if it's based on my seeing that flame which is that fire; Pan's belief that Rufo's here is "best grade" by the canon of evidence if based on Pan's smelling that dog who is Rufo.

A formula for factual knowledge "of".

Resume'. The main results of our investigation so far can be summarized in this formula (take a real deep breath): A reporter's ascription of factual knowledge k_1 of kind K_1 of information c_1 of kind C_1 is true just in case there are facts k_2 of kind K_2 and c_2 of kind C_2, where k_1 and k_2 are respectively states of S that are conditions for S's utilization of c_1 and c_2, S has "position" on c_1 and c_2, and k_2 specifically is knowledge-"that which" of c_2; c_1 and c_2 may be identical or distinct, but *kinds* C_1 and C_2 are distinct and figure together in some established explanatory pattern. k_1 and k_2

are respectively "of" c_1 and c_2 and c_2 is "evidence for" c_1 and a "reason" for S to believe c_1.

The objectives of "Rational Epistemology".

Rational epistemology, as I envisage it, should supply (i) a classification of K_1's, e.g. as *recognition, noticing, seeing, understanding*; (ii) a codification of "positions", as e.g. *direct perception, testing, general experience, education* and (iii) a codification of evidence ("logic"). Some writings in the field suggest that rational epistemology should also take notice of the "subjective certainty" of a knowing subject's belief, that perhaps to be gauged by its betting propensities or by the "resilience" of its information-utilizing behavior.

We believe that such an investigation would teach that best grade theoretical factual knowledge of information is by way of a direct verification of the relevant formulation of that information.

This envisaged rational epistemology should, among other things, afford means for the assessment of ascribed knowledge. I have already urged that a subject will have best grade knowledge of information F by the combined application of the canons of access and evidence if he has gained that knowledge by a direct verification of F. But that is actually misstated. The "direct verification" is not strictly of F but always rather of a truth that formulates F.

So we amend the rule to read that knowledge is "best grade" by the combined canons of access and evidence only if by direct verification of the relevant formulation of the known fact. Such knowledge is available only to subjects who have access to theoretical knowledge of truths. I have no rule for best grade "mere" practical knowledge.

There's a "gap" between evidence and truth.

Before proceeding to other matters, I feel constrained to consider an issue that has bothered our subject. One may support a statement that S knows F by arguing that S has position on F-like information, that S is ready to use F, that the balance of evidence favors F and that S has knowledge-"that which" on a telling bit of evidence, and yet F not be so. The senses may be dulled, techniques be misapplied, education stultify and one's trust in others be abused. Yet surely there must be some connection between the satisfaction of a canon and truth-conferring fact! For one thing, every justification by a canon must itself be judged before the bar of truth. Since the identification of information is relative to a formulation and that formulation is true only when true to fact, we want to establish a bridge between knowledge-components and truth.

There are two things I want to say on this matter, one rather simply having to do with verification and the other, a "peircian" note, having to do with the perfection of available evidence.

Verification, if actualized, closes the gap.

A subject having the position of direct verification cannot be wrong. Best grade knowledge by the canon of access provides a sure bridge to truth. Of course, if the subject thinks he has verified the fact when he hasn't, then he thinks he knows but maybe he doesn't; he also does not occupy the credited position of direct verification.

Justified belief cannot be false.

The more elusive "peircian" point I wish to make goes back to our notion that a subject knows only if he believes as he should or ought (p. 206). Now one should or ought to do what is right, and similarly one should or ought to believe what's true. However, the reporter's justification of the subject's belief must be something other than a flat claim to truth. The respondent, rather, provides reasons for thinking that the subject's belief is true. Indeed, I believe that terms like "ought" and "should", when used with reference to action, thought or emotion, always incorporate a "quantifier" over all *reasons* (not just those subject knows as reasons), and are then also used to say that the balance of those reasons favors this line of action, feeling or belief. Now reasons within a field , separately or collectively taken, are facts different from the facts they are reasons "for". Still, it strike me as sense defying to suppose that the "balance of reasons" could favor holding a false belief anymore than that a balance of reasons could favor doing a disastrous deed. That is obvious if the balance of evidence includes (as it sometimes but not always does) the fact itself. Justified belief cannot be false. Now, it may be that we can never demonstrate an "ought"--for belief or for action--but still it may be so, that this subject ought to believe such and such, in which case our subject could know, though perhaps no one could know that he did.

We shall now be concerned with the certainty of knowledge of truths.

18. I now take leave of matters of "How?" and "Why?"--of access and evidence--to broach an issue that dominated such classical epistemologies as those of Aristotle and Locke and has recently been restored to philosophy in Wittgenstein's essay "On Certainty".

"Certain Knowledge" is the theme--not the "subjective certainty" of a knowing subject's belief (which may be greater or less), but the greater or less certainty of known truth. The issue arises only for our theoretical knowledge of truths. Knowledge of certain truth is certain knowledge. (Notice that the certainty of knowledge is not gauged by the quality of a knower's access: my knowledge of most mathematical certainties is less good than that of mathematicians.) It was, I think, Aristotle's view that "knowledge of reasoned fact" had high relative degree of certainty but less perhaps than "intuitive knowledge" of "first truths"; again, Locke seems to have held that sensitive knowledge is less certain than demonstrative which is itself less certain than "intuitive"; Wittgenstein wished to explain the certainty of some beliefs and propositions of commonsense which G. E. Moore had certified as true.

> Wittgenstein did not mistake Moore's message, as a number of recent writers have. Moore, in his "Defense of Common Sense", was not concerned to combat skepticism; none of the beliefs he instanced as certainly true were in regard to our knowledge of anything.

Certain knowledge, I repeat is only of truths; it is never "merely practical". Known truths commonly pertain to some field of inquiry such as Combinatorial Topology (Mathematics), General Relativity (Physics), Soil Structures (Civil Engineering), Medicine, Marine Biology (Natural History), European History, Literary Criticism or Commonsense, and the knowledge of these truths may be accordingly codified. There surely are these fields of inquiry, named within uncertain boundaries. Some known truths are matters of plain commonsense, e.g. that the sun rises in the east, that the fall is normally cooler than summer, and that creatures may take offense when scolded. Others are matters of "education", e.g. that when it's fall in Patagonia it's springtime in the Rockies. Still others come to be known as a result of controlled observation and by statistical processing: this kind of knowledge, where "method" matters much, is of a kind often dignified as "scientific", and includes medicine and such things as the length in seconds of a year calculated between two successive summer solstices, the "half-lives" of radioactive materials, and the normal temperatures of mammalian species. Differently, *science* itself may (or may....

not) provide knowledge of why (e.g.) radio-active substances have roughly constant half-lives. *Mathematics* is something else again: it exists at least in part to certify the consequences of commonsense, science, statistics and other modes of inquiry in regard to such things as magnitude, order and transformation. "School subjects" such as history, anthropology, geography, and natural history are all distinctive fields of inquiry, with their own peculiarities and disciplines. These many styles of inquiry interact and fuse in manifold ways, e.g. mathematics is used everywhere, and historical data may be appropriated to any discipline. A fully worked-out rational epistemology would provide a "code of inquiry" for the classification and comparison of these different fields--a task I leave to others.

Known truths commonly pertain to fields of inquiry and may be accordingly codified, e.g. as mathematical or historical. Rational epistemology should provide a "code of inquiry" for these intermingling classifications.

Truths are graded for certainty according to how "deeply entrenched" they are within a field of inquiry. Disproof of a "most certain truth" would effectively undercut and overturn the field of inquiry, and to doubt such truths would be to doubt the *bona fide*'s of that field of inquiry rather as nowadays many doubt Astrology and Theology.

Truths are graded for certainty by their depth of entrenchment within a field of inquiry. Most certain truths are propositions whose truth is required for the very existence of a field of inquiry.

An index to certainty of truth is the ease with which we would cite it as an entrenched point of comparison by saying, "I know that as well as I know (e.g.) that 2+2=4, that Sunday follows Saturday, that the sun rises in the east, my own name". A short, easily extended list will illustrate this conception of certainty.

(i) $2+2=4$ but not $169=13^2$.

(ii) $log\ 1=0$ and $log_{10}10=1$, but not $0.3011 < log_{10}2 < 0.3012$

(iii) *That a triangle divides a plane into an "inside" and an "outside"* but not *That you cannot connect three points of the inside each to three points of the outside without at least one crossed connection.*

(iv) *That Sunday follows Saturday,* but not *That the calendar must be set back a day sometime before completing a westward circumnavigation of the earth.*

(v) *That a non-spinning body falls together with its parts, all at the same speed* but not *That like situated bodies are all subject to the same gravitational acceleration.*

(vi) *That Africa is south of Europe* but not *That Cape Horn is south of Good Hope.*

(vii) *That the sun rises in the east* (as I've heard it doesn't on Venus), but not *That the stars do.*

(viii) *That fall follows summer,* but not *That it's usually cooler in the fall than in the summer* (as it isn't in Berkeley, Calif.)

(ix) *That Charlemagne flourished after Julius Caesar* but not *That he flourished around 800 AD.*

(x) *My own name,* but not *Yours.*

These examples illustrate the concerns and certainties of Arithmetic, Chronography, Geography, History and so on, in regard to objects of different kinds: some of them are truths of Mathematics, others of Science, others of History, Natural History, Commonsense, and the last is peculiar to myself. Of each contrasting pair, the first member is "more certain" than the second: I find all of the first members to be equally certain. First truths are, in the jargon of the trade, both "intuitive" and "transcendental" in the sense that doubting them would stultify inquiry. The examples argue that certainty isn't logical necessity, for both members of some pairs would be reckoned necessary by most readers of this text and, for other pairs, both instances are contingent. Nor is the criterion of certainty here one of general familiarity: No one could do logarithms without knowing that log 1 (any base) = 0; but lots of competent calculators haven't a clue about logarithms; similarly, no one could pursue the study of the history of Western Europe without knowing of Charlemagne and Caesar, but a Chinese engineer could be ignorant of all that.

A feature of the "most certain" examples, I claimed, is that, If **This** weren't so, then a field of inquiry wouldn't exist at all--Arithmetic, calculating by logarithms, keeping a calendar, Geography, or Western History. But what of your own name? Surely, you needn't know it, for you might forget it in a moment of amnesia, and besides lots of us have names we've never known. Yes, but then, if you didn't know "your name", there would be lots of everyday things you couldn't do. In certain social contexts, people are preferentially known "by name"--one name or another; for that to be so, each nominee must be able to introduce himself "by name", which implies that he know his own name. In a squash competition, I tell my previously unmet opponent that I'm Shwayder; now he knows who I am. (Cf. identification by social security number, which matters much in bureaucratic contexts, but is "most certainly" determined in some file somewhere.) "Knowing my own name" plays much the same role in this system of social communication as does "2+2=4" in systems of calculation.

My general formula, then, is that "first" or "most certain truths" are propositions whose truth is required by the very existence of a field of inquiry and investigation. In that sense, they are "most deeply entrenched" in that field of inquiry. They are the "intuitive certainties" from which demonstration begins and which are otherwise assumed when field workers set off to find out what's so.

A worker within a field of inquiry must know all sorts of truths in order to get on with his work. This knowledge can be derivatively graded for certainty according to the certainty of the truths it is of. The most evident features of such a fieldworker's knowledge of certainties is that that will be pervasively on call in his investigative activities across the field of inquiry in question.

A fieldworker's knowledge of certain truths--his "certain knowledge"--is pervasively utilized in his fieldwork.

Now, as we saw, all knowledge is "practical". A wittgensteinian theme, which rings right to my ear, is that the certainty of a known truth is also measured by how pervasively knowledge of that truth affects the theorizing behavior of a subject working over that field: the behavior of an accountant will be affected several times every working day by his knowledge that 2+2=4; an amnesiac who forgets his name is crippled for many ordinary human activities.

Understanding of first or most certain truths accessed by direct verification is "best of all".

19. *On Understanding as a perfection of knowledge.* Knowledge, we saw, is of many "kinds", e.g. it may be observational, learned, noticing or understanding. Is any of these "best"? We can't actually say, for we don't know what they all are. But consider: My hearsay knowledge of some theorem of topology is punk in comparison with that of those mathematicians who really understand such things. Fido may know the shortest way home just as well as I do, but my knowledge is so far better by my understanding that two sides of a triangle are together longer than the third. Einstein's knowledge of the Lorentz Contraction and of the Equivalence Principle was better (though no more secure) than that of Michelson and Eoetvoes. This point about the superiority of understanding is the lasting truth of Rationalism.

> While Einstein's understanding of the Lorentz Contraction was probably no better than Poincarés, Einstein's knowledge has been given superior credit because (I suggest) directed more toward established physical fact. Einstein's knowledge of the Equivalence Principle was suspect until predicted phenomena were observed. Cross-cutting assessments of knowledge are predicted by our account.

If that is so, then, combining the several canons of access, evidence and certainty, we are ready to say what knowledge--knowledge of truths, anyway--is the very best of all. Briefly: Understanding of first truths that are also "accessed" by direct verification seems qualified as being best knowledge according to the combined canons of access, evidence, certainty and kind. Instance: Our knowledge of the answer to the question of which number comes first in the sequence of natural numbers. I think that Leibniz believed that God knows everything rather in the way that we know this truth: *1 is the first natural number.*

"Commonsense" is the most viable field of knowledge.

20. *On the "primacy" of "commonsense".* We have yet to say what "field" of knowledge of truths is "best". Mathematics, Physics, Medicine, Natural History or Farmer Jones' Commonsense knowledge of the weather? The Vice Chancellor for Academic Affairs had better not express his preference in public. No one of these fields is unequivocally best, and it is doubtful whether there is any way of assessing our knowledge of truths according to a canon of inquiry. Yet, in our actual human lives, they fit together in ways that give some of them "primacy" over others. What philosophers[11] call "commonsense" seldom

attains to that best grade "understanding" we so much treasure; but it was there before Study and is bound to survive when more formal inquiries fall into disuse.

History might never have progressed beyond legend, or physics arisen; the rationality of the courtroom may decline into social control by proclamation rather as Economic Forecasting has replaced Astrology, but the child's sense of justice will survive. Commonsense, however much modified under the influence of science, could never be completely discredited. It is the ultimate epistemic viability or (changing the figure) the field of inquiry whose epistemic rights are utterly unalienable. The bounty of other cultivated fields descends from natives primitively rooted in commonsense: our commonsense of how-many and how-much provides a "natural foundation" for mathematics, our everyday understanding of part and whole an axiom for Galileo's demonstration of the law of free fall, and the understanding of everyday consociation for anthropology; medical inquiry begins with our common notice of dysfunction; Rational Epistemology, if it ever arrives, cannot, even under tutelage from Statistics and Logic, be expected to rise more than a scant half-step above the Commonsense of knowledge and belief.

NOTES

[1]For argument in support of this point, see Ramsey's *On Truth*, edited and posthumously published by Nicholas Rescher and Ulrich Majer (Dordrect, 1990), esp. in Chap. IV entitled "Knowledge and Opinion".

[2]The formula suggests a resemblance, which a reader remarked on, between my doctrine and F. I. Dretske's book, *Knowledge and the Flow of Information* (Cambridge, Mass., 1982). There are certainly affinities but also important differences. I once thought to include a comparison, but have now decided that the following commentary together with occasional remarks later on would do just as well for readers who are informed enough to have an established interest in some such comparison.

[3]Dretske (*ibid*, p.215 et. seq.) speaks of simple primitive concepts, that are identical when referentially equivalent.

[4]Not all *knowledge "that which"* is "*tigein*". A reader mentions someone's knowledge of *that treatment which is best for rheumatic fever* as a contrasting case. The subject is being credited with knowledge of what treatment is best for rheumatic fever. One can also hear the contrasting message: subject has very good knowledge of that treatment which, as it happens but unknown to him, is also best for rheumatic fever. It may be that there are difficult techniques known only to experts, first developed as therapies for (say) acne that turn out to be good for certain cancers.

[5]E.g. G. E. M. Anscombe at #32 of her little classic, *Intention*.

[6]From Austin's frequently reprinted contribution to an *Arist. Soc. Symp.* (1946) on "Other Minds".

[7]As does Dretske, *ibid*, pp. 229f. Plato did too as, I believe, also did Aristotle, implicitly.

[8]Due to D. Riggs in a seminar. I am assisted in my reply by observations due to J. Harker, D. Jordan, J. McHarry and R. Schubert.

[9]See p. 84 of the D. H. Mellor edition of *Philosophical Papers* (Cambridge, 1990).

[10]With thanks to Melnick.

[11]Such philosophers as G. E. Moore. T. Griffin reminds me that *everyday* commonsense has scarcely anything to do with the existence of other minds, but rather covers such things as realizing, when changing a tire, that you had better loosen the lugs before jacking up the wheel (his example).

CHAPTER 3

STATEMENTS AND THEIR CRITERIA

1. *General Exposition.*

Among the conditions of success indicated in an act of assertion are that there exist two sometime applicable procedures, one for verifying and the other for falsifying the statement that would be produced in the assertion, were it to succeed.

Statements are both verifiable and falsifiable by application of procedures which must be known to anyone who is prepared to assert or to understand assertions of those statements. Procedural action is deliberate and self-conscious

My choice of "procedures" is perfectly natural in this context, but not without consequence or challenge. Though I doubt not my capacity to recognize procedures for what they are, I have been pretty much defeated in a struggle to say what they are. There are procedures to follow for assembling machines and for trouble-shooting them, for checking out an airplane before takeoff, for conducting meetings, for applying for leave, for changing fonts or deleting words in word-processing and, in the non-human world, for nest building by birds[1], but not (e.g.) dancing by bees. These examples suffice to show that procedures may be investigative or non-investigative, staged or single-stepped, and they may or may not require conceptualizing capacities; they may or may not be routine, but in either case are to be distinguished from what we have come to call "routines" which may or may not be routine, e.g. dance routines or superstitious routines of various kinds. In applying a staged procedure, the agent works through an ordered succession of positions. The working-through may or may not be "recursive". It won't be if one or another position cannot be attained without the subject having to make non-prescribed adaptations, e.g. a form-filer changes a line or an ingenious robin sticks a twig off-angle into the wattle. The only generalization I am willing to hazard is that the procedure following activity consists of acts that are both *deliberate*, in the sense of being attentive, and *self-conscious*, in the understanding that the agent must, in doing what it does, know what it is doing, in respect to some but not just any "what", e.g. our bird knows that it's inserting twigs in the wattle but not that it's following a procedure. This generalization, as it seems to me, carries over to one-step procedures too, e.g. the one-stepped procedure I just followed in bringing the cursor on my computer screen to the end of a line. I am stumped to provide good formulas for either *deliberate* or *self-conscious*. In any event, these stipulations fall short, for they do not distinguish procedures from routines, which also seem to be self-conscious and deliberate. [A suggested[2] *differentia* is that routines, unlike procedures, consist, in their actualizations, entirely of acts whose success lies in the very movements that constitute them (some call these "basic actions"; I've dubbed them "endotychistic--see p. 65, fn. 1); realization of procedures, according to this suggestion, requires action "on" things, e.g. computers or twigs.]

These procedures are methods of proof.

If an assertion succeeds and a statement is produced, then a successful application of its indicated verification procedure proves that the statement is true and a successful application of its indicated falsification procedure proves that the statement is false; otherwise said, it belongs to the sense of an assertion to indicate procedures for resolving questions over whether the statement is true and over whether it is false. Now such procedural resolutions are usually left untried. It is more to be noticed that such procedural proofs, attempted and even taken to completion, might never be successfully applicable on either side of the question, and the produced statement be neither true nor false. Still, assertionally indicated verification and falsification procedures must both be sometime applicable, if the statement exists. They must, moreover, be familiar and available to anyone prepared to make the statement, or to understand any assertion of that statement.

You may *object* that a statement could be asserted after the fact, when it could, in principle, no longer be verified or falsified, e.g. a statement to the effect that Socrates chatted with Crito. There is indeed a problem here, one which will occupy us in Chapter 22, over how we do conceive what is no longer so. For now, it suffices to reply, simply, that a 20th-Century assertion of the statement would have to indicate that the verification and falsification could have been only *previously* achieved, e.g. in Athens sometime during the 5th Century B.C. That would be known to anyone who could assert or understand an assertion of the statement. Those of us who are capable of such assertions also *now* know how to apply the indicated procedures, e.g. by listening into a conversation between Socrates and Crito. We also know that all occasions for applying the procedures are buried deep in the past. So, while the procedures, by assertional indications, cannot now be applied, they are sometime applicable and are (now or anytime) familiar and available to anyone prepared to assert or to understand an assertion of the statement. (The "could" of our "could be applied" is the "timeless" "could" of "Gold could have isotopes".) Two assertions of the same statement, one before and the other after the fact, must alike indicate common procedures applicable on the same occasions. *Tense* is a device for indicating ranges of occasions of verification and falsification relative to occasions of assertion. One who understands an assertion couched in the past or in the future tenses must know that procedures for the projected statement were or would become applicable only on prior or later occasions.

The two assertions "mean" the same thing; they have the same "speaker meaning". To secure this, the speakers in question would of course respectively work with expressions having different "word meanings", where it is also possible that strings of expressions with unaltered word meanings may, on different occasions, eventuate in utterances with different speaker meanings, e.g. "The moon is new". Speaker meaning is resolvable into indicated conditions of success, e.g. that these procedures are applicable on these ranges of occasions; words, by having the meanings they do, enable speakers to indicate these conditions or perhaps other ones.

Not both procedures are successfully applicable.

What has so far been stipulated is subject to only one general restriction: the verification and falsification procedures for a statement are not both successfully applicable. The successful application of one procedure on any occasion rules out the anytime successful application of the other. This restriction secures the important principle that no statement is both true and false. This principle does not forbid the existence of contradictions of the *p and not-p* kind: such statements are never true and may be simply false. The principle certainly does not forbid people from contradicting themselves, by asserting statements contradictory to others they previously asserted. Our principle stipulates simply that not both of the indicated procedures for a statement are successfully applicable.

The restriction is not trivial and may be controversial. I believe it is agreeable with common understandings. Take a case of fission--a topic which will occupy us in Chapts. 9, #5 & 17, #8 (pp. 405, 436): A sectioned window is separated into its two window-parts; here we can successfully apply procedures both for identifying and for distinguishing part-windows from the original whole window; those procedures, it might seem, would respectively serve for the verification and the falsification of a statement that either part was the same as the original whole; I doubt, however, that there is any such statement, for (as it seems to me), a person familiar with the situation, if asked *which* of the two part-windows was the original window, simply wouldn't know what to say; he implicitly knows that there is no statement to provide a "right" answer to the "Which?" question, as there might be such an answer had this been a case of trimming a once larger window to fit into a smaller casement.

The conditions for and the mechanisms by which the successful applicability of one procedure excludes the anytime successful applicability of another, while practically familiar to anyone who comprehends an assertion, are not always analytically obvious to a theorist's eye. The satisfaction of those conditions and the implementation of the mechanisms may be metaphysically charged. An instance arises in connection with the ancient issue of the exclusion of contraries. The atmosphere around us exhibits a great spread of temperature and pressure. Procedures for proving out all these conditions are likely right now successfully applicable. Obviously, to secure an exclusion of contraries here we have to say *where* it's -20°C. Aristotle argued that there is no exclusion of contraries unless there are individual substances. The case of the atmosphere around us makes me think that contrariness anyway needs individuals, and that there are no statements predicative of features selected from ranges of contraries unless objects can be individuated.

Plato anticipated Aristotle at *Parmenides* part II, Hypothesis VII, where he argued that without "unities" we could have only the appearance and not the fact of featuredness. (I don't see why, by Plato's Theory of Forms, an individual should not stand in participation-relationship to contrary-determining Forms, and that makes me suspect that the Theory may be incompatible with Hypothesis VII.) The problem of contraries survived Aristotle's examination to plague the philosophies of Leibniz and Wittgenstein. Kant's answer was a major criticism of Leibniz. See p. 338, also p. 366 fn#39.

To summarize: Any statement is both verifiable and falsifiable in that there exist sometime applicable procedures for proving and for disproving the statement; neither of the procedures need ever be successfully applicable, but if either ever is, the other never is. Both of the assertionally indicated procedures must exist whenever the statement does and, therefore, whenever the statement is successfully assertible; the sometime occasions of application may be at almost any temporal remove from the assertional occasions upon which they may be indicated.

Specimens

Three Examples. Consider, for purposes of illustration, a statement to the effect that there are at a particular time a certain number of apples on a particular table: To verify such a statement, one must, at the time in question, first get himself into a position where he may both touch and see every displaceable object resting on the top surface of the table; having sequestered the apples among those things, he would thereupon proceed to touch them in turn while concurrently reciting in order a sequence of numerals; his terminating with a pre-selected numeral would verify the statement. His finishing the count at any other position would, in this instance, imply that the falsification procedure for the statement in question could be successfully applied. Again, one might wish to test a statement that there was a body atop the table. Once in proximity to the table, he would verify the statement by reaching out his hand and displacing something touched. Failure in that performance would imply nothing about the falsifiability of the statement. Falsification in this instance would require that he place a body (e.g., his hand) somewhere into every region bottomed by the top of the table without ever displacing anything else in the process. Finally, a controversial instance: The statement is to the effect that *if* a particular apple on the table is a jonathan it still isn't ripe. The statement would (roughly said) be verified by showing that the object in question was a green hard jonathan; it would be falsified by showing it to be a red less-firm jonathan. The terms of the assertion make advance allowance for the apple being a golden delicious, in which case both of the described procedures are applicable but unsuccessfully so; I submit that in this event the assertion succeeds in its own terms and produces a statement that is neither true nor false.

The availability of procedures is a condition for the existence of statements, and the character of the procedures determines the character of the statements

A statement would not exist failing the availability of the indicated procedures. Such procedures once available may be lost, in which event a onetime existing statement may later cease to exist. Then the statement would also cease to be assertible because unknown procedures are not indicatable. A onetime existing statement may otherwise cease to exist because of the passing of authorities or otherwise because of the loss of that knowledge, indicated among the other cited conditions of success for assertion (pp. 88-91). The statement would never have existed if either of the indicated procedures were *always* inapplicable. Assertional indications that fix the occasions of procedure-applicability may, of course, take place almost any time. This condition, that both procedures should be sometime applicable, is unaffected by the vicissitudes of human knowledge and the erosions of time, but brings in all manner of other contingencies that bear upon both the "form" and the "content" of the statement itself. These conditions of applicability fix the character of the statement as a somewhat "abstract object". I add that the "truth value" of the statement, if it has one, owes entirely to which of the procedures is successfully applicable, if either is. Altogether, then, the sometime applicability of verification and falsification procedures is, of all those conditions of success indicated in assertion, the most salient for the analysis of statements and is the thread that will guide us through what ensues in this treatise. The applicability of procedures also ties into the fund of knowledge that backs the statement, for the successful application of the verification procedure of a statement, if it is applicable at the time of assertion, affords the most fundamental way of gaining knowledge of the information formulated in that statement (p. 208).

The relevant conditions of application for procedures are "circumstantial", not subject-centered

The "able" of "applicable" conveys the sense of "can" and is *schematic* over conditions for applying procedures (see pp.151 et. seq, 129f.). That covers too much; specifically, we must, at least for the most part, exclude conditions on the existence and capacities of procedure-applying subjects. I mark the distinction by speaking of "circumstantial" and "subject-centered" conditions of application. Geologists are now familiar with procedures applicable for proving out the occurrence of earthquakes in precambrian times. The procedures did not exist in those ancient days; but presumably conditions for their applicability did then obtain, e.g. there were proximities from which the self-displacement of the earth could be observed. I resist putting this in terms of "possible applications" by "possible observers". The alluded to circumstantial conditions, if they obtained at all, were perfectly "actual" at the time; if they did not obtain, then presently available procedures were not then applicable and there does not now exist a statement that an earthquake then occurred. If the statement does now exist (as I believe), it will cease to exist for reason other than the inapplicability of procedures when, because of the ultimate extinction of our species, those procedures themselves no longer exist. Knowing subjects may indeed enter into the circumstantial conditions of procedure-application when their procedures are applicable in respect of themselves; again, we shall argue that delimiting procedures circumstantially presuppose the sometime existence of testers; but these are matters of special notice; absent such special notice, the assertionally-indicated conditions of procedure-application are "circumstantial", not "subject-centered".

We now confine ourselves to verification. Verification procedures are "tests".

The relationships between verification and falsification affect the determination of the statement's forms, and will be brought under examination in Chapter 14 on "Form". For now, however, it is enough to know that two procedures should be applicable for every statement and not both be successfully applicable. That stipulation is symmetric between verification and falsification, and we may, accordingly, gain economical full purchase by confining our words to verification.

Verification procedures are distinguished as kinds of action by being methods of proof. Putting name to this distinction, I propose to call them "tests". This verbal proposal is not entirely free of stipulation. Tests, in ordinary speech, are usually staged and well articulated. The so-called Nitrate Test learned by students of chemistry and comparison tests for series convergence learned by students of the calculus are examples. That everyday sense of a test is too restrictive for us. All that one needs to verify that I am scrabbling away is to look at me. The reformed idea of a test must be construed to include acts of observing, examining, looking at, for, and through, comparing, counting, measuring, as well as such tests, properly taken, as titrating. Test operations may be slight, brief and, so to speak, "natural"; they may come to nothing more than taking a look. A test in present understanding is a kind of action characterizable in considerable part in terms definable by lists of conditions for doing and conditions of success. Not any kind of action, of course, is a test in our broader sense, and I shall presently set out to specify the distinctive features of tests. But first a caveat: The procedures of which we have so far been speaking are all of them supposed to be tests. However, not all tests are to be reckoned among those procedures, but only those that factor into what we shall presently dignify as fundamental criteria for statements.

Tests are action-kinds. We call their instances "applications".

We call completed acts of a test-kind "applications of the test". The application of a test is action and a happening occurring at a place over a definite stretch of time, initiated and carried forward to completion by some animal. As already observed, the happening or process need not be kinematically prolonged, neatly articulated or nicely staged--it may be a mere "looking"; nevertheless, a test application when completed is something an agent did and meant to do.

Tests are kinds of investigative action with perception being part of their measures of success.

Tests as action kinds are (partially) definable by a listing of conditions of success. Other conditions (for doing) must also be satisfied if a test-application is to be completed by a particular subject on a particular occasion. Any test application fully carried out is either successful or unsuccessful, and we may accordingly speak about successful or unsuccessful test applications. (NB: Failure *in* applying a test must be contrasted with failure *to* apply the test and with failure to complete the application. The question whether an application is successful or unsuccessful cannot be answered unless the act is done, i.e. the application made and completed.)

Tests are kinds of investigative behavior, and an application of a test succeeds only if a perception is induced in the agent. That is a first distinctive feature of testing. A test is a kind of trying to find out. Testing is a species of what ethologists have called investigative behavior. In our scheme, that means that testing is a kind of behavior whose measure (*not* condition) of success is the occurrence of a perception. Every species of testing will accordingly always have as characteristic conditions (for doing) that the animal should be specifically perceptually sensitive, e.g. be sighted. We can then always go on to specify as conditions of success that there should be present in the environment objects such as sounds, smells, sights, etc which can be heard, smelled, or seen; a source of light, etc. The requirement that the successful application of a test should lead to a perception rules out as tests those action-kinds in which the agent means only to move himself in a particular way. Acts of standing up, say, or of jumping a creek do not qualify as test applications (Here recall the "suggestion" advanced on p. 218 for the definition of *procedure*).

I shall hold that "individuation" is a kind of testing, and shall propose that bodies are individuated by displacement. By what has just been stipulated, mere self-displacement by standing up, say, or by lifting an arm "directly" does not suffice to individuate one's self or one's arm as a body. In order to bring off the individuation of a body, the tester must also perceive the displacement of something.

Perception: A change in a subject's body, classified as being "of" a "perceptum" other than the subject itself, which change is induced by a material phenomenon involving the perceptum and which change may or may not induce subjects' "direct awareness" of the perceptum.

A first digression on perception[3]. Perception is notoriously "problematic". I do not know that anyone can say, in so many non-circular words, under what conditions perception occurs. One can imagine cases of which it would be hard to decide whether happenings should be classified as perceptions or as hallucinations. (A classical instance due to A. Gans is that of a man who every night regularly at 9:30 looks at his watch from which he gathers the usually mistaken impression that it reads "8:30". At 9:30 one night a companion sets the watch to read "8:30", and the subject gathers the usual impression: Is this a case of perception or hallucination?) Nevertheless, one can recognize perception for what it is, and investigate its mechanisms. A perception is a happening in a subject induced by the environment; it regularly but not invariably brings subject into a state of knowledge--a condition of readiness for the utilization of information available from the environment (see pp. 182f); that accession of information to the subject is most often (but not invariably) manifested in the subject's response to that environment. Now that is hedged, and also vague! I hope the following observations will serve to keep us on course.

First: Perception is a change in a subject's body induced by some concurrent material phenomenon.

Second: The inducing phenomenon is either itself a change in quality, e.g. temperature, pressure, color or a gradient of such qualities which subject's body moves across.

Third: *What* subject can then be said to perceive is rarely if ever the inducing phenomenon itself. The change in pressure resultant upon subject's moving his finger in and out of a bowl of water induces him to perceive, not that change, but the top surface of the water in the bowl. What is perceived--call it the "*perceptum*" is only a feature or aspect of the inducing phenomenon. This explains why we can be truly said to hear silence, see darkness and why Fido can look to see that mistress is away before he sacks the garbage can, for in all these cases there is a perception inducing phenomenon operating. The observation is supported by what the physiologists have discovered, that the inducing phenomenon influences what is perceived at a sensitive surface by inducing prior or concomitant changes in adjacent surfaces. In any case, it is obvious that the sometime static "object of perception" or "perceptum" cannot always be equated with the dynamic phenomenon that induces it.

Fourth: Different inducing phenomena may be perceptually equivalent. Changes within different mixes of radiation notoriously induce identical color perceptions. The existence of such "perceptual equivalence classes" of phenomena and the difficulty of finding physical rules to define them does not imply that the induced perceptions are "subjective".

Fifth: Those changes in a subject which are perceptions are preferentially classified as perceptions either by reference to a physical classification of the phenomena whose changes induce the perception or by reference to that part of the subject's surface at which the changes are induced or both. Color perception is either *of light* or *through eyes* or a combination of both. In cases of conflict, I believe we would give preference to the physical classification. I am told that bulls' eyes are not suited to discriminate color and that bulls therefore do not have color perception. Still, a physiologist dedicated to the traditions of the bull-ring might investigate to determine whether bulls might not perceive color through their horns.

Sixth: Perception, when it is characterized, as it commonly is, within a "perceptual mode" (e.g., as sight, hearing, touch), is always spoken of as if "of" something. What it is "of", the *perceptum*, is fixed in the observer's formulation. There should always be a "clarified" "that-which" formulation to fix the object, e.g. we say that Tom saw that bird which was in the bush. Commoner formulations may be "opaque" over substitutions of references to what the perception is "of", and the observer must then take caution not to put his perception into the eyes of his subject. Different formulations of the same seen thing yield different characterizations of perception. I say that my cat sees a bird, which means at least that Tom is visually aware that a bird is there. Suppose that the bird Tom and I both see is a sparrow. If it was correct to say that Tom saw a bird; it might or might not have been right to say that he saw a sparrow. Here "see" is "opaque".

Seventh: The perception, which is a change in the subject, may or may not be "voluntarily" sought by subject in action. It appears, however, that some kind of activity on subject's part usually contributes to the inducement of perception.

Eighth: Perception may or may not be "awareness" (see pp. 183f). However, if not, as when one is wakened by a sound or blankly sees the passing scene, we like to think that this subject could recover an awareness from its previous perception. While *perceptual awareness*, as we may now dub it, is distinct from perception *per se*, it does seem to me to be the "proof" of the perception in rather the way in which a scintillation on a screen is a proof of the presence of an electron.

Ninth: Perceptual awareness is knowledge and may or may not then be subclassified, e.g. as *notice* or *recognition*. Such knowledge may be attended by any or no degree of self-consciousness.

Tenth: Not all awareness of things is perceptual. Recollection, for example, isn't. Perceptual awareness must be "direct" in the sense that

(i) It is an "aspect" of that change or gradient which induces the perception,

(ii) A report of perception should not by itself imply awareness of anything else,

(iii) Any other awareness of the same thing is subordinated to the direct awareness, if that is available, in the sense that "direct awareness" is a check against the other; but not conversely. So, for example, "hearsay" is indirect relative to a witness.

> Differently, the visual awareness of roughness is subordinated to the direct tactual perception of same, where in such cases one might also be said visually to perceive the light reflected from the surface; the subordination of the awareness is reversed in the case of color.

Eleventh: Not all "direct awareness" is perceptual We must, in particular, distinguish perceptual awareness, which may or may not be of self, from *sensation*, which is a subject's "propriocentic" direct awareness of itself. (An observer reports that you and I both see my left thumbnail: in neither case is it implied that the thumbnail is an actual part of me. Contrastingly, only I can feel the pain from the hammer blow my nail once suffered, and *my* sensation must and can be of *me* alone.) The kind of sensation that will matter most for us is a subject's "kinesthetic" direct awareness of the movements of its own body. (The direct awareness of relative movement is by tactual or visual perception; kinesthetic awareness of *relative* movement is only "indirect".) Perception and sensation can occur independently of each other. It remains that sensation is a more "primitive" kind of direct awareness than is perceptual awareness and provides what Gibson styled a "basic orienting system" or "frame of reference" for distinguishing and coordinating perceptions of different kinds. A subject comes to know that he sees through kinesthetic awareness of eye movements; some physiologists have plausibly hypothesized that we can be truly said both to see and feel shape because these different perceptions trigger the same behavior, resulting in the same kinesthetic sensations. Some kinds of direct awareness such as of relative distance from self require both perception and kinesthesis. It is likely, as we shall presently argue, that successful testing should always involve both. [NB: These observations do *not*, as I once believed, solve the problem of how a subject coordinates the presentations of different perceptual and sensation modes without (as is ruled out) a common mode of perception or sensation. By what sense, or in what other manner do I coordinate the presentations of sight, touch and kinesthesis?]

Testing is proficient behavior.

Though testing may be as slight as taking a deliberate look, not all such investigative behavior is testing. A cat's scan of the garden outside the window is not a test for birds. Garfield might, however, test a proffered goody for an unwanted pill. That's something he would have learned to do and at which he may improve; it's a skill, and qualifies as proficient action by some of the marks we have given, e.g. a condition for this cat doing that kind of action is that he be capable of making a possibly non-test-like investigation to establish essential information (see pp. 42). Again, a counting child may *test* by counting; a "counting"-horse cannot: the horse unlike the child is not investigatively competent to locate where it has gotten to in a counting sequence.

Testing is self-conscious.

Though testing may be as slight as taking a deliberate look, not all such investigative behavior is testing. We are back to the matter of testing being procedural, where (as it seems) procedural action is always self-conscious in respect to some "*what* the subject is doing" (p. 218). A testing subject--even our pill wary cat--must know what he is looking for and be able to know he hasn't found it if he fails. *(Now*, Garfield, finding no pill, proceeds to eat its bit of mullet.) Any test application envisages alternative "possibilities". The condition of self-consciousness is automatically satisfied when testing occurs in the service of assertion--a kind of action that self-consciously indicates the character of the test now applied (p. 52f.). Since, as in our cat-case, testing may occur unentangled with language, I would like to find a lesser condition of self-consciousness for testing. That the behavior should simply be "skilled" is plainly too little, for skills may be exercised non-self-consciously, as when a dog catches a frisbee. That action should conform to "enabling rules" may be sufficient for it to be self-conscious;, but this condition, though lesser than the possession of language, is still excessive for our purposes: Garfield's pill-seeking behavior is not that sophisticated, but it is self-conscious. (For further discussion, see my *Stratification*, pp. 266f). Our examples of self-conscious activity seem clear enough to vindicate our sense that self-consciousness is a fact; but a formula for self-consciousness eludes me: so (per earlier anticipations) I must disappointedly resign the problem while appealing to the fact.

Testing, in summary, is self-conscious procedural exercise of investigative skill or know-how. My theory holds that any subject's capacity to conceptualize--itself a network of skills--must be fastened down to a repertory of procedural competency. Now a

Defending the thesis that testing proficiencies are conditions resident in agents who apply tests. Testing is "naturalized" in behavior.

subject's actual skilled behavior, taken in sequence, may be fitted into indefinitely various patterns of procedural competence or none at all. (I think that even follows from our earlier conclusion that the presence of know-how is not a "condition of success" characteristic of action-species, but rather a condition for doing proficient action. pp. 37f). Consequently, there is no proving that a deed witnesses one skill or another or any skill at all. The observation raises a problem which Kripke, in his exposition of Wittgenstein's later philosophy, has called a "skeptical paradox". Children learn to count and we know that they do; yet there seems to be no way of knowing that some here-and-now recitation is just this kind of counting. Are we then to conclude that there is no "fact" resident in the child that is its knowing how to count, and that the whole fact of the matter is the child's more or less appropriate behavior in the human context of its other activities? Allow that words of skill have "dispositional meanings", implicating the occurrence of fairly specific behavior under pretty open conditions: We have already argued that such dispositional ascriptions commonly imply the *presence* not the absence of a (usually unnamed) something in the subject (e.g. pp. 38ff.). Some may now wish to argue against this: Since skills are acquired, the anytime presence of a skill in the subject would entail that there have been a *first* skilled deed done by this subject *from* this present skill--a first word read or batch counted; but that is absurd. But why should there be this entailment to a "first", anymore here than for virtuous actions (See Aristotle's *Nic Ethics*, II.4)? Consider: cars are broken in by being driven; the sticky valves of a new motor get better seated with driving, which gives us reason *not* to suppose that there must have been a first smooth valve flip but no reason at all to suppose that these moving valves are not part of the motor mechanism. Similarly, I as yet see no reason to suppose that our virtues and the testing skills we acquire as a foundation for conceptualization are not "facts in us", or that testing is not naturalized in behavior.

Perceptions that occur as the measure of success of testing are self-conscious

I have not defined *testing*. I hope I have said enough about it to make the notion serviceable for the heavy employment we are going to be making of it. Before going on to adapt the notion of a test to our account of assertion and statements, I want to register one last claim in regard to testing in general. Subjects may perceive things without knowing that they do. Now, as it seems to me, if a perception is generated as a result of a deliberate test, then the subject must also know that it has gained the perception it does. It must know that it finds what it was deliberately and self-consciously seeking. Even our wary pill-testing cat will know that it tastes the pill it spits.

Tests applied to verify or to falsify statements are called "criteria".

A subject may successfully apply a test and say nothing in response; indeed, subject may have never learned to make such reports. Furthermore, when an utterance is elicited and concepts exercised, the performance need not be an assertion. Again, a test applicable (e.g.) to verify a statement that an object weighs five pounds might be otherwise applied, not to verify a statement, but to resolve a question about the object's weight. Again, one may test whether it's windy by putting his hand into the weather, no particular spot in mind, and issue a non-assertional "feature placing" (Strawson) report that it's blowing out; in this instance, the successful test application does not serve to verify a statement. On the other side of the matter, a statement might be established, not by successful testing, but in some other way, say by demonstration. Now, the weather-testing example at least raises the question of whether every test *could* serve for verifying statements.

> I later propose that statement verification procedures must have a certain minimum complexity involving the agent's being able to localize the application more narrowly than is required by our description of the above mentioned test for the state of the weather. "Idealists" hold that we can never do better than tests for the weather: Idealism won't sanction even so much as a directed response, let alone the individuation of bodies or the making of statements, and disallows any separation of subject and environment.

The variety illustrated by these cases portends endless qualifications which I elect to bypass by simply appropriating the now fashionable term "criterion" to denominate tests that are applied to verify or to falsify statements.

"Criteria" are "hybrids" of testing into proving and disproving statements.

A criterion in this confined usage is a kind of action that is a hybrid of testing into statement-proving or disproving. The Original Purpose is to establish that a statement is true or that it is false; the Original Way is to apply a test (by which one might come to know); the hybrid result is statement-verification or -falsification. Once switched away from side-railing mention of falsification, we can now speak of applying a criterion as a kind of action that takes its conditions of success from both testing and proving. Chief among the conditions of success for proving is that something have been said, e.g. in assertion, to raise the question whether it is so. Proving super-adds its conditions for doing to that of testing, notably that all the other conditions for the existence of the statement identified in assertion are satisfied. A (verification) criterion is an action kind having the general character of a test, whose measure of success is the perception of a statement-formulated fact. (A falsification criterion seeks the perception of a fact formulable in other statements.)

Kant, in the A-edition version of his "transcendental deduction of the categories", moved from the schematized categories of the imagination to their realization in judgement; he seems to have reversed the transition in the B-edition. I speculate that he may have done so from a realization that his test-like procedures need not be "criteria". The shift, as it turns out, mattered little, since Kant was exclusively concerned with intellectual representation and, at that level, the concept determining role of procedures is the same whether their categorial aspects be located in the imagination or in judgement.

We call a situation in which a criterion for a statement can be routinely applied to gain a trivial proof of the statement a "verifying situation" for the statement.

The verification procedures appropriate to a statement may be various, depending upon the placement of testers. Occasionally we may assert or otherwise come to consider a statement in circumstances where we can routinely determine that the statement is true. Thus one may or may not be in a position where, by merely reaching out to touch an object, he may verify a statement that the object is rough textured. Again, one may routinely determine the properties of a number by calculating. If one is in position where he need merely reach out or calculate or measure to secure verification or falsification, the result, though perhaps not obvious, is (as mathematicians say) "trivial". The test, though possibly complicated and prolonged, is routine. Let us call a situation where we could thus routinely apply a criterion and trivially establish the truth of a statement a verifying situation for the statement.

There may be numerous ways of getting into a verifying situation, whence one may proceed to apply a test in a routine way. The application of a criterion may be thought of as a two-stage affair carrying one from a situation in which one asserts or otherwise considers a statement into a verifying situation, and thence to a successful or unsuccessful termination. Consider an earlier example, roughly set out: Someone may assert that there are seven apples on the table in the kitchen. If another understands the assertion, he knows that one would test the produced statement by entering the kitchen and approaching sufficiently close to the table to see and touch what is on it, thus getting into a verifying situation; then he would count the apples he sees there; if he counts in the regular way to "seven" and not further, the test is successful; otherwise not. I stress the two stages: first, putting oneself into position to touch the apples; second, counting the apples in a routine, straight-forward way. Again, any idiot can show by Mathematical Induction that the first n natural numbers sum to $n(n+1)/2$; but Gauss displayed precocious auguries of genius when, by himself, he found the recursion formula, presumably by folding up the sequence in just the right way. Generalizing: the application of a criterion for a statement is normally a two-stage operation of first entering a verifying situation and then routinely proceeding from there. The first-stage of getting into the verifying situation often calls for ingenuity, imagination and good luck; the second-stage operations may demand nothing more than time and sweat.

Criteria applicable in verifying situations are called "fundamental criteria". These are parcelled out one pair to a statement. Fundamental criteria are the only ones whose applicability must be assertionally indicated as conditions of success.

The second stage of the application of a criterion enjoys a status different from the rest of the test. No matter how subject gets himself into position to touch the apples, actual verification that there are seven requires that he should count. Counting is characteristic of the statement; the precedents are not. Generally: It is only the applicabilities of these second-stage procedures that are to be reckoned among the conditions of success indicated in an assertion of a statement, and they alone satisfy the stipulations entered in the earlier paragraphs of this chapter. I call the second-stage of a criterion that is applicable within verifying situations of a statement *a fundamental criterion* for the statement. Looking, counting, and measuring are familiar sorts of fundamental criteria. My theory demands that there be a pair of such routinely applicable fundamental criteria for every statement and anyone who could understand an assertion of a statement should know-how to apply those tests to completion and be capable of knowing whether the applications succeeded or failed.

> The question arises whether these procedures are "algorithms" in the sense explicated within the theory of recursive functions. Some are, and all of them are meant to be "like" algorithms; but most of them fall short of the strict definition. That is because they cannot always be realized in an alphabet of symbols as marks on paper. Squints and ear-cockings are just as eligible as calculations. They are like algorithms in that they are "terminating" and specifications can be written down for bringing them to completion. There must be a "yes" or "no" answer to the question of whether an application is *completed* on any appropriate occasion. However, there may be no routine "yes" or "no" answer to the question which might have prompted this activity; there may be no routine available for determining whether either of the fundamental criteria for a statement is successfully applicable; furthermore, indeed, it may be that neither is.

Since one may get into a verifying situation in any number of ways, there may be any number of criteria appropriate to a given statement. However, once in a verifying situation, the number of second-stage continuations is diminished. I propose to cut them down to a single pair, one for verification and another for falsification. To test how-many, for example, one must *count*; counting is the fundamental verification criterion kind for statements of the "how-many" form; so, if you wish to verify the statement that there are seven apples, you must count to *seven*. This may cover many variants; it may not matter whether you count this apple first or that; we still think of these as versions of a single test. This policy of criterion individuation results from the consideration that the theoretical identification of criteria should be made relative to our actual ways of thinking about the world, for only thus can an accurate analysis of conceptualization be achieved. It might or might not matter whether a test that consisted of reaching out to touch were further specified to foot or hand; if it actually made a difference to our conception of cardinality whether we counted from right to left or the reverse, as in certain systems of arithmetic it may make a difference whether we expand $(a+b)(c+d)$ as $a(c+d) + b(c+d)$ or as $(a+b)c + (a+b)d$, that fact would affect our identification of the criterion.

Take it, then, that, while we may apply any number of criteria either to verify or to falsify a given statement, the characteristic terminating stages of these tests have a uniform specification, and pairs of fundamental verification and falsification criteria are uniquely assigned to statements.

Statements are theoretically identified by specifying a pair of fundamental criteria.

Anyone who understood an assertion that 6 was a perfect number would have to know that this statement would be proven out by summing 6 up out of its proper factors plus l, and shown false with any other result; similarly, one who understood an assertion that there were seven apples on a table would know that this would be verified by counting those apples to seven and falsified by any other count. I propose that statements are to be theoretically identified by specifying pairs of fundamental criteria. I shall expand that proposal, by way of a general theory of testing, into a general theory of statements. I believe that this "criterion of identity for statements" is more discriminating than any other in the field.

The analysis requires no reference to a language, and is not "paraphrastic". Still, the requirement that we can describe the circumstances under which the verification fundamental criterion for a statement could be successfully applied resembles the familiar principle of formal semantics that everything formulable in the object language is also formulable in the meta-language.

Summary for verification. To summarize what has so far been said about verification: A statement is verifiable in principle: the verification would be secured by the subject's application of a criterion carrying the subject into a verifying situation, thence, by application of a fundamental criterion, to a successful termination. If the criterion is actually applied and with success, subject proves the statement. In order to know what the fundamental criterion for a statement is, we must know what occasions are verifying situations, how to operate within those situations and what will count as success. This knowledge is implicit in anyone's knowledge-how to assert the statement. Our theory of statements will seek to represent all this knowledge solely in terms of occasions for the application of the tests.

Since fundamental criteria are always so called with reference to statements, that fact requires no further mention. I shall accordingly now speak indifferently of "test" or "criterion" while implicitly limiting myself to tests hybridizable into fundamental criteria.

A test is represented by a set of occasions of applicability together with a subset of occasions of unsuccessful applicability. The included subset is determined by formulations of conditions of success for the application of the test.

I now prepare ground for a uniform representation of tests. In order to identify a test one would have to know this much at least, first, on what occasions the test could be applied and, second, under what circumstances or on what occasions of application it could not be successfully applied. (The negative formulation reflects our first approach to the matter of the conditions of success for action. See p. 24) Actually, nothing more is needed for the purposes of the ensuing analysis. We may, accordingly, think of a test as being represented by a specification of these two sets of occasions--of applicability and of unsuccessful applicability, where the latter is a subset of the former. A specification of the set of occasions of unsuccessful applicability is a kind of cashing out by reversal of those conditions of success for the test that attach to the circumstances (these conditions are non-subject-centered). We normally gain a specification of occasions of unsuccessful applicability by writing down conditions of success, e.g. one could successfully count an assemblage to *seven* only on occasions where there were at least six objects present. The specification of the wider set of applicability occasions may be thought of as cashing out circumstantial (non-subject-centered) conditions (for doing) characteristic of the testing purpose. Here again the actual specification will normally come in reverse, e.g. one cannot make a measurement on occasions where no standard is available. (Caveat: These conditions for *doing* for testing are conditions of *success*--not for doing--for assertions of the asserted and testable statement. See pp.300f. below.)

So, in summary of these thoughts and in anticipation of a formalism: Tests may be uniformly represented with a specification of the set of occasions on which the test could be applied together with a specification of the subset of those occasions on which the test could not be successfully applied. These two specifications may respectively be taken to represent the circumstantial conditions for doing characteristic of the test kind and the circumstantial necessary conditions of failure.

I now stipulate as a principle or axiom of *testing*, hence of criteria, that, on every applicability occasion, a test is either successfully applicable or unsuccessfully applicable but never both. If this axiom seems nothing more than logic, it is still worth setting it out in so many words because of difficulties over "could" and "able", . It could be that one condition for successful application of a test on an occasion was satisfied and another not; observers might then want to say both that it could be done and that it couldn't. Our axiom comes down to the stipulation that, if there were any circumstantial condition that would make it impossible to apply the test successfully on an occasion, then the test would be unsuccessfully applicable on that occasion; otherwise, it would be successfully applicable. "Impossible"? Our theory advisedly makes no appeal to modal notions, and there is no way of introducing them into the representation. The problem is easily overcome in this instance. We simply stipulate that the set of occasions of applicability of a test is the union of two disjoint sets of occasions, these being the sets of occasions of successful and unsuccessful applicability respectively.

A test is either successfully or unsuccessfully applicable on every applicability occasion, but never both.

We shall say that a test is "qualified" just when the representing set of occasions of applicability is non-empty, "qualified", that is, to represent a statement. If an assertional indication is of an unqualified, inapplicable criterion, there can be no corresponding statement and the assertion must fail. We may now tidy up our representation of qualified tests: The set of occasions of applicability is simply the non-empty union of a representing disjoint pair of sets of occasions of successful and of unsuccessful applicability.

A test is "qualified" to represent statements just in case it is representable by a disjoint pair of sets of occasions whose union is non-empty.

A statement, we held, is representable by a pair of tests. Since a test may in its turn be represented by a pair of sets of occasions, it comes out that a statement may be represented by a quadruple of sets of occasions, where the unions of the first and second entries and of the third and fourth entries are respectively non-empty and where the non-emptiness of either the first or the third requires the emptiness of the other. A statement may be pictured thus:

A representation of statements as a quadruple of sets of occasions.

$$
\left<
\begin{array}{l}
\left.\begin{array}{l}
<{}^{+}\omega_{V_s}\ (= \mathrm{Set}_1), \\
\qquad \cup \\
{}^{-}\omega_{V_s}\ (= \mathrm{Set}_2)>
\end{array}\right. \neq 0 \ = \ V_s \\
\\
\left.\begin{array}{l}
<{}^{+}\omega_{F_s}\ (= \mathrm{Set}_3), \\
\qquad \cup \\
{}^{-}\omega_{F_s}\ (= \mathrm{Set}_4)>
\end{array}\right. \neq 0 \ = \ F_s
\end{array}
\right> = \ S \ [\mathrm{Set}_1 \neq 0 \rightarrow \mathrm{Set}_3 = 0]
$$

Statements admit of five "values" determined by the emptiness and non-emptiness possibilities of their representing quadruples of sets of occasions.

Tests may be successfully applicable on some occasions of application and unsuccessfully applicable on others. The test for the existence of a book on a table might be unsuccessfully applicable over some parts of the surface of the table and successfully applicable over others; again, illumination may set bounds on the proximities from which one might visually discern the title of that book. Some criteria may be successfully applicable on no occasions of application and others be successfully applicable on all occasions of application. So either or neither but not both of two test-representing sets of occasions may be empty, and (by our stipulations) there are five possibilities of emptiness--non-emptiness relationships among statement-representing quadruples of sets of occasions[4]. I shall appeal to these possibilities respectively to give definitions of statement-untruth and necessity. Consider, for example, that if a test for finding a book here and now on the table at which I am writing were everywhere unsuccessfully applicable, the statement that there was a book there would be untrue. On the other hand, the statement that the sheet of paper I am writing on is larger than its bottom half, which seems true of necessity, would be verified by folding the bottom half up over a part of the whole; that test, surely, is successfully applicable wherever it is applicable at all.

Riposted objections arising from the concreteness of occasions and the contingent determination of statements.

At this point I anticipate that some readers may wish to enter objections, stemming from the consideration that statement-determining occasions of testing are "concrete" and, accordingly, subject to various pretty accidental sorts of contingency. My account, no doubt of it, is laced with contingency, and I am sure that that will make it hard for many theorists of intensions to swallow. I like it! My brief replies to these objections will convey a sense of some of the more distinctive (or, perhaps, "idiosyncratic") features of my theorizing.

The first objection was once made in a seminar: The verification test for a statement that a certain object was 10.34 cm. long could be applied only on occasions where the object to be measured was present, some kind of measuring instrument available, etc. Suppose the statement was true: could the test in the circumstances be unsuccessfully applicable? If not, then we would have the unwelcome result that the statement about the object's length was necessary. My answer to this is that the test might not be successfully applicable on all of its verification occasions simply because there was not enough light on some of those occasions for the discrimination of decimilimeters, even though the tester thought there was. He could, in the circumstances, have shown that the thing was 10 cm. long, but not that it was 10.34 cm. long. But then I am also prepared to allow that if, *per accidens* there was enough light, then the statement would be necessary *per accidens*.

Second objection: Suppose it were totally dark on all those occasions: Are you then going to say that it is untrue that the object was 10.34 cm. long? I reply, Not quite; but rather, that the statement I think you have in mind doesn't exist, although some other statement might exist or perhaps some other true "constative product". Notice in this connection that in fact the simple visual length of the object couldn't be known.

Third objection: "Visual length", you may wonder? Why shouldn't we use calipers instead? Yes; and I believe that the use of calipers defines a more fundamental idea of length than does sighting. But then these are different if related ideas of length, and the statements they contribute to are also different. I can imagine a new version of the first objection, arising from the possibilities that the calipers might or might be capable of spanning the object on all of the indicated occasions of verification. But now I think enough has been said to project my replies to such objections.

Our way of confining the testing of statements to limited occasions sets this account apart from others.

Different tests are applicable on different sets of occasions, and, for us, statements--our most basic truth-vehicle--are distinguished by the limitations on their occasions of verification and falsification. So much is obvious from what we have said. The point is worth noticing, however, for it distinguishes our method from standard logic and other successors to the liebnizian philosophy. We interpret those theories to hold that all statements, hence all tests, are uniformly definable over a single cosmic totality of occasions: there is no occasion on which any test could not be applied.

Occasions are time-specified, closed and connected regions of space with their contents that are tactually circumscribable and partitionable.

We now consider the postponed question of what these occasions are. The representation cannot tell us, but we must know anyway in order to apply the theory. It is evident from my reply to the above registered objection that I take occasions in a very physical way indeed. Every occasion, in present understanding, is a closed, connected region of space, with its contents, specified to a time, small enough to be tactually circumscribable by some animal at the time, and large enough to be partitioned with probes (the resulting parts themselves not necessarily occasions).

Remarks:

"closed, connected regions of space": Every such region is bounded and pairs of places within the region may be reciprocally reached or enclosed along a tactually traceable path lying entirely within the region.

"small enough": I don't know how big or how small an animal can be. I shall for purposes of illustration sometimes assume that The Earth occupies a region too large to be an occasion. But I allow that there may, somewhere in the cosmos, be a spider of humean proportions with tentacles long and strong enough to reach across that much space.

"large enough": This is to assure us that some animal can operate within the region. Again, I can't say how small an animal might be or what the lower threshold of tactual perception is.

"specified to a time": This is not a pretense, but simply another kind of openness. The time of a test occasion is the time at which some animal could make a tactual probe. But that is not *a* time, you may say, for any act of that sort must take *some* time. I respond that the probe could be made as quickly as you wish. My idea of a time could be brought under the familiar definition of convergence, now relativized to probing: For any temporal interval i, there is an interval i', smaller than i such that a probe could be made over interval i'. Now, if that in fact is not so, if there is a smallest probing-i, then I would accept that as "a time".

"tactually circumscribable": Only touch among our human sense modalities affords controlling means for surveying limited regions. This grubby fact is of great consequence for our "conceptual order".

"contents": This condition is "circumstantial", not "subject-centered". Although our *conception* of an occasion is explained relative to our conception of a tactually sensitive animal, it is not generally required that such a creature should actually be present on such an occasion. The stipulated contents may include light, sound, wafted chemicals, and various *tangibilia*. Note well that *what* a tester is said to see on an occasion of testing need not be part of the occasion, though perhaps the instruments he peers through and light emitted or reflected from the seen object will be.

"probe": Ideally, this would be a tactually sensitive part of the tester; however, I *may* (reluctantly) have to allow that we sometimes need the use of such instruments as probers with small points and perhaps even wire-like grids, as extensions upon our bodily selves.

We may now, for purposes of illustration, go back to the apple examples used on pp. 221f. One example was a statement to the effect that are some apples on a table. The occasions for both verifying and for falsifying this statement are ones included within a particular room at a particular time, themselves all enclosing a particular table at the time in question. There must also be enough radiation in the room for the somehow detection of *apples*. Of these differently centered and variously sized places at a time those ones that also include at least one apple are occasions of successful applicability for the fundamental verification criterion and those ones that do not are occasions of unsuccessful applicability for that test. Application of the fundamental falsification criterion would require exhaustive sort-testing of objects selected from the tabletop at the time in question, that presuming a complete resolution of the tabletop into exclusive surface parts no larger than the diameter of a smallest apple; there may be numerous such partitionings of the relevant region bottomed by the tabletop, hence numerous occasions of application for the procedure; these application occasions are all of them those of the verification test further limited by the condition of an accomplished circumscription-cum-partitioning of a region bottomed by the tabletop; of those occasions (still insufficiently fixed by the just-given recipe), any one that contained an apple would be one for the unsuccessful applicability of the test, and the others (whatever they are exactly) are occasions for successful applicability.

A statement to the effect that there were seven apples on aforementioned table would be verified and be falsified by exhaustively assigning the apples on the table to an adequate set of counters. The occasions for applying either test are subject to the circumscription-cum-partitioning condition of the falsification criterion of the previous example, but differently constrained by the required presence of apples on the table and by the need for a set of counters, in this instance numbering at least eight. These several constraints require as a circumstantial condition the actual presence of a tester with counters in hand. The occasions of successful applicability for the verification test are restricted to those on which there are seven apples on the table; occasions of successful falsification comprise those on which the table holds any other number of apples. Because there are so many constraints on these tests and so little left open, it may come out that, if any occasion is one of successful applicability for either test, then that test is successfully applicable on all of its occasions of applicability. It also appears that one of the tests must indeed be successfully applicable. (Such numerical statements, it seems, are "leibnizian": either true or false and, if true, necessary.)

The third example was a statement to the effect that if the apple on the table is a jonathan, it must be unripe. The occasions for both verifying and for falsifying this statement are those of our first "existential" statement further limited by the condition that there be exactly one apple on the table; the occasions of successful applicability for both tests are further conditioned by the apple's being a jonathan; the occasions of successful verifiability are still further constrained by the condition that enough parts of the thing be hard, green and tart, with more or less opposite constraints for falsification. If our description of the tests is correct, then neither would be *successfully* applicable if the apple were not a jonathan. This statement is at best "vague" because of the indetermination of *enough*, and it may be neither true nor false for that reason or perhaps because the apple's a pippin and the statement neither successfully verifiable nor successfully falsifiable.

Occasions are the "ultimate possibilities" envisaged by our "conceptual scheme".

I now resume the abstract discussion of the relations between occasions and statements. Occasions on which either criterion for a statement is successfully applicable are "actualities" in relation to that statement. Other occasions of application are ones on which the statement "might have been true" or "false" (as the case may be). These occasions taken altogether comprise a "space" or "*Spielraum* of possibilities" for the statement. Possibilities for other statements are, of course, also places-at-times. Occasions, for us, are the ultimate constituents of all those possibilities envisaged by our conceptual scheme. These occasions are bounded in space and themselves--like aristotelian "now's--temporal boundaries for "events"; they are no less concrete than space and time themselves. (Still, as Kant was wont to remind us, places and times are not phenomena directly sensible in themselves.)

Our apparatus affords means for formulating conditions on the applicability of tests. We never have to say that a test is successfully applied.

Our theory affords means for formulating conditions on the applicability and successful applicability of tests in terms of the applicability and successful applicability of other tests. Thus occasions for applying tests that are constrained by the condition that there be a single apple on a table are also occasions for the successful applicability of a test that at once selects an apple from the tabletop and exhaustively shows that other objects on the table are not apples. There is ever so much more than that kind of thing to be said about tests and testing which we shall never find need to say. Specifically, we shall never have to say that a test was (actually) successfully applied.

Remarks on skepticism.

Because there is nothing in our account of conceptualization to require, even less to assure that the underlying tests should ever be successfully applied to yield verifying perceptions, it is compatible with and therefore opens no escape from the kind of global skepticism scouted by Descartes in his hypothesis of the malevolent demon[5]. So far as this theory goes, we could, as in the contemporary physicalistic version of The Demon Hypothesis[6], be slightly pulsating brains in vats interconnected through a central computer and conceptually booted by a *Deus ex Machina*. Our theory, like Husserl's "transcendental point of view", is an account of the constitution of sense; the present line of argument makes me doubt that Husserl provided that refutation of cartesian skepticism which he claimed. I am equally dubious of every manner of "paradigm" case argument including those that would require concepts to have their referents be their causes. The Demon Hypothesis suggests the possibility of our having been brought to concepts by causes other than those which our concept-underlying repertory of testing is meant to get us back to.

It remains that our account does require the sometime existence of spatial objects, and that for reasons reminiscent of Kant's celebrated "Refutation of Idealism": The exercise of concepts implicates the applicability of tests that may in turn presuppose the successful applicability of tests for the existence of bodies. However, as I have said, there is nothing here to require the anytime achieved successful application of any test and so far nothing to warrant our claim to know of the existence of an object that would be proven out by the successful application of a test.

I am, however, ready to claim that our "epiphenomenal behaviorism" offers the prospect of local solutions to local skeptical challenges. Verification is a method of proof which, if *successful*, establishes the truth of a statement beyond a shadow of a doubt; still no method of proof is proof against error. Mistaken application may occur on any occasion and the subject come to think he has verified a statement when he has not. One may accordingly be skeptical about any application of any test without having to be skeptical about testing in general. Such mistaken test applications and misjudgments as may occur are always of a specific sort which can be identified after the fact and against which advance precautions could have been taken. The forever possibility of error, which follows from the fact that we are concerned with tests, no more provides grounds for thinking that what the statement formulates is irremediably doubtful than does the generalized possibility of challenging a delegate's credentials provide grounds for refusing him entrance. What cannot ever be doubted is the general principle that a successful application of the test establishes the truth of the statement. If, after application of a test, a doubt remained, nothing better could be done than to apply the test again. One could not doubt that the truth of a statement could be established in that way without raising doubt about what was said. "Transcendental arguments" for that conclusion can be framed by considering circumstances in which one exhibits doubt in the very course of demonstrating a successful application of the test. If one says, "I doubt that I am talking", or "I doubt that I exist", or "I doubt that there are in the following sequence one, two, three, four, *five* numerals", a conceptualizing observer could share those doubts only by supposing that the subject is doubting something other than what he says he does.

> *Instance*: Might he not be trying to see whether he's awake? Yes, and of course he succeeds. The observer can have no doubt. Mightn't the subject? Not if he is applying a test to himself which he could apply to others, for the application of a test must be self-conscious. But precisely what he wishes to know is whether he is conscious at all! But if he has done anything then he is conscious!

We don't deny that the doubting subject has his doubts. It's just that the observer could not grasp the expressed sense of such a doubt without knowing it was "ill-founded".

> Evidently our "epiphenomenalistic" tactic for meeting local skeptical challenges that is reminiscent of Descartes' use of his "Cogito" to parry the Malevolent Demon. I think we are better grounded than was he, for at the place where he entered his elegant one-step argument that the "Cogito" does the job, he seems to have forgotten his "Dream Hypothesis" and to have neglected the possibility that he was only dreaming that he doubted, and I don't see that his earlier "austinian" reply to the Dream Hypothesis counts against the possibility of dreaming doubts.

Tests either are or are not applied on every occasion of applicability. Tests needn't be applied on any occasion of their applicability. We formulate this requirement as a "rule of inference": if the applicability of one test entails the application of another, infer that the two tests are distinct.

It seems reasonable to stipulate that a test either is or is not applied to completion on any appropriate occasion of applicability. Our theory requires that we sometimes be able to say which, with nothing added about actual success or failure. We may now put in place an additional principle of testing: A test needn't be applied on any of its occasions of applicability. The rationale of this is that otherwise the tester couldn't know whether a test were applicable on an occasion unless it was applied to completion, hence certainly applicable. The principle seems plausible. I know of objections to it, to be taken up on pp. 279ff. below. For now, we are faced with a tricky task of giving it an acceptable formulation. "Needn't" is the sticking point. Let's restrict ourselves now to *qualified* tests. *viz* ones that are sometime applicable: these are the only ones fit for representing statements, and the condition of applicability gives us something to work with. Now a test might in fact be applied on every occasion of applicability, so there is no way of capturing the wanted sense of "needn't" by restricting the domain of actual application. Nor can we unblushingly say that there is no occasion on which the test is *necessarily* applied, for our theory of testing is concerned only with actual conditions on testing and it cannot matter whether those conditions are formulated in necessary truths or in truths *simpliciter*. Technicians will know that the appearance of "necessarily" in the formulation of the principle would imperil that formulation with "opacity". Here, I suggest, we may borrow a page from the book of logic. Logic seeks to elicit truths in regard to what necessarily follows from truths that may or may not be necessary. While we may (perhaps) allow that the truths of logic are themselves necessarily true, they do not affirm that of themselves. How then do they tell us about necessary consequence? Well, in some systems anyway, by being derived from the first truths of logic in accordance with stipulated rules of inference. Every system of logic must somewhere appeal to some rule of inference or other, and the method of logic puts rules in place of modality. I submit that we can gain our principle that tests need not be applied on any occasion of applicability in just that way, by introducing a rule of inference for our theory of testing. My proposed rule is this: If, from the assumption that a qualified test t_1 is applied on an occasion, one infers in n steps that a qualified test t_2 is applied on that occasion, infer at the $n+1st$ step that t_1 and t_2 are distinct. This rule allows that, while the actual application of a test may be required for the applicability of another, it is not required for the applicability of itself.

Tests are not to be identified with their representations.

We have been setting the background for a uniform *representation* of tests and of statements in terms of tests. It should be clear that a test is not to be identified *with* or "reduced to" those ranges of test occasions by specification of which the test, in my scheme of representation, is identified. We could, after all, just as well have used *other* representations for tests, e.g. as pairs of sets of occasions of applicability with a subset of occasions of unsuccessful applicability. A test is not "something more", it is something quite different from the sets of occasions on which it is applied. Nonetheless, all of what we shall wish to say about tests can be captured in the occasion specifications I contemplate. The application of a test might, for example, require movements of a certain kind, which I think can be accommodated by demanding merely that the occasion should leave "enough" free room. We might effectively define the whole range of "look-see" tests by merely requiring that the occasion be minimally illuminated and that there be room enough for a tester to get visual perspective on the scene.

Summary.

My presentation has proceeded in a systematically backwards direction from statements to occasions, and a summary reversal might be useful at this point. An occasion is a temporally specified closed region of space with its contents. A test may be represented by a specification of two disjoint sets of occasions. An occasion belongs to the successfully applicable set when the contents are such that a testing animal could achieve a self-conscious perception of specified fact on that occasion; only when that is not so does an occasion belong to the unsuccessfully applicable set. Tests that are also procedures for establishing the truth or falseness of statements are *criteria.* A test is qualified as a criterion only when the union of its two representing sets of occasions is non-empty. The verifying criteria for a given statement are alike in their terminations, as are also the criteria for falsifying a statement. The two terminating procedures, which are characteristic of the statement, are called *fundamental criteria* for the statement. A statement is verified by the application of one fundamental criterion and falsified by application of the other. The only general restriction on a pair of qualified fundamental criteria is that not both are successfully applicable. If A successfully applies the fundamental verification criterion for a statement *s*, then A is authorized to claim to know the truth of *s*. The accomplished verification is an ungainsayable basis for such a claim to know, its relevance is unquestionable, and it cannot be challenged on other grounds.

The so-far expounded account of statements in relation to their criteria implicates a "philosophy" against which various objections have been directed. Before resuming our explication of the representation of tests and statements I should like to stop over to review the philosophy by comparing it with some others and to formulate and respond to the objections I anticipate.

A "conceptual order" is an interconnection of representative skills based upon a repertory of investigative skills.

An apparatus of conventional representation--what I style as a conceptual order--is a network of proficiency for undertaking utterance. Among the strands in this network, for our human selves, are capacities to realize predicative, referential, form and mood indicating, applicative and other such uses of words. (See pp. 300 & 363, fn. 20.) These skills are conspicuously, but not exclusively, manifest in the assertional representation of fact, and we shall (in #4) consider how a general semantic theory can be projected from that central position. A guiding theme of this investigation has been that our conceptual order is based upon *another* repertory of investigative skills. These skills issue in tests that also figure as the fundamental criteria for the statemental products of (successful) assertion. Conceptualization, in our view, requires some such basis. Our network of representation must all around be fastened down to the procedures by which we "get to" the represented world and by which the statements we produce in our assertional representations are proven out as true or false. The apparatus of testing we've been scouting and shall develop in Part II of this work is, I believe, the actual and adequate basis of our own conceptual order.

The working out of this way of looking at a conceptual order results in a "philosophy" that overlaps but also differs from several others.

The thesis that conceptualization requires some such downpinning eventuates in a "philosophy" that is at once "empiricist", "rationalistic", "verificationalist" and "pragmatist", but with important differences from other doctrines that march under those banners. My philosophy is also "kantian" to an extent that allows for what I believe are useful comparisons. Our opposition includes the "noumenal descriptivism" of Leibniz and theories of meaning that would base our capacities to make predicative representations upon an otherwise unsecured nexus of relationships, perceptual or semantic, between ourselves or our words and the objects of our representations. What follows in the next sixteen pages is a sounding of the schools, a philosophical interlude, in which I shall review these various affinities and oppositions, starting with the oppositions. In pursuing these many comparisons, I shall frequently be anticipating the developments of Parts II and III, especially when ticking off points of comparison between my doctrine and those of Kant and Arthur Melnick.

In comparison with Leibniz.

Ideas, according to Leibniz, are powers or capacities for representing forms in nature (see *Discourse on Metaphysics*, XXVI). So far so good. Ideas or concepts "guide us to their referents". They are not to be thought of as impressions stencilled on the mind by our experience with instances of which they are the ideas. Support for this may be drawn from some remarks Frege made in criticism of the correspondence theory of truth. He observed that in order to determine the fidelity of a picture, we must antecedently know what object is being pictured (his example was Cologne Cathedral); but you don't need that to determine the truth of whatever it is that's going to be true; if there were such a correspondent, the truth-vehicle itself would have to show you what that object was. Leibniz and Frege agreed on this, that the truth of a representation is a non-relational feature of that representation. The point at issue for right now is that representation "guides you to its referent", and that ideas or concepts , which are capacities for representation, are in no wise "images" of what they may allow us to represent. Leibniz went further to maintain that these capacities have *no* experiential connection with the forms they represent and must, with greater or less apperceptivity, be innately present for all eternity to every representing spirit. "Empiricists" will complain that our representational capacities in actual fact are at least conditioned by our experience of the world. More to the present point, I endorse Kant's argued protest[7] that Leibniz' ideas are, save for possibilities pure and noumenal, *objectless* and hence, taken by themselves, for human conceptualizers anyway, unqualified as representations. Concepts are representations of objects, and collapse without means for getting to those objects; qualified representations need what Kant called "forms of intuition" by which they may be referred to what they represent. Leibniz was persuasive in his arguments for the conclusion that our representative capacities cannot be derived from unguided experience alone; still, we also need, what he lacked, guiding operations, by which those capacities can be, not "taken from", but "brought to" experience. Conceptual skills cannot stand without procedural support[8].

Our doctrine is free of the "under-determination" that besets all those semantic theories that would seek to explain meanings in terms of relationships between expressions and objects, sets of objects, and sets of sets of objects

Those theories that would base our representative capacities solely upon relationships that obtain between ourselves or our words and represented phenomena, including the "contact" empiricisms of Aristotle and Russell (see pp. 111, 114), theories of habituation by stimulus and response, and various developments of "model-theoretic semantics" leave those representations unacceptably underdetermined. (As Quine, perhaps in a spirit of doctrinal self-examination has argued, e.g. in Chapter 2 of *Word and Object*; I adduce considerations of "intentionality" to argue the same point, at pp. 256f. below.) These doctrines all leave the notion of an identifiable object of reference and then of sets constructed from such objects so far unfixed. (See pp. 367 below). Contrastingly, the investigative repertory we invoke comprises procedures for individuating, separating and identifying objects of reference and for delimiting collections of such objects. and it allows for finer discriminations of meaning than what is afforded by the sense-functions of model-theory.

Our doctrine is "empiricist" with a difference: Comparisons with Locke, Lewis, Ayer and Quine.

The empiricist opposition comprises the "contact empiricism" of Aristotle and Russell, the "phenomenalism" of C. I. Lewis and A. J. Ayer, the "verificationalism" of Vienna and the theory of language Locke scouted in Part III of his *Essay*. Leibniz would have accused them all of treating concepts as images. We spent a lot of words on Aristotle and Russell in Appendix A. Let me now briefly consider the others just mentioned, starting with Locke. These writers all seek a foundation for thought and language in perception and sensation. Our theory of testing is also at least a theory of what such experience contributes to conceptualization. In actual fact, only subjects capable of self-conscious perception could learn to apply tests. However, we take much more seriously than did Locke and his successors what I regard as the "real" problem about the "origin of our ideas". Locke's "ideas" were items preferentially identifiable as "constituents of propositions"[9]. Locke also held, of course, that ideas originate in non-conceptual brute experience. The "real" problem of origins is to account for the transition in behavior from the promptings of experience to our perception of relationships among ideas. Locke's whole thought on this issue comes out, unsatisfyingly, in his confused appeal to *abstraction*. He used abstraction, ostensibly, to explain how some ideas become general. But the theatre of this exposition was Book III, and there is no doubting that he connected abstraction with our coming to the use of language. Locke invoked abstraction exclusively in relation to predicative ideas; in exempting particles from abstraction he showed blindness to what Liebniz appreciated, that the kind of generality he really needed was that in which *all* words, including particles and singular terms, may be repeatedly employed to signify the same idea. This (as Leibniz knew) is a generality of " knowing-how to"; "predicative" generality is but that special case of knowing how to apply words in a single sense (to signify a single abstract idea) to the different objects comprising the "extension" of the idea. My account of the transition from brute experience to conceptualization is part of a theory of behavioral competence, issuing finally in a theory of language as behavior. To this we annex our anti-leibnizian requirement that those linguistic skills in reference to which ideas may be identified as "constituents of propositions" (roughly, "assertion") are themselves bottomed on other, (investigative) skills of a kind that needn't yet be fully conventional. I propose to explain our "ideas" by reference to the features of those underlying procedures. I differ from Locke and other traditional "empiricists" by stressing these behavioral foundations; that, I claim, enables me to explain, what should have mystified the empiricists, how our ideas come to have conceptual structure. Conceptual structures are defined by the different ways in which subjects may seek experience by applications of tests. The inquiry issues in a theory of meaning, adequate, I believe, to the analysis of all variety of thought from the most concrete to the most abstract, from the commonest to the most recondite, but preferable to other empiricist theories by being more deliberately set out and for being disencumbered of mischievous suggestions of privacy and clarified of obscuring speculative psychology. I can with good conscience steer clear of those phenomenological barricades at which skepticism may present a challenge and our progress be delayed by inquiries which, from my point of view, are irrelevant. I also bypass questions in the theory of knowledge over confirmation and evidence, and need consider only those methods of proof by which our words are given their senses. Here my theory differs from those of Lewis and Ayer according to which meaning must either repose on incorrigible reports of one's own experience, or be infinitely diluted by the endless verification of all the implications of what we say, yielding at most a questionably converging confirmation[10]. Finally (to echo what was noticed before), we readily avoid a kind of difficulty best known from the writing of Quine: Quine, taking verification to be merely a relationship between experience and language, easily shows that our language is underdetermined by our experiences. The intervention and accretion of procedures, these being plural in respect of both phenomena and perception, resolves such underdetermination.

Our "verificationalism" is mild and expansive and resolutely "non-descriptivist"; it works into an actual theory of conceptualization.

Our inquiry stands aside from epistemological issues over what is basic in experience and what is necessary for the confirmation of hypothesis or judgement in science or everyday life. There is accordingly nothing in my account corresponding to the troubled "Protokolsaetze" of Vienna, the "expressive judgements" of C. I. Lewis or (of more recent vintage) the "occasion sentences" of Quine. These are doubtful instrumentalities by which a passive intellect is supposed to record its most basic experiences. It's not my current business to champion or challenge epistemological priorities. Our account argues against the existence of corresponding priorities among statements, though it does so from the hypothesis that there is indeed an order of dependence among tests (Chapt. 5. #1, p. 377). Our account also allows for statement possibilities--all of them equally basic--unconsidered by other verificationalisms, while at the same time making advance allowance for such unverifiable or unfalsifiable products of non-assertional constative language as generalizations and conditionals. The philosophy that emerges is epistemologically mild, logically unconfining and uncommitted to "positivism" or any other ideology. More to the point of present comparisons, the conceptualizing intellect, as we conceive him, is active not passive. We hold that only those prepared for inquiry and ready to make an active search for fact can have language for formulating facts, and it is this activity that imposes a conceptual structure upon "the amorphous materials of experience". My brand of verificationalism is akin to the active "operationalisms" of Peirce and Bridgeman and to the Waismann's "Thesen"[11]. I certainly find these writings suggestive and congenial but still altogether programmatic. My program issues in an actual theory, replete with detail and girded by refutable hypotheses that generate problems and predictions, and has applications beyond those prompted by an ideological devotion to science.

Comparisons with Wittgenstein, with particular attention to "Criterion".

Wittgenstein's *Tractatus* was "descriptivist" in pretty much the leibnizian manner. I interpret his later writings, especially those in the Philosophy of Mathematics, to be arguing for my thesis that our system of conceptualization is grounded on skills such as counting and measuring. My own thinking was conditioned by a reading of those works. Wittgenstein, in his later work, also frequently talked, loosely (as he allowed), of *verification* and *criteria*. I believe that this use of "criterion" threw him back into a kind of "descriptivism", as I shall now try to explain.

Wittgenstein in the writings at question talked as if he thought that criteria were "for" concepts, and furthermore that a criterion for a concept was a consideration bearing on the concept's realization as a predicable of objects. Two among a handful of examples were these: a criterion for saying that a person has angina is the presence in him of a certain bacillus; again, a governing consideration for our saying that a person *understands* something is how he behaves when questioned. Here now is my summary of what I can glean from Wittgenstein's discussion of those examples: First: While criteria are for concepts, not all concepts have criteria, e.g. while *understanding red* does *red* itself does not. Second: a criterion for a concept is something one may *perceive* in the course of determining that something falls under the concept in question; a criterion in this familiar understanding is not a test but a phenomenal feature of something. Third: the presence of this phenomenal feature is of itself never a condition either necessary or sufficient for the presence of what its for. Nonetheless, fourth, a criterion for a concept is connected in "grammar" or in understanding--"logically", if you wish--with what it's for. Now the presence of a bacterium in the blood stream, in Wittgenstein's first (badly selected) example, is not a feature of angina, nor is behavior a feature of understanding. I conclude from this, as a fifth observation, that a criterial feature is not a feature of what its for. What then? It appears to me, more as a conclusion than as an observation, that the criterial feature is "of" the underlying phenomenal basis in which the concept is realized, located or "instantiated": our pathologies and understandings are in us, and their criteria--blood composition and behavior--are features of ourselves. A criterion "for" something is a feature of the "medium" that "instantiates" that something. I believe that this conclusion fits our observations. I shall later on call this "instantiating medium" a "location". Objects of most any kind are shown to exist, individuated, separated and identified "in locations": A locating phenomena is a phenomenon of a kind different from the located objects. Now objects may be severally located, allowing at least for the prospect that the located object may have various criteria, none of them necessary and perhaps no one of them ever sufficient for the presence of the located object. What does seem to remain inescapably "logical" in the picture is that these objects are observed by observing their locations. I repeat that the criterial features we may then perceive the location to have are various and are variously related as evidence in support of what they are criteria for: some are "recognizing"-features (to be discussed in Chapter 13); some come down from theory and are specified in an aristotelian-style "definition" or formula; some are even demonstrable (e.g. a criterion for a number's being divisible by 3 is that the sum of the numbers represented by the digits of the original representing arabic numeral be itself divisible by 3). Now, since, in testing (our sense) for the presence of the located object, we seek out *those* features of the location--that's why I call this a "descriptivism"-- the tendency of our theory of testing suggests that the evidentiary "relevance" of the criteria is also "logical". Despite these efforts to impart some kind of structure to Wittgenstein's notion of a criterion, it does finally seem to me to be a soft notion that is sure to be "problematic" and of little service for an inquiry such as ours.

Fending off the "verificationalist fallacy": a description of the verification procedure for a statement is, for us, a part of the "analysis" of that statement and in no wise to be taken either as a variant assertion of the analyzed statement or as a description of such an assertion.

"A statement that the cat is on the mat would be verified by showing that the cat is on the mat". That, some will sneer, is no better than a stilted stutter. The complaint misconstrues my doctrine. It consequentially neglects falsification, which, for us, matters no less than verification. In any event, the description of a verification, for us, is always to be taken as the description of a procedure for getting to the facts and never as a description of these facts themselves, or, otherwise of the "truth conditions" asserted with the producing of any statement that the described procedure would serve to verify. Our doctrine is addressed to the analysis of statements; while it is adaptable to the task of giving the meaning of assertions, it provides no recipes for paraphrasing those assertions. A description of verification, I repeat, describes one way of getting to the facts; it makes no reference at all to those facts themselves. If an asserted statement is true and its verifying procedure successfully applicable, there is then indeed a fact asserted; but that very same fact might instead have been gotten to by following other procedures definitive of other statements. The notion that a description of verification is itself an assertion of the stated fact or otherwise of the "truth conditions" for a statement or (differently) a description of some such assertion are different versions of the "verificationalist fallacy". I hope it is now obvious that our doctrine is free of that contamination.

> Our account does indeed provide paraphrases for certain things sayable *about* statements, e.g. to say that a statement is true is, for us, to say that the statement is successfully verifiable.

Our philosophy is a kind of "pragmatism"

I have no quarrel with the "descriptivist" assumption that a conceptual order requires that we stand in *semantic* relationship to objects. Our claim is that those relationships could not be semantic properly said unless based upon our investigative activities and, additionally, that the activities we shall adduce are adequate for the burden. The resulting doctrine is a kind of pragmatism[1], one so far unencumbered with morals, whose theoretical thrust is parallel with the "operationalism" of Peirce and initially at right angles to the "descriptivist" "passive pragmatism" of Dewey and Quine. (It may be swung closer to parallel with the Dewey-Quine notion that our conceptual apparatus is also an "instrument" that may be brought back to experience as a control.)

Considerations of "intentionality" favor our position in its opposition to the claims of a procedurally unsupported semantics.

Our kind of "pragmatism" gains support from considerations familiar from the "*Intentionalitaet*" philosophers, going back to Brentano, considerations that are interestingly but uncertainly revived in Quine's allowance that his theory of stimulus meaning eventuates in radical semantic underdetermination.

Any theory of meaning must stand ready to certify such relationships as obtain between a name and its nominatum or a predicate and its extension. Now *any* actual relationship can be variously described. But then, when we ponder the matter, it would seem that, in giving these *different* descriptions of the same *one* relationship, we are dealing with *different* meanings. Take the naming relationship as a case in point: Any relationship that obtains between "Shwayder" and me also stands between "Shwayder" and the author of this sentence. But surely the latter description of the relationship would have been better taken to identify the meaning of "The author of this sentence" than the different meaning of "Shwayder". What the observation almost shows is that *no* identification of the relationship between a word and its extension suffices to determine the meaning of the word.

The second supporting consideration of intentionality is that meaning may be fully realized in words and yet there be no corresponding object for the words to relate to. That is suspectedly so with thc meaning of "Homer" (= the meaning of "'omeros").

"But", you may wonder, "What's the option?" The examples prefigure the answer. It is the meaning of "Homer" that sends us in search of Homer. The *Sinn*, as Frege could put it, shows the way to its *Bedeutung* (see pp. 304 et. seq.). Meanings indicate a way of getting to an "object": they may subsist without an actual quarry, but not without a procedure to make the search. So the realization of the meaning minimally indicates a procedure. We believe that nothing more is needed and that we may identify the meaning by describing the procedure.

Our approach yields a doctrine of "objects as referents" and is a "transcendental idealism".

My inquiry into the foundations of our conceptual order issues in an analysis of our notion of objects conceived of as objects of thought or as "referents". Objects most fully thus taken, as referents, are conceived of in relation to an identifying reference. All *notions* of course owe their existence to thinkers, but equally "of course" we contrive all manner of referential notions that do *not* fix their referents in relation to thought. The notion of a *referent*, however, does additionally implicate the conception of a thinker. This characterization of the metaphysical phase of our enterprise as being of the conception of an object conceived of as referent, qualifies the resulting philosophy for being what Kant called a "transcendental idealism".

Our metaphysics is the study of Being *qua* Referent, those words to recall Aristotle's differentiation of studies as of Being "qua" something or other. The "qua" implicates conceptualization, e.g. as having attributes of extension. Aristotle's characterization of First Philosophy as the study of Being *qua* Being led him to the conclusion that the inquiry is primarily the study of Being *qua* substance, where (he argued) substances are individuals characterizable as being of some unique species. Our *referents* need not be so fully-formed.

Comparisons with Kant and with Melnick.

In comparison with Kant[1]. The Kantian metaphysics is the study of our notion[12] of reality conceived of as "phenomena". He argued that the study could not coherently escape that definable limitation. He did indeed allow that we have a conception of reality unconceived and he conceded the possibility of there being conceptualizations of reality other than our own. However, we have no way of saying what unconceived reality is nor can we even imagine how representations other than our own might run.

> Perhaps Spinoza's mention of infinite "attributes" other than thought and extension also anticipates the possibility of other representations. But who of us is to say what they are?

"Phenomenon", for Kant, meant reality conceived according to subjective conditions or "forms" of space and time. These subjective conditions or forms are, for Kant, activities--ones that, in his scheme, play much the same role as procedures for identifying bodies do in mine. It may at first blush seem that my doctrine is less stringent than is Kant's, since we do allow for reality not yet actually tested, whereas, for Kant, every representation (in the sense of a represented object) is a phenomenon. He, however, hedged exactly as I would, by saying that phenomena may be encountered "in the possible advance of experience" (*First Critique,* A-493, B-521). He also denied the berkeleian dictum that perceivable phenomena could not exist except as actually perceived.

Kant's transcendental idealism and my own both stand opposed to certain other forms of idealism, alike the noumenal idealism of Leibniz and "phenomenalism" (what Kant called "empirical idealism", e.g. *ibid.* B-519).

> Kant viewed Leibniz' "noumena" as candidate objects of the Divine Scheme of Conceptualization, but for Kant (and me) these are no better than pure possibilities, mere representings in the imagination so far unqualified for the status of represented reality.
>
> Against empirical idealism, Kant compellingly argued that our sense of succession is grounded upon our experience of *external* phenomena.

Absolute Idealism, I believe, presents an unanswerable challenge to both of us. That doctrine, in denying that individuation is ever actually successful, gainsays any legitimate separation of represented reality from our representing selves. I see no way of proving that the body-individuating procedure I shall later describe is ever actually successful.

Kant certified three components to his "transcendental idealism": (i) the unity of apperception, (ii) the possibility of experience and (iii) the need for reality. To these we must add the conclusion of his most important argument, as explained in the resolutions of the two Mathematical Antinomies, *viz* that no phenomenon met with in space and time is one and the same with any "thing in itself" and (by contraposition) no thing in itself is identical with anything met with in space and time (to which I add) *under any description*. Phenomena are not to *dinge an sich* as The Evening Star is to Venus or as iron is to Element #23; there is not even a correspondence between them. This, it should be stressed, is not inconsistent with Kant's (and my) view that phenomena are to be met with in space and time by application of procedures that "take us to reality".

(i) is matched in my account by the demand for self-consciously and uniformly applicable (general) procedures. Kant similarly appealed to the actual or imagined application of unified procedures both for the schematization of the categories and for the representation in intuition of the possibility of singular existence (e.g A-105,140f; B-155, 179ff.).

> Melnick argues that this insistence upon activity was Kant's decisive departure from Leibniz and is what distinguishes his system off from most of its successors.

(ii) is matched in my account by the stipulation that these procedures should be investigative or "tests". (iii) is matched in my account by the idea that these procedures are "ways of getting to the world"; only when so understood can they stand as foundations for our representations that something is so. I agree with Kant that objectual representations are not taken from objects; rather, the procedures "give us" objects and their attributes. What corresponds, in my scheme, to Kant's negative conclusion from the "Mathematical Antinomies" is simply that there is no sense trying to conceive particular unconceptualized realities. I agree with Kant that the procedures by which we establish the objective existence of phenomena take us to reality *en bloc*.

> I remark on the side that the contemporary appeal to exportation of reference for purposes of clarifying otherwise "opaque contexts" still leaves us with references to identifiable objects or phenomenal substances.

I have been astonished to find so many parallels[11] between my philosophy and Kant's[13]. There are differences too. Of these the most obvious are ultimately the least important. Some of what may seem to be merely small points of detail turn out to matter quite a lot.

The first, most obvious point of difference is methodological. Kant's approach to his subject is "subject-centered" and may perhaps be dubbed "introspective" or "phenomenological". My "conceptual epiphenomenalism" is a "behaviorism". Still, Kant was clearly theorizing about *our* scheme of representation and he never doubted that we humans are all pretty much the same in this. From our side, the difference is neutralized by our conclusion that a conceptualizing observer must at once know-how to do what he may observe in the behavior of a conceptualizing subject and then be a self-conscious observer of his own conceptualizing activities (pp. 52f.). The difference in "point of view" would matter for pinning down the merely practical knowledge of a non-conceptualizing subject. However, Kant was exclusively concerned with apperceptive conceptualizing activity, and so have I been in the foregoing comparisons. Transcendental Idealism is not a philosophy for dogs, cats and babes in arms. It would of course be foolish to think that creatures do not respond to and have practical knowledge of reality unconceptualized by themselves. And perhaps there is a niche for this in the kantian system after all: I am only slightly embarrassed at interpreting his undifferentiated sensory manifold as the behavior of an observed such subject, made whole, *viz* "unified", in the reports of a conceptualizing observer.

A second but vanishing doctrinal difference I once thought mattered has to do with causation. Kant docketed causation as a "category" and hence as a constituent of our conception of represented, phenomenal reality. I thought this a mistake because causation is inseparable from explanation which reaches beyond the mere constitution of concepts (pp. 171ff.). However, I am now pretty much convinced that Kant always took causation to be "succession...subject to a rule" (B-183) and never explanation. Successiveness is, in my system, defined by the equivalent of a kantian schematized category--what I call a "proto-criterion". So, finally, there is no difference here[14].

A third somewhat mattering difference between Kant's system and mine is that I offer detail in regard to the underlying procedures that he pretty much neglected, both for the differentiation of predicables ("Concepts", See Chapter 13, pp. 408 et. seq.) and (much more importantly) for the resolution of "intuition". What, in my account, corresponds to kantian intuition factors into a nest of procedures for establishing the existence, individuality, separateness and identity of bodies, and for delimiting bunches of bodies, those procedures in a consequential order of increasing dependence. I stress "order of dependence" or (as I often dub it) "stratification". Some similar detail emerges with Kant's deduction of the categories, where he argues that intuition must have these several wrinkles if we are to predicate sensible features of objects and in the "Analogies", which strikes me as being the kantian counterpart to a theory of identity and distinctness. His "schematization of the categories", as I understand it, by which those notions are brought under rule (*viz* intuition), suffices for introducing corresponding predicables, e.g. predicables of being somewhere at a certain time. His presentation, however, does not impose an order of dependence upon the factors brought together in intuition. We shall presently get a glimpse of why that matters.

A fourth difference has to do with what space and time are. I agree with Kant that these "forms" are not directly observable phenomena. I balk at his identification of them with the very activities by which phenomena are elicited and made observable. I would myself prefer to place them as "constructions" from bodies and other things met with in space and time, hence as phenomena all right, but ones that are only indirectly observable. Kant sometimes seems to accede to some such view (A 215, B 262). On the other side, I agree with Kant that these places and durations incorporate an ineliminable "subjective factor" and here my line of analysis converges upon his. That is because places are proven out and individuated by the use of delimitation procedures. These operations alone among my basic repertory of test-kinds are circumstantially conditioned by the sometime existence of a tester. (See the remarks on pp. 242, about our apple examples and, p. 406 for detail about what will be coming at #1 of Chapter 12 in Part II.)

A fifth real difference has to do with the "unity" of space and time. Since the procedures Kant identifies with space and time are uniform hence "unified", his doctrine of the unity of space and time seems to follow in train. Allow that. Still, as I see it, bodies and other things met with in space and time are referred to places and are given dates only through the use of identifiable bodies as "frames of reference". Since two different such frames of reference may be inaccessible to each other, we allow for the fragmentation of Kant's unified forms into reciprocally inaccessible spaces and durations. Kant perhaps could have allowed as much, but he didn't; our alertness to such possibilities owes to our placement of both identifying and delimiting procedures as annexations onto a repertory of (progressively) less dependent procedures for securing separation, individuation and for proving existence.

As a sixth difference, we countenance, what Kant seemingly denied, that particular substances may come into or pass out of existence by coalescence and dissolution. Kant argued that the eternality of substance was a necessary condition for the unity of time (B229). I agree that the conceptualization of temporal order is based on procedures for identifying bodies. However, our account of this procedure leaves room for unsuccessful applicability by cause of the recent coalescence or dispersion of individual bodies and we accordingly dispute his thesis of the eternality of substance. We allow, however, that Kant was certainly on to something true in finding the connection he noticed between the two disputed theses of unity and eternality. (The weakest sufficient condition I can find for the temporal relation of earlier-and-later being uniformly asymmetric is the eternal existence of at least one identifiable body--see p. 405).

My differences with Kant over the unity of space and time and over the eternality of substance pivot on the question whether his "forms" are to be taken simply or (as I maintain) broken down into procedures for proving existence, individuality, separation and identity, as I maintain. Stratification is crucial. I am unable to prove that we cannot get by with unfactored forms. Kant may be right after all. His doctrine would then have the advantage over mine, not just for generating those spectacular metaphysical consequences of the unity of space and time and the eternality of substance, but also for penetrating more deeply into the foundations of our conceptual order. Melnick has made a strong case for the kantian claim; we shall presently consider that; but first:

Certain passages of the *Analytic* once made me think that Kant believed that the (active) individuation of bodies always implicated an intuitive non-active individuation of "the rest" as a kind of amorphous but undeniably individual muchness (an "indefinite dyad") and that he believed that so much is required to ground our conception of generality (e.g. B-40, 48, 136b). That if true would force me to the terrible concession that activity is not sufficient grounding for representation. I was glad to find that later, in the *Dialectic*, Kant unequivocally repudiated the non-active intuition of individual "muchnesses" (e.g. at B-457) and I hazard to think that he would have dismissed any such appearance as a "transcendental illusion". It remains that we do indeed seem to have this concept of "the rest" for which I shall be challenged to find a procedural basis (see response to Objection 2 below).

Compliance as an Alternative: Melnick's Theory. Arthur Melnick, to assist his exposition of Kant's *First Critique*, has worked up a general system of philosophy which, to me, is at once congenial and challenging[15]. I stop over to compare his offering, as I understand it, with my own. Melnick, in well defined stages, introduces a "K-language" of directives issued to secure compliances on the part of an emerging conceptualizer. These compliances, in their natural order of introduction, are realizations of Kant's forms of intuition; they correspond, in my account, to those test procedures I claim underlie our conceptualization of objects as identifiable referents. Melnick's compliances are brought in as responses to language, itself not yet "fact stating", and thus in relation to but at a certain remove from representations *per se*. Melnick has no need for a theory of language in order to secure an elucidation of representational capacities answering to the basic notions of First Philosophy. That must be right for Kant. (Melnick does indeed have a theory of language, one which is not very different from my own; he lays greater stress than I would on the idea of language being a kind of substitute for investigation.)

One of Melnick's "K-directives" could be issued in some such terms as these: "Take steps counting to five as you go; stop, turn to the right, take steps, counting to three as you go." Call the ensuing compliance a "positioning". Positionings are proficient acts; they needn't always be self-conscious and they needn't seek findings; hence, they needn't be tests in my understanding. Positionings usually do result in reactive perceptions ("sensibility") and, when meshed with other positionings of oneself or another subject, may come to implicate self-consciousness and possibilities for error equivalent to falsity.

A positioning may be indefinitely prolonged by simple directive, e.g. by simply saying "Take steps". Such positionings reach beyond any of my tests. From the other side, my tests could be domesticated to Melnick's compliances by countermanding a subject's activity when that subject is observed to have an experience or by directing subject to self-countermand upon its having an experience. All of this gives

Melnick's *positionings* a theoretical advantage over my *testing*. The advantage is seemingly increased upon further consideration.

In my exposition, a theorist's (the *We* of p. 18) analysis of reports on a subject are framed in the language of "occasions". Occasions are places at times; I must accordingly be able to refer to places at times as if they were objects. You may complain that my effort to account for a subject's conception of an object is therefore circular and flagrantly so for bringing in the non-sensible etherealities of space and time (see Objection 10 below). I reply that the circularity charge would hold up only if, inadvertently, I had invested the subject with those capacities to refer to places and times from the outset. I take great caution in that matter. I do, however, acknowledge a responsibility to recount at some point the procedures that underlie our perhaps too easy talk of occasions. These procedures all involve the use of separable bodies as probes. I require, of course, that nascent conceptualizing subjects must move out into occasions if they are ever to apply tests at all; one may protest on that account that the testing of occasions must be mastered before one can come to make body separating tests. I reply that we must take care to distinguish testing "in" or "on" occasions from testing those occasions themselves. After all these defenses, I must still concede that my distinctions between theorist, observer and subject are rather stilted and may be artificially rigid. Melnick's approach, if it works, more smoothly accommodates the acquisition of concepts of space and time than does mine. An attractive feature of Melnick's system, if it works, is that the subject, as it seems, is directly given a sense of place and time without need for objects to fill places or to clock changes. A subject, stepping out, counting, and now and then changing directions, simply moves into place.

Melnick's system, if it works, also penetrates more deeply into the foundations of conceptualization than does my own. That is so both because it operates more simply and more broadly (see above) and, as we just noticed, because it gives us conceptions of space and time that lie below the level of our conception of an identifiable body.

For all these reasons, Melnick's system, *if it works*, seems superior to what we shall develop in this treatise.

To defend the integrity of my own enterprise, I shall now argue that Melnick's system in fact does *not* work within the bounds he sets. It is of the last importance that his subjects should be knowingly able to change directions. That knowledge affords the principle of exclusion that Kant rightly faulted Leibniz for lacking (for more, see p. 338 below). A sense of *turning* is not enough to assure that knowledge, for one may turn back into an initial direction. It looks to me as if a conceptualizing subject, in order to control its knowledge of the directional aspect of its positioning, must either be able to sight on an identifiable reference body or have the use of the equivalent of a transit or a protractor. It therefore looks to me that Melnick's conceptualizing subject could not undertake its various tasks of positioning without being able to individuate and separate bodies. I have every confidence that Melnick's representation is faithful to Kant. But then, to repeat, I believe that my resolution of Kant's forms of space and time into various orders of testing is vindicated by the state of our conceptual order and is an advance upon the kantian philosophy. The progressive accession of these procedures eventuates in our finally coming to methods for identifying bodies and delimiting regions, thence to a capacity to "individuate" places and durations.

Replies to objections.

I am familiar with numerous objections to the system of philosophy sketched in this section. These objections adduce different sometimes conflicting reasons for doubting that my fundamental criteria can be as fundamental as we claim. Broadly classified, these objections argue that (A) Our methods are inadequate for a discriminating representation of available forms of human conceptualization; (B) The underlying doctrine is systematically ineffective and perhaps even incoherent; (C) Our brand of verificationalism is epistemologically myopic. I do little more in this place than list these objections under classifications A, B, and C, make a brief reply, and illustrate the situation, as I see it, with one or two examples. I hope, in this way, without supposing that I shall have settled anything once and for all, to show that my theory has survival potential.

A. Replies to objections charging that our methods are inadequate for representing available forms of human conceptualization

Identity is untestable.

1. Hume argued that a supposed fact of identity was over time; but (echoing Augustine) any experience is at a time and can never be

of what takes place over time. It would seem to follow that facts of bodily identity are not to be experienced and therefore that statements of bodily identity are untestable. No test, by the terms of our theory, is applicable except on occasions at times, and it would then seem that there can be no test for bodily identity "over time".

The objection is serious business, for identification is of great moment to our conceptual order, and I'll keep coming back to this matter (see esp. #2 of Chapt. 9, #1 of Chapt. 11 and #8 of Chapt. 17, pp. 401, 403, 428f.). For now, I hope that the following will suffice as a *reply*: A subject may, in the course of testing, *have* been testing and know that he has. ("Have been" signalizes a present tense of perfect aspect.) So I may now have been tracking a body and know that I have. A test for identity is applicable and then perhaps also successfully applicable on an occasion of having tracked. One may then come to observe a fact of body identity, and verify a statement of bodily identity, on that very occasion. Such a test application does indeed "presuppose" that subject have come from an earlier occasion on which the identified body could have been individuated. It would, however, be plainly excessive to demand that a testing subject should ever have to be observing the satisfaction of all the "presuppositions" for its anytime application of any test. Now, abstracting from actual testing and looking solely at the occasions for applying an identifying procedure: any such occasion is conditioned by the fact that there is in it an individual body that has continuously existed as an individual from an earlier occasion.

Totality is untestable.

2. Our sense of *totality* (everything, everything-else) is that of an undelimited hence untestable range of cases.

In first reply: I shall be able to describe procedures for delimiting "small" regions (Chapt. 12); these tests are sufficient for purposes of verifying restricted universal statements of the "all of these are" form.

But then we are also able to propound general truths that reach beyond such statements. These comprise such sundry items as the truths (if such they be) that all creatures with kidneys have hearts, that π is transcendental (*viz.* no polynomial equation is satisfied by π). Our one contribution toward the analysis of such general truths was that they should entail every corresponding universal statement (p. 141). Nothing so far observed explains how *that* part of our sense of generality is founded on testing procedures, and the objection still stands.

Let me try another approach. When we do secure individuation or restricted delimitation by application of test, we get the idea of separating a "this here" from "all the rest", which latter is conceived as a kind of "indefinite muchness". I believe but cannot prove that this idea of "indefinite muchness" comprises the whole of our idea of generality not provided for by restricted delimitations. But now, I argue, this idea of "the rest" is that of what is *not* here to hand, and our idea of unrestricted generality is accordingly a *negative* notion. But surely (and here I invoke the supporting authority of most philosophers from Plato to Kant) our sense of what is not must be based upon some revelation of what there is. And surely our sense of "the rest" is grounded upon capacity to go on indefinitely making separations and restricted delimitations. That is the gist of my *second reply.* I hear the howls: "indefinitely", indeed! Is that notion of "indefinitely going on" a tractable one for us? "Knowing how to go on indefinitely" is a skill that every reader of this treatise will have long had. It is paradigmatically manifest in our ability to count, not just to 5 or 69 but "indefinitely". Counting affords a representation of all other indefinite procedures and does, I believe, make a foundation for our sense of indefinite generality. So let us consider *counting*. Particular counts, e.g. to 5 or 69, are skilled activities that can be recast as tests. *Counting simpliciter* or "indefinitely" is also a skilled activity that requires that subject know how to make particular counts; but it is not just a skill in making some particular count. The question now is whether it is anything more than skill in making count*s* and accordingly can be conceived as a genus of testing. One is pulled both ways: From the one side it seems that knowing how to count comprises nothing more than knowing how to make particular counts; from the other side, knowing how to count involves the notoriously ineluctable "and so on". I would like to be able to show that one could come to this idea of "and so on" just in coming to know how to make particular counts. So, as a backup to my second reply, let us consider how one would come to counting. A subject who knows how to make a particular count knows how to count various bunches of things to that place. Let us now get our subject to count his counters with something put in front. That skill, though it requires reflective capacities, is something that would be no less visible to an observer than would be any other skill: we see the subject doing the deed. What we observe is not just skill in making a particular count, e.g. going from 5 to 6, but skill in counting *simpliciter*. I hope that this counting of counters illustrates how a subject can come to or be brought to *counting*. Having that in hand, he can also learn the method of what is called "Mathematical Induction", which is a two-stage "finite task", that can be used as a method of testing: test the "first" thing and then the "n+1st", e.g. we prove that the first n natural numbers sum to n(n+1)/2 by showing it holds for 1 and then, adding n+1 to n(n+1)/2 and reducing the fraction, we get (n+1)(n+2)/2. Use of Mathematical Induction is a bona fide test that seems to sum up the "and so on" element of counting. These ruminations give me confidence that our whole sense of *generality* is founded on "knowing how to go on" and that that kind of operation can be recast as a test. There are new objections. We can conceive totalities that cannot be delimited by application of a recursion formula, notoriously the totality of real numbers. Still, I am impressed that a proof that there is such a totality is gained by a reflection on the use of recursive methods of expanding real numbers and of ticking off and switching the digits in those expansions. The theorem shows that, no matter how the ticking-off is done, there is always something left in "the rest". These observations do, I believe, make a case for thinking that our actual notion of generality, even when used in relation to non-denumerable sets by mathematicians, requires for foundation nothing more than testing. So, in summary of what I pray and believe to be true: we have a notion of generality based entirely on testing which is incorporable into various "things said" that cannot themselves be proven out by testing.

"Natural kinds" are untestable.

3. "There are no tests for showing that something is an apple, a dog or gold. Yet all of us are ready to spot such things for what they are. Generally: This brand of verificationalism is inadequate for explaining our references to and predications of 'natural kinds'" (with thanks to Tim McCarthy). The challenge here, as old as Aristotle and as new as Hilary Putnam, is a serious one. While common-notions of color and magnitude might perhaps be dealt with by the use of exemplars, the look-see's by which we distinguish cats from dogs or iron from gold are, as procedures, uniform across those differences and do not of themselves afford a

basis for theoretically securing the distinctions we need among these various names and predicables. So far as testing goes, we simply look and see that that's a dog or a cat; that's all there is to it. These "kinds" do have "essences", and that will matter a great deal when we finally come to discuss species and other *sorts* in chapters 13 and 16. However, for reasons well known from Locke's *Essay*, the appeal to essence is unsufficing for distinguishing kinds: we do not test for kinds by running through a list of essential specifications.

The gist of my *reply* is that the objection is really an exorbitant demand. It requires that the indications realized in a natural kind reference or predication must actually *identify* the kind. That this is exorbitant is evident in the consideration that there are always plural such identifications. It is enough that the speaker give indication that there be *some such* identification of the kind. Let me now fall back a bit in order to get a better run at this answer. We must, by my theory, be able to mark any semantic distinction by a distinction in testing. Now, as I believe (following Aristotle), natural kinds have names and their predication is systematically dependent upon the availability of those names. I can do something to prepare amplification on my reply to the objection by saying something about how I view the semantics of naming. Now there are myriad kinds of names--of individuals, sorts and stuffs, among other things. In giving the meaning of a name we must somewhere say that it is indeed of an individual, sort, stuff or whatever. That kind of difference can, I claim, be resolved by appeal to tests. The explanation of the meaning might also tell us that the name is (e.g.) a personal name or a chemical name; further, that it is (e.g.) a given-name or the name of a metal. The possible add-on's are all but endless, e.g. we can say that it is a male given-name after such and such saint. All of this, I claim, can be elucidated in terms of tests. An attentive reader will have by now noticed that we have still said nothing about what it is to be a name, where surely an expression's being a name must be part of its semantics. Now certainly part of what is indicated by the use of a name and must be mentioned in any explanation of what a subject would mean in using a name, however much or little else we have to say, is that the use implies that there should be an answer to the question "Which?". That implication of naming can be elucidated in terms of my theory of testing, *viz* that the use of a name indicates that tests presuppose the successful applicability of an identifying procedure or analogue thereof. However, what the use of a name does not and cannot indicate is what the "which"-answer should be. An answer to a "Which?" prompted by someone's use of "Newton" might be the author of the *Principia*, the first Englishman to demonstrate the fundamental theorem of the calculus or the person credited with having first reconstituted white light from is spectral components--all of these are equally good answers, and no one of them is given absolute preference by the imagined speaker's use of "Newton"; again, iron is the commonest metal on earth, the core of most magnets, element #23 and the stuff smelted from Mesabi ore; dogs are domestic canines, a cross between wolves and jackals and man's best friend. A theory of speaker meaning should be able to explain why natural kinds are identifiable for what they are; but what we have just adduced all but demonstrates that a theory of meaning should not try to give those identifications; so no theory of meaning should be able to say anything about what those differences are that prompted the objection. I believe that my theory of testing can be made serviceable for the admissible semantic chores, e.g. of explaining what it is to be a name of a sort with or without essences; I am not vanquished but rather confirmed by the observation that it cannot tell us what the sort is.

"Vagueness" is untestable.

4. "'Vague' statements cannot be tested." Far from acquiescing in that judgement, I claim that our method of analysis secures several different kinds of statement-"vagueness", to include *indefiniteness, imprecision* and *indetermination.* To say that there is something here (in the corner) is to say something "more definite" than that there is something here (in the room). To say that there were six students present is "more precise" than to say that there were six or seven, or that there were several. To say that there were several *may* be undetermined in truth value when to say that there were four might simply be true or false. I claim that these differences can be explained in terms of our presentation and that we can describe procedures for verifying and falsifying statements that are "vague" or "slack" in these several ways, all without appeal to the highly suspect or notion of degrees of truth (see #7 below).

Meaningful assertions may produce no verifiable statement.

B. Replies to objections that our doctrine is systematically defective.

5. Some may be moved to object that an assertion may be perfectly meaningful but never verifiable. I *reply* with applause: one may assert something, even though an indicated fundamental criterion is never applicable. Meaning includes an indication of conditions for verification; actual verifiability is something else again. Failed assertions are nothing if not meaningful. Indeed, we may assert and fail to make a statement, but never find that out. Our ancestors may have lapsed in that way when they spoke about Adam and Eve. We commonly adapt to the fact or even the prospect of failure by adjustments of meaning, asserting testable statements in place of others we might otherwise have made. I say "Your brother is here", and you ask which one; I adjust my meaning by annexing a description, or I may say "if". Logic books are full of instruction in the use of devices by which meaning-indicated conditions of failure may be obviated, to subrogate a statement we might otherwise had made with one that just could turn out to be "vacuously" true or false.

Distinct statements may have identical representations as pairs of sets of occasions for testing.

6. The next objection is directed against my way of representing tests by pairs of sets of occasions. There are statements of obviously distinct forms that are respectively successfully and unsuccessfully both verifiable and falsifiable certainly by different tests but still on exactly the same occasions. An example: Statements to the effect that there are six and that there were several students present are different; no doubt of that. The tests for verifying them are different too: one requires counting and the other doesn't. Yet, it seems, those two tests may be applicable and successfully applicable on exactly the same (actual) occasions. Here it should be well noted that the terms of this theory do not permit us to appeal to "possible" occasions on which the "several" statements could be shown true and the "six" statements be shown false.

This objection from excessive concreteness has cost me more grief than any other (see pp. 239 above). My whole *reply* is that I have yet to meet with a case of distinct statements I cannot resolve by appeal to actual occasions. Consider the cited example: While the "six"-asserted and "several"-asserted *facts* may be the same, the indicated occasions for coming onto that fact are not. There will be occasions on which a given assemblage is observable to be *several* and yet be uncountable, perhaps because objects separable from one direction are not so from another. Distinctions among logically equivalent statements are readily managed in this way, e.g. we verify a statement that all these hounds are beagles by perusing a collection of canines with a beagle-seeker's eye; on the other hand, we test the equivalent statement that none of these hounds is not a beagle by surveying as hound-hunters a sequestered part of an indiscriminate crowd of judges, beaglers, spaniels and other brutes for being non-beagles.

Statements are testable in variant ways

7. The next objection is simply that there may be more than one way of verifying a statement. I have stipulated to the contrary only in the matter of *fundamental* criteria. To defend the stipulation I must appeal to the representation of tests by sets of occasions, and then *reply* that different tests bring different occasions of testing hence different "possibilities" for truth and falsity into play, which seems to me to be conclusive for saying that the represented statements are distinct. Sometimes, when there seem to be two tests for the same statement, one procedure is clearly fundamental and dominates the other. Bank clerks may weigh the number of coins in a stack, but counting is the check. Sometimes different concepts hence different statements are involved. The determination of color by spectrographic analysis of emitted or reflected light and the determination of color by looking or by comparing to exemplars are of different (though related) concepts of color; what are shown to be different mixtures of different colors by the first sort of test might come out as the same color by the second. Sometimes what are thought to be the same or equivalent tests turn out to give variant results. The difference between perimeter length and surface size was an important discovery of the ancient world. The arguments by which this was shown made it clear that different concepts (in this case magnitudes) were being used and different statements being made.

Statements may have variant "formalizations".

8. Different formalizations of one and the same theory incorporating the very same statements may bring in different concepts based on different operations or tests (thanks to D. Riggs). My *reply* is, simply, that an equivalence between two formalizations must be demonstrated, which is enough to argue that they are (only) equivalent, not identical. This sort of mathematical reformulation is analogous to the assimilation in physics of the principles of geometric optics to the theory of wave propagation.

The use of instruments for testing displaces the subject-matter.

9. Some may want to object that our appeal to tests on cloud chambers as a way of explaining what is meant and said about electrons displaces the subject-matter from electrons to cloud-chamber tracks. I *reply*, simply: tests for the identity of electrons secured by observing cloud chamber tracks are not tests for the identity or of anything else about tracks. There could be those tests too; and their results are "presupposed" when we turn to the matter of electron testing. But surely one test presupposing another doesn't yet throw the good standing of the first one into question. I can't forebear from remarking that the idea that observability and testability diminish with magnitude is dubious. At the time this was first written out, the testing of electrons had apparently become commonplace, whereas then (and maybe still) there were no procedures available for identifying particular molecules. Tests for identifying colors elude our everyday representations, but now may be provided against the background of Quantum Mechanics. Here a theory of the very small assists our observations of the gross.

The representation is circular.

10. We have already noticed (p. 263) the objection that the account is circular because it would try to explain our notion of a referable object by mention of tests that are applied on or to objects, namely occasions. I have learned from experience that the impression of circularity will, as this work goes on, increase not lessen in some readers. A brief *reply* is that the account needn't be circular so long as our descriptions of how a nascent conceptualizer applies tests on those occasions, to which of course *we* must make reference in our theoretical exposition, does not presuppose that *he* be able to make such references. The defense is not a disproof. We shall be continuingly in danger of speciously crediting our imagined subject with capacities so far unexplained. So I must beware. What poses the danger also bears an advantage: When we do, as we should and shall, finally turn to explain a subject's capacity to refer to occasions or those other speculative capacities we bring with us to this enterprise, we can use our own practice as a check against the proposals we shall advance, and, oppositely, use the theory as a gauge on the practice. We may be in for some surprises here, for there are indeed reasons for suspecting that our everyday ways of thinking and talking about occasions (places-at-times) is incoherent. It will turn out (at #4 of Chapt. 23) that, in the terms of this theory, occasions are not referable objects strictly taken; it also turns out that our seeming references to them are

criterially underfulfilled. This attenuation of occasions as objects will leave the skeptic with less to pull on.

Occasions have no testable status as objects.

11. The final objection under the heading of systematic defectiveness is on the other side from that of excessive concreteness. The objection is that these occasions or places-at-times with which we work from the outset are dubious commodities because unobservable by the senses. (Kant used this point as an argument in support of his conclusion that space and time are subjective factors and forms of activity see pp. 260.) My *reply* is that, while places-at-times, are not sensible in themselves, statements about them are made testable by the use of perceivable enduring space-fillers. We shall go into this when, finally, we do come to explain our conceptions of space and time (chiefly at #4 of Chapter 23, p. 443).

C. Replies to objections that our verificationalism is epistemologically short-sighted.

The account is applicable only to trivialities.

12. Only trivialities, it may now be protested, are testable by our recipes: Any idiot may "verify" that a certain function is the integral of another by differentiating the first into the second; but it may take real mathematics to find the solution in the first place. I *reply*: While statements may indeed be only trivially asserted on occasions of testing, those very same statements could also be non-trivially asserted on other occasions. Surely the objector cannot be claiming that we have different statements once the matter is set up for checking by calculation? *Assertions*, not statements, are trivialized by the circumstances of utterance. The "real mathematics" seeks those trivializing circumstances. Certainly one cannot coherently question the applicability of the basic rules of multiplication to the verification of the truths of arithmetic, or that one proves that F(x) is the solution of a differential equation by differentiation; a rainbow is a sort of thing which can be looked at and seen; lengths are measured and weights determined by weighing. None of this eliminates the need for ingenuity, reasoning and gifted guessing, e.g. by detectives looking for evidence. Advances in the established fields of mathematics and science often consist precisely in finding new ways of getting into verification situations where known fundamental criteria can be routinely applied.

The doctrine downplays other, better ways of knowing truths.

13. We hold that statements are proven and disproved by application of the fundamental criteria that define them. It may be protested that there are other better more certain ways of establishing the truth of statements, e.g. by mathematical demonstration. Agreed! There are lots of ways of knowing most anything; of those, verification is fundamental, not "best"--not by the canon of certainty, anyway (pp. 213f.). Mathematical demonstration illustrates both the claim and the concession, for (I believe) such demonstrations may always be taken to show that a certain test is successfully or unsuccessfully applicable in certain circumstances, and the argument then provides a kind of "understanding" and certainty of knowledge not available from actual testing.

Note well that the issue here is not one of "subjective certainty". One may be more certain of what he has been told than of what he has seen because he distrusts himself more than he distrusts others, but that cannot jeopardize the epistemic superiority of observation over hearsay.

The doctrine is incompetent to account for mathematical knowledge.

14. It may be riposted, as a follow-up objection , that what I have just held in regard to mathematical demonstration is, at best, an account of the use of mathematics and not of mathematical knowledge in and of itself. The least part of my *reply* to this objection that I do not account for mathematical knowledge in and of itself, is that formulations of our knowledge that such and such *is a theorem* are verified by exhibiting demonstrations; such *exhibiting* passes muster as a kind of testing. More to the point: equations and other such mathematical truths are often enough demonstrated by calculations, which are tests. Other general mathematical truths cover statements that are verified and falsified by such operations, e.g. the truth that π is transcendental (see pp. 266f). My opinion, succinctly put, is that mathematical knowledge is "of" operations, whether those operations be mathematical or not. Mathematical knowledge is produced by such demonstrations, e.g. Cantor's demonstrations that operations of denumeration can be successfully applied to certain sets and cannot be to others. If this "philosophy" is as natural and plausible as I find it to be, that finding also reflects credit on our representation of statements

This "philosophy of mathematics" is, I believe, a liberalized version of Hilbert's: Hilbert held that mathematics is ultimately about operations of excising initial segments from strings of strokes. I think that holds up for the various theories of number; I would annex procedures of connecting, circumscribing and superimposing figures to Hilbert's more austere geometric number theory.

This theory of conceptualization is undermined by its contingency.

15. The next objection challenges the epistemic credentials of this very theory. An inquiry, such as this one is, into the nature of conceptualization, seeks *a priori* knowledge of a body of necessary truths. However, the objection runs, our use of these procedures or those is only human history; no system of conceptual truth can rest upon so contingent a foundation. I won't dally to consider whether those same "contingent facts" could not also be given necessary formulations or to argue for the possibility of *a priori* knowledge in respect of contingent formulations (e.g. that "Saul Kripke" names Saul Kripke). Let me make my *reply* be a simple opposition to a modal principle espoused by Leibniz: Our claim to have a priori knowledge in regard to a system of necessary truths about statements is compatible with those statements being contingent existents. I explain as follows: The necessity of conceptual truths owes to nothing other than that those truths are about possibilities for the truth and falsity of statements and other products of conceptualizing thought. These possibilities, in the case of statements, coincide with test-occasions. Such possibilities must be meaning-indicated in assertions of those statements and, accordingly, we may come to know these truths "a-priori" simply by reflection on what we mean. There is nothing in this story to contradict (what I believe) that statements themselves exist contingent to our activities. To the observation that this would eventuate in the consequence that necessary truths about statements would have contingent entailments, *viz* of the existence of those statements, I say Amen! I have seen no non-circular arguments to convince me that this rule, that necessary truths cannot have contingent entailments, isn't a false principle (for further discussion, see pp. 347f. below).

This reply joins with the earlier one in regard to mathematical knowledge. Theories of "natural mathematics", certainly to include logic and number-theory, are commonly reckoned to be bodies of necessary truths of which we have *a-priori* knowledge. I explain that sense of what "natural mathematics" is along lines drawn just above:

> Such mathematical truths owe their alleged necessity to their having for subject-matter other conceptual techniques, e.g. counting for arithmetic. Mathematical truths are "about" how we think; they speak to what our forms of conceptualization allow and forbid; they are therefore "about" what could or could not be otherwise, as we actually do conceive things. Still, we might have thought otherwise than we do, and then our mathematics, if we had any, would have been different too.

Statements may survive the repudiation of theories underlying their testing procedures.

16. The use of particular procedures for testing perfectly good statements may seem to depend upon the acceptability of underlying theories, which may be refuted or otherwise discredited or perhaps simply abandoned. My brief *reply* is that that only seems to be so. Measurements on the Piltdown skull tell us nothing about the physiognomy of prehistoric man. The measured lost weight of a charred log is no index to how much phlogiston the thing took on. In the Piltdown business, some anthropologists were gulled into "presupposing" that they were dealing with an authentic skull. What is interestingly brought out by this objection is that the "presuppositions" of a statement might, for one reason or another, be more "problematic" or less certain than the statement itself. What must remain troubling to my account is that a theory whose truth is presupposed by some statement is usually itself unamenable to direct validation by test. Still a theory is commonly validated by its falsifiable predictions. Such statements as do survive the shipwreck of a theory are tested by procedures that don't draw upon the theory in any essential way. We once did and could again test for things being blue without drawing upon those spectrographic theories which even philosophers have heard about.

The possibility of discovering test procedures is incompatible with this being an a-priori theory of conceptual truth

17. Since tests can be discovered to provide answers to questions previously raised, the existence of those questions and of their answers could not have depended on the prior provision of the tests. *Reply*: The discovery of a test might just happen (as perhaps with magnetism or of the relation between the mass and luminosity of stars), or it might resolve a question raised by an hypothesis in a theory (electrons). But an hypothesis about what should or might exist is not yet a statement that it does. Since the Rutherford experiments and the invention of cloud chambers, physicists can now make statements about the presence, trajectories, etc of

electrons, where previously they could only conjecture and hypothesize.

18. Your doctrine is too static to accommodate those kinds of conceptual change that go along with the advance of science. *Reply*: We provide for concepts of all sorts--old and new, and with many options--and incidentally an apparatus for doing autopsies on defunct concepts. Doubtless there will be relationships, sometimes hard to fathom and often much in need of elucidation, between the old concepts and the new ones. That is a kind of problem our account can live and help with. We may for one thing want to say that these different concepts afford different representations of the "same phenomena". That does raise the interesting question (to which we shall make an approach) of how the phenomenon itself is to be identified, by preference.

This doctrine is incompatible with conceptual change.

19. Finally, some skeptical readers may complain that I have given no directions for verifying and for falsifying the statements of my own theory (Hugh Chandler). I *reply* that there are few if any statements to be found here or in any other theory. My proposals and my conclusions are for the most part generalizations. These generalizations are to be checked-out, not by testing, but by referrals to testable statements in regard to our own conceptualizing activities. A carpet-cutter, if he can, will immediately set about to measure a room whose dimensions are in question, and might go blank if asked why one should make measurements here. Again, a person who wondered about the truth of a certain something said would, I believe, be satisfied with the observation of a verification. If I am wrong about that (as may be), then I am wrong in my analysis of the concept of truth, as set out in #5 below. This theory like any other is to be tested by cases. I have made lots of missteps along the way, and am aware that many of my proposals are vulnerable to refutation, e.g. about the grounding of our sense of generality upon procedures that may also serve as criteria for statements (see reply to objection No. 2 above). So (to repeat), the principles of my theory are not testable statements, but the total doctrine is nonetheless up for grabs and out for checking.

The account is non-self-applicable.

Other objections of a more particular character will be noticed as our work scrolls forward. My replies, here and later, are too perfunctory and undemonstrated to satisfy everyone; and there is also the forever threat of undisclosed pitfalls. Still I hope to have transmitted to skeptical readers a share of my own guarded confidence. I now proceed to more systematic development.

2. A REPRESENTATION OF TESTS.

Tests may be represented by disjoint pairs of sets of occasions

We have been heading toward a representation of tests, taking sight only on their occasions of applicability, of successful applicability and of actual application. We have already, in the foregoing, said a lot more about tests than that. So, in what follows, we'll be using only part of what we can say about tests for purposes of representing them.

> At risk of being tedious, I must stress that this is to be a "representation" and not a "reduction". (Cf. the "representation of a point by an ordered triple of numbers or--closer to our case--the representation of a line segment by a pair of end points.) This cannot be "reduction" for a good reason, buttressed by an argument. The reason is, simply, that tests are not pairs of sets of occasions, but action-kinds, and we just noticed that there is much not provided for here that can be said about tests other than that they are applicable or successfully or unsuccessfully applicable on such and such occasions. The buttressing argument is that, if we can represent a test t as a pair of sets of occasions of successful and unsuccessful applicability respectively, we could, without loss or addition, just as well but differently have represented it as a pair of sets of occasions of applicability and unsuccessful applicability. Indeed, that's the place we started from (p. 236).

A first principle or axiom of testing is that every applicability occasion for a test is also an occasion on which that test is either successfully or unsuccessfully applicable, but not both.

To gain a second principle of testing, we stipulate that a test either is applied or is not applied to completion on any of its applicability occasions. It must be one or the other and it can't be both. (For present purposes, incomplete application is thrown in with no application at all.) Speaking formally, we shall postulate that there should exist for every test an "application function" which provides, for every applicability occasion of the test, that it is or is not applied on that occasion.

Every applicability occasion for a test is one on which the test is exclusively either successfully applicable or unsuccessfully applicable and one on which the test either is or is not applied to completion.

I speak of "postulate" rather than of "axiom" in connection with application because here we are, not so much eliciting the gist of principles suggested by the very idea of a test, as stipulating a condition for the use of our theory. The postulate stipulates that we be able to say something about tests that we might otherwise have to leave unsaid. Now (to resume an old theme, p. 244f.) we shall never have to be able to say that a test is *successfully* applied. This relieves our theory from the defeasible contingency of the actual establishment of fact by test. Mere application doesn't require that. Why then should we have to say even that a test is applied? Because certain "concepts" are defined by tests whose application involves the application of other tests, as with the conjoined and otherwise "composed" operations to be taken up just below. An interesting instance that will matter in our later analysis of predicables is that of the subordinate features of objects of a sort or a stuff: in testing for something's being a tiger, I must also apply a test for one or other subsidiary feature--no one in particular--e.g. having to do with stripes, whiskers, body proportions or something.

A "rule of inference" to capture the idea that no test need be actually applied on any occasion of applicability

A third principle of testing is that tests may be left unapplied on any occasion of applicability. Otherwise put, this principle holds that if the applicability of a test t_1 requires the application of a test t_2, then t_2 must be distinct from t_1. I find this principle obvious beyond question, for, otherwise, one must wonder how one could set about to apply a test on an occasion if the occasion were not fixed in advance of the application. The principle resists easy formulation and, despite its abstract plausibility, I know of at least two counterexamples that have been alleged against it. Let me first deal as best I can with the alleged counterexamples and then proceed to the task of providing it with a formulation.

First alleged counterexample: A test to show that a point lies at the intersection of two lines presupposes the existence of that point. But the point is actualized only by the construction. So here the applicability of an intersection-construction test presupposes the actual application of some such test.

Reply. An intersection-construction could be unsuccessfully attempted, as when the lines don't meet. So here there is an occasion for applying the test on which the point is not actualized by the test. The objection confuses the "actualization" of a point with its "existence". Points like other mathematical things exist as "possibilities", in this case for bounding line segments. Generally, there is no way to *establish* the existence of a possibility other than to actualize it. However, a geometric point may exist unproven by actual construction. In sum, we need say nothing about intersections that is not normal and unproblematic, *viz* that we cannot establish the existence of a point except by applying a test.

Second counterexample[16]: *"Double-slit experiments"* in Quantum Mechanics illustrate the possibility of cases where the existence of the tested phenomenon intrinsically involves the actual application of the test.

Analysis. The "experiment", often cited by Bohr and others as a theoretical certainty[17], to illustrate the consequences of wave-particle duality in Quantum Mechanics, also seems to witness the disturbing idea that the character of an observed quantum phenomenon is, within the bounds of indeterminacy, a function of the observation itself. Here is the situation as I understand it: Electrons coming from a source through a slit are recorded to show a certain distribution pattern on a screen. When we record electrons coming through a pair of slits, the resulting pattern may show the peaks and hollows of an interference phenomenon and not a simple "addition" or superposition of the patterns separately recorded. However, if we attach detectors at the slits in order to determine from which direction an electron registered on the screen has come, we now find that the recorded combined pattern of the electrons that are also detected at the slits is a simple addition of the two patterns. The test for the interference pattern involves the passage of electrons undetected at the slits. It seems, then, that an attempt to observe the actual passage of the electrons changes the nature of the observed phenomenon, and so the determination of occasions of successful applicability for the test depends upon whether that test is actually applied.

Reply: Now, finally, my "reply". We have described *two* different tests both of which terminate in observations of the screen. One of those also requires as a condition for application that detectors be placed at the slits: The tests for direction could not be applied unless that condition were satisfied. (Actual application of the test would require the coordinate observation of blips on the detectors and blips on the screen.) The applicability of *this* test, which requires the presence of detectors at the slits, does not require *its* actual application, *viz* counting the blips on the screen. The test for the interference phenomenon requires no detectors at the slits at all. Neither test envisaged in the "double-slit experiment" would require its own actual application. (This answer does not, of course, talk to the main theoretical issue for physics, over whether there *could* be a way of tracing all features of the interference pattern to electrons that separately come from the directions of the two slits and does not touch on the competing explanations of why there should not be.)

I come now to the tricky matter of formulating the principle. It was "modal" in our first statement, for we said that the test "may" be left unapplied on all occasions. Such formulations, we know, create "non-extensional" or "opaque contexts", allowing for the unacceptable possibility that a condition of the world might satisfy the stipulation under one characterization but not under another. Now a test could be applied on any occasion and, indeed, on all of its occasion of applicability. We cannot appeal to how things actually are in order to say how they may or needn't be. So how *are* we to say that a test needn't be applied on any occasion? Accustomed formalizations of logic suggest a way round this impasse: such formalizations always require "rules of inference", which also sum up a sense of a proposition having to be true if something else is true. Now the "something else" in our case is that the test be simply applicable; and we must then *not* be able to conclude that the very test is actually applied. With those thoughts in mind, let me now propose that test theory, thought of as a kind of logic, is subject to the following rule of inference: If, on the hypothesis that a test t_1 is applicable, we conclude in n-steps that a test t_2 is applied, then, at the n+1st step, conclude that $t_1 \neq t_2$

I now proceed to introduce some terminology for talking about tests, all definable within the resources of the proposed representation.

A test is qualified when its set of occasions of applicability is non-empty.

Assertionally indicated conditions for applying a test may actually fail to be satisfied, in which circumstance there are no occasions for applying that test; so, for example, if there was no Homer (as some believe), there is no occasion for testing for "his" being blind. When a test is applicable, this being formally understood to mean that the set of occasions for its applicability is non-empty, we shall say that the test is "qualified", *qualified*, of course, for the representation of statements (p. 237).

A test whose representing pair of sets of occasions when taken in reverse also represent a test is "commutative"

We have stipulated that every test may be represented by a disjoint pair of sets of occasions. But surely not every disjoint pairs of sets of occasions need represent a test. Some do and some don't. The question takes on character when we ask it of the reversal of a test representing pair. The first member of a representing pair is the set of occasions upon which the test could be successfully applied and the second the set of occasions upon which the test is unsuccessfully applicable. Consider tests for existence: The test for the existence of a body (in a room, say) may be successfully applicable on some occasions (e.g. one's containing a table) and unsuccessfully applicable on other "empty" occasions. Suppose we were to "reverse" our sense of success and unsuccess here, interchanging the two sets of occasions in the representation: What would we be testing for then? Certainly not for the non-existence of anything in the room, which can be done only on occasions where the whole room is brought under examination. It is not obvious what else it could be either. So perhaps nothing. In contrast, any occasion upon which a test for an object weighing 5 lbs. would be successfully applicable would also be an occasion upon which a test for it's not weighing 5 lbs. would be unsuccessfully applicable and conversely. If we reverse the two sets of occasions representing either test, we represent the other test. It will be useful to dignify the difference with a name: If the reversal of a pair representing a test t_1 also represents a test, t_2, we say that t_1 is "commutative" and that t_2 is the "commutation of" t_1. Obviously, the commutation of a commutative test is commutative, and a commutative test is the same as the commutation of its commutation.

One test will be said to be "kinematically like" another when the occasions for applying the first are included among the occasions for applying the second.

Tests may stand in various relations which we might hope to define in terms of our representation. As an important candidate instance, consider that one test may or may not consist of the same kinds of movements as does another. Tests of reaching out to displace something and of reaching out to touch that thing are in this respect physically alike and together physically different from tests of looking over to see the color of the thing. I shall say that the test of reaching to displace is "kinematically like" the test of reaching to touch. Since there may be constraints on the movement of the first test not binding on the second, the relation is not symmetric. Is there any way to capture, within the toils of our representation this idea that one test is physically or kinematically like another? Consider: I might have applied the same tests, reaching to displace or to touch, not with my hand, but with my foot or my head, using a foot perhaps because something present on the occasion may have blocked the passage of my hand, or simply because I wished to use a foot. A test may be applied except when circumstantially blocked, and a test is kinematically limited only by the physical character of the occasions of application. I conclude that the kinematic constraints upon the possibilities for applying a test are determined by the specification of occasions of applicability. It appears to fall out, then, that a test t_2 is kinematically like a test t_1 just in cases where t_2 can be applied on occasions on which t_1 can be applied. So let that be our definition of one test being kinematically like another. I sense that this definition gives the right results. Tests which are physically like others, but asymmetrically so, are kinematically coordinate with the others, e.g. reaching out to displace and reaching out to touch. I may reach out to touch even if nothing is there, but I cannot reach out to displace what is there if nothing is there[18]. The occasions for applying the latter test are included in the occasions of applying the former, but not conversely. The latter is kinematically like the former, but not conversely. In contrast, tests which are "physically different" have exclusive or overlapping occasions of applicability. So, for example, acts of reaching to touch and acts of sighting to see: an act of sighting to see can be performed even when I am circumstantially blocked from moving my head toward the place I look to, hence there are occasions for looking which are not occasions for probing, and look-see tests are therefore kinematically unlike probe-touch tests.

A test kinematically like another is "included in" that other, if the other is successfully applicable when the first is, and is actually applied whenever the first is.

A test physically or kinematically like another might if successful yield a different result from that of the other. Tests for material existence and non-existence are cases in point. Sometimes, however, a test, t_1, physically like t_2, also seeks a similar result. Testing to tell whether a displaceable chunk is hard is "just like" testing to tell whether anything is hard--another chunk or a wall or the earth--except that a particular displaceable something must be at hand. The chunk-testing test, t_1, is physically like any other test for hardness, t_2, and also, if successful, leads to a like perception. Whenever the test for chunk-hardness is successfully applicable, then so too would be the general test for hardness. Furthermore, if a test for chunk-hardness is actually applied, so too is a test for hardness *simpliciter*. The one test is a constrained version of the other and is, so to speak, included in the other. We can readily capture this idea of being *just like* or *included in* by sole appeal to the terms of our representation: First, t_1 is included in t_2 just when t_1 is kinematically like t_2; second, the occasions for successfully applying t_1 are included among the occasions for successfully applying t_2 and, third, t_2 is applied on occasions on which t_1 is applied.

One test may "presuppose" another in the sense that the first may require for its "qualification" the successful applicability of the other.

A test t_2, included in t_1, may differ from t_1 only in that t_2 is subject to additional constraints as regards application and successful application. The additional condition on a t_2 included in a t_1 may be, for example, that a certain displaceable object be present on the occasion, or that something else, perhaps, have existed somewhere, sometime. The applicability of t_2 in that way "presupposes" the presence or the existence of something. That existence or presence may be the kind of thing which could be proved out by the successful application of some other test, t_3. However, the application of t_2 in such circumstances could proceed provided only that t_3 could be successfully applied. I think we have now, within our representation, defined a notion answering to what philosophers have sometimes had in mind when they have spoken of "presupposition": one statement "presupposes" another if the verification test for the first bears the just-described kind of "presupposition" relation to the verification test of the other. So, more formally: One *test* "presupposes" another second test if the first is included in a third test, subject to the additional constraint that the first is applicable only if the second is successfully applicable. We say that the first test presupposes the second over the third.

A first test "strongly presupposes" another if the first is applicable only on occasions on which the other is successfully applicable.

We may test for the electric charge on a body by looking not at it, but at a remote gauge. We test for the shape of a body, however, by applying our instruments to the body itself. Both tests require that there be a body there to test, hence presuppose a procedure for individuating the thing. There is a difference, however. The shape-test is applied only on occasions where the presupposed individuation can actually be brought off, whereas the electric-charge test is applied elsewhere. Here I shall say that the shape test *strongly* presupposes individuation. More generally and somewhat more formally: t_1's presupposition of t_3 over t_2 is "strong" when the occasions for applying t_1 are included within the occasions for successfully applying t_3.

Tests may presuppose themselves.

A verification test for a statement to the effect that some particular old grey mare is female would consist in making observations for gender upon the mare in question, on condition that it was indeed a mare. The test therefore presupposes what it tests for, that the creature is a female of its species. This possibility in no way gainsays the sense of our earlier rule that a test cannot require its own anytime (actual) application.

A presupposition of a test that includes all tests with the same presupposition, weak or strong, is "minimal".

Different tests may presuppose the same test over a common test in which they are included. Reaching out to determine the position of a body and reaching out to examine its texture, while certainly distinct, are both "just like" tests for determining the existence of a body and both equally presuppose (successful) individuation. Though neither of these tests seems to be simply a further restriction upon the other, both look to be further restrictions upon some still other test. We can suppose that there at least *may* be a test, distinct from bodily individuation itself which minimally (in this case strongly) presupposes body-individuation over reaching out to touch. A like possibility is conceivable for weak presupposition. Generally, then, a test t_1 *minimally presupposes*, weakly or strongly, a test t_2 over t_3, when t_1 is distinct from t_2, and when every other test that (weakly or strongly) presupposes t_2 over t_3, is included in t_1.

If a test is applied and successfully applied by applying and successfully applying two other tests, we say that the first test is a "strong conjunction" of one of the other two onto the other.

We sometimes wish to string two procedures in tandem into a single dependent test. Thus a doctor may first take the patient's pulse and then listen to his heart. It is common to record the results of making such a test as separate items of information, although it may make no practical difference which of the constitutive tests is first applied; still, different orders of application make for visibly different procedures. Let us symbolize a test, t_3, whose application consists of first applying test t_1 and then another test t_2, as "t_2t_1"; we call t_3 a conjunction of t_2 onto t_1. The application of t_2t_1 obviously must consist of a terminal application of t_2 on an occasion appropriate to t_2. t_2t_1, moreover, is successfully applied on that occasion only if t_2 is successfully applied. Briefly, t_2t_1 is included in t_2. It is clear also that t_2t_1 is successfully applicable on an appropriate occasion only if t_1 is successfully applicable on some occasion. We capture that requirement by saying that t_2t_1 presupposes t_1 and (to assure uniqueness) that the presupposition is "minimal". The presupposition may be weak or strong. Finally, any actual application of t_2t_1 must additionally actually involve the application of t_1 as well as the application of t_2. I believe with that further condition we have gotten what we need for one kind of "strong" test conjunction.

Although t_2t_1 and t_1t_2 are usually different tests, it is evident that if either is successfully applicable so too is the other.

"Strong" test conjunction will, I hope, be serviceable for the definition of such conjunctive predicables as *being red and juicy* (≠ *being juicy and red*). It is, however, much too "strong" to afford verification criteria for conjunctive *statements*: a "strong" test conjunction of two verification tests of the conjuncts presupposes the successful applicability of one of those tests; if the presupposed test were verification criterion for a false statement-conjunct, the presupposition would not be satisfied and the conjunctive test would be inapplicable; so here the falseness of the one conjunct would make the conjunctive statement non-existent.

The above argument shows the need for an additional "weak" test conjunction. Here it seems enough to stipulate that the test is applicable whenever either of the conjuncts is applicable and is successfully applicable whenever one or other of the two "strong" conjunctive tests is successfully applicable. I call this "weak" test conjunction, and will symbolize it with a wedge, thus:$t_2 \wedge t_1$.

If a test is applicable whenever either of two other tests is and is successfully applicable whenever either strong conjunction of the other two is successfully applicable, we say that the first is a "weak conjunction" of the other two.

A hostess looks at her guest list and thinks that if one of the invited couples doesn't come another won't either; she then tells her husband that either three couples will come or five will. He verifies his wife's statement after guests have arrived by applying a disjunction of a three-count with a five-count to the assembled set of couples. Here successful applicability of either "disjunct" eventuates in the failure of the other. It would not be that way in a test for something being either a radio or a phonograph. We must now seek a general formula for the definition of a *disjunctive* test from two component tests. A disjunction of two tests, t_1 and t_2, and which we may symbolize as "$t_1 \vee t_2$" is a test which may indifferently terminate with the application of either t_1 or t_2, and which is successfully applicable in case either of the components is successfully applicable.

If a test is applicable on occasions appropriate to either of two other tests and is successfully applicable if either of the other two is, we say that the first is a disjunction of the other two.

That suffices as a definition. One must, however, take caution over occasions of *unsuccessful* applicability for the disjunctive test: they are not always simply the *unions* of the occasions of unsuccessful applicability of the component tests, for if any one of those occasions were also an occasion for the successful applicability of the other component, we would be in violation of the disjointedness rule on testing, that a test not be both successfully and unsuccessfully applicable on any occasion (p. 278). To avoid the off chance of a mishap in this matter, I warily propose to define disjunctive testing by separately specifying the sets of occasions for successful and for unsuccessful application. We must limit the occasions for unsuccessful application of the disjunction to occasions of unsuccessfully applying either t_1 or t_2 which are not occasions for successfully applying the other. That is one source of difficulty tediously but easily obviated. Another is to distinguish a disjunctive test from both of its components taken either separately or together. To that end, I propose that we "unify" the application of the disjunction by demanding that it would consist at least in the application of *both* components, *viz* the disjunction is not applied unless both disjuncts are.

Notice, incidentally, that, since the definitions of both weak test conjunction and of test disjunction are symmetric as between their "components, it follows that $t_1 \wedge t_2 = t_2 \wedge t_1$ and that $t_1 \vee t_2 = t_2 \vee t_1$

"Ostensive" tests require the sometime application of some test.

Some of our tests in actual use require an adjustment to or use of a part of the world given to hand. So with the use of bodies as "frames of reference" for locating things in space; again, tests for substantial predicables require a sometime familiarity with named instances. There is no applying such procedures unless something or other is displayed or ostended. Even when mistakes are made, perhaps because the ostended object was faulty, subject must have a sense of preparations in place. All that is needed for expressing that thought, I think, is that subject have actually applied some kind of test on an "appropriate" occasion. We call "ostensive" those tests whose qualification requires that another test have been sometime applied.

Might not perhaps all tests be "ostensive"? If so, I'm in trouble. I want certain procedures to be "most basic"--a hard demand if "most basic" tests required the actual application of other presumably less basic tests. Fortunately, our candidate for "most basic" among tests are probes-to-touch. Since these seem to require no instance of anything, some of them also seem to be safely "non-ostensive". It bears repeating at this point that all tests, including "ostensive" ones, are pathways "to" not "from" the world, which in no manner jeopardizes the true opinion that testing subjects must be *practically* responsive to their surroundings.

I have now fashioned a bare bones apparatus for dealing with tests. The explanations have been brief, and the formulations succinct. Details have been eschewed: so, for example, a fastidious reader may wish to confirm that disjunctive and conjunctive tests are defined whenever their components are. The apparatus will be greatly expanded when, in Part II, we introduce the idea of a test that is, not only applicable "on" occasions, but also applicable "to" what I shall call "locations".

3. A REPRESENTATION OF STATEMENTS.

Chief among the conditions of success indicated in an assertion are that fundamental criteria can be applied. That such a pair of tests be applicable or "qualified" is also the chief condition for the existence of a statement.

Statements are producible in successful assertions, and questions concerning the existence of statements arise most fundamentally and preferentially in connection with assertions understood. A statement would be proved to exist by showing that somesuch assertion succeeds. Since assertion is a kind of utterance, an assertion does succeed whenever all the indicated "understood" conditions of success are satisfied. (p. 55f.) These dicta allow for the possibility of unasserted statements. But statements must anyway be assertible, and are proven to exist by establishing the success of an assertion. Just as a prospectus may canvass questions about a partnership for which the registration papers have not yet been formalized, so too questions about the existence of unasserted statements can be raised, but then always with an eye to the possibility of assertion, and such a question would be settled by showing that all the conditions of success that would be indicated in such an assertion are satisfied. Chief among the assertionally indicated necessary conditions for the existence of a statement are that a pair of verification and falsification tests are both sometime applicable or "qualified" and that not both are successfully applicable.

A method for showing that a statement does not exist is to show that some condition for applying one or other of the tests is not satisfied.

The first mentioned rule, that both tests be qualified, will be our main instrument for settling questions about the existence of statements. By showing that assertionally indicated constraints on the application of a test are *not* met, we can show that the test cannot be applied and that the projected statement does *not* exist. (The assertion of course is still "meaningful" though unsuccessful.) There can be no comparable demonstration that a statement does exist, and for two reasons. First, the applicability of tests is not a sufficient condition for the existence of statements. Second, we have no method for showing that *all* the conditions for applying a test are satisfied.

Another way of proving the non-existence of a statement is to show that both tests are successfully applicable.

Statements can also be shown not to exist by instancing circumstances in which both of the assertionally indicated procedures could be successfully applied. Consider the example (p. 220) of the identification of a onetime two-paneled sectioned window. Looking at a pile of window panels, a superintendent asks one of his workers, "Which is the window that came out of that casement?" The worker may tell his boss the answer, even if all the windows in question were stripped of their handles and hinges, and indeed even if the mentioned window panel had been unhinged from a companion panel since destroyed. But suppose that both of the onetime companion panels were together there in the same pile, and that all of this was known to the worker. I submit that then he wouldn't know what to say in answer to the question, simply because in this case there is no right "statement of identity" to do the job. More theoretically said, we establish window-identity by continuously tracking a window into a window where it is separated from some other thing. We prove window-distinctness by separating a window from some other thing into which a window was tracked. We verify a statement of window-identity by applying a window-identifying test; we falsify the statement by applying a window-distinguishing test. We can identify a window separated from its handle and distinguish it from other windows. In the case of separated windows one can successfully apply both a window-identifying test and a window-distinguishing test for each of the parts with the original double window. Pending clarifications, there can be no statement of identity in relation to either part with the original whole, *viz* no answer to the question over *which* of these two windows was the original whole. The original double-paneled window, which is both identifiable with and distinguishable as a window from each of its window parts, cannot be successfully asserted to be identical with or distinct from either part, at least not as a window. (Had the superintendent "clarified" the matter by demanding to know which of these single panels was the double-paneled window that was before in the casement, then there would *clearly* have been no answering "statement of identity" to be provided.) I have assumed that "statements of identity", when they exist, answer "Which?" If there is no answer in the described circumstances, then that is evidence that there is no corresponding statement, which conclusion supports our thesis that statements cannot be both successfully verified and successfully falsified. (The reader may confirm a similar lapse in the existence of statements of the identity of unsorted hunks.)

A consequence: Statements require identifiable particulars.

We shall later (in Chap 9, #4, p. 401) extend arguments from Plato and Aristotle that there is no contrariety without individuals (see above p. 220), to hold that identity is also needed if we are ever to establish that the successful applicability of one test excludes the successful applicability of another. The upshot of this is that we could not find pairs of tests of which the successful applicability of the one excluded the successful applicability of the other unless a presupposition of identifiable objects were met. The existence of statements implicates the disputable "metaphysical" consequence that there are identifiable particulars.

Contradictions are possible but never true.

> The second rule for the existence of statements, that not both of the indicated procedures should be successfully applicable, is the test-theoretic counterpart to one version of The Law of Contradiction, namely that no statement is both true and false. It does not prohibit contradictions--neither the assertion of contradictions nor contradictory statements *per se*. Suppose we have an (existing) statement *s*, and another statement *not-s* whose verification and falsification tests are those of *s* taken in reverse. If (reasonably) *s and not-s* is a statement to be verified by a weak conjunction of the verification tests for *s* and *not-s* and falsified by a disjunction of their respective falsification tests, then the verification test for *s and not-s* is applicable but not successfully applicable; the falsification test is applicable on the same occasions as is the verification test and successfully applicable if either the verification or the falsification test for *s* is successfully applicable. If either *s* or *not-s* is true, then the contradictory conjunction is simply false. If *s* is neither true nor false, then *s and not-s* is also neither.

This version of The Principle of Contradiction generates no companion Principle of Excluded Middle.

Our theory does not demand that either of the indicated tests be successfully applicable on any occasion. A statement may exist and in actual fact be neither true nor false. The envisaged possibility must be sharply set off from cases of meaningful but unsuccessful assertions which produce nothing in the way of statements true or false, because they produce no statements at all. We allow for the successful assertion and actual existence of statements which, it happens, are neither true nor false. I believe that this liberty, while unorthodox as theory, leaves the meanings of "true" and "false" intact[19]. I would feel more comfortable in this (precedented) departure from the orthodox opinion that the

Principles of Contradiction and Excluded Middle are intertwined laws of thought, if I could exhibit statements in actual production that are in fact neither true nor false. A prospective kind of case is that of conditional statements (to be distinguished from conditional assertions and from conditionals; see, p. 146f.). Our third apple example (pp. 222) was a candidate. Here is another: I spot a dog chasing a rabbit apparently into a hollow log. I station myself at the opposite end, and assert that if the rabbit is in there, he'll come out here. If the rabbit appears from the hole, what I said was true; if instead he made his way out through a hatch centrally leading into the warren, what I said was false. But what if he never went into the log, only seeming to, when he really scooted over the top? By saying "if" I meant to leave that option open. Why should I allow myself to be brow-beaten into a "true" or "false"? I might have said something else, of course, e.g. that either he didn't go in at all or else he's coming out here; but that is not what I suppose myself to have said. Other prospects are indeterminate statements produced in vaguely worded assertions: I say that there are several people in a room occupied by four. Is that true or false? I'd say "neither".

Candidates for statements neither true nor false have been advanced by Mathematical Intuitionists as villains. They have in effect argued that *reductio* demonstrations of the unsuccessful applicability of falsifying tests do not establish the successful applicability of verifying tests. I agree. However, since these theorems concern non-denumerable sets for which there are available no delimiting procedures, necessary in my scheme for the definition of the wanted universal statements, these examples leave open the question whether there are any mathematical statements, strictly taken, which do not conform to Excluded Middle. Alternatively (and I tend to this opinion myself): It is an old idea that talk of infinities is just fancy for talking, in an indeterminate manner, about undelimited "other cases". The matter is left very "undetermined" indeed, when there's no indication of how the "and so on" is to proceed, and ever less determined when we are given no indication of how those "and so on"s are themselves to go on. If that thought can survive the rigors of Cantor, Dedekind and Frege, then we may allow that some statements about infinities are simply "vague" and may be neither true nor false for pretty much the same reasons that "several"-statements may be neither true nor false.

Goedelian incompleteness results suggest artificial examples. Suppose we have a formula F that is read to say that F is not formally derivable. A statement D_F that F is formally derivable would be verified by making a formal derivation of F. Call that test V_{DF}. Similarly, a statement $D_{\neg F}$ that the formal negation of F is derivable would be verified by application of test (call it) $V_{D\neg F}$. Now *if* the formal system to which F and its negation belong is consistent (which may be true even if unprovable), then not both V_{DF} and $V_{D\neg F}$ are successfully applicable, and we *may* take them as a pair of criteria defining a statement D_G (which, if it exists at all, is distinct from the statement represented by the formula F). By Goedel's argument, neither V_{DF} nor $V_{D\neg F}$ can be successfully applied and D_G is neither true nor false. I cannot formulate the constructed statement in so many words, and am glad to leave the question of whether there are such examples in actual mathematical production for the experts to decide.

Two assertions produce the same statement if they both indicate the same pair of qualified fundamental criteria.

We now move from considerations of statement existence to considerations of statement identity. A question of statement identity arises most fundamentally and preferentially when we ask whether two supposed successful assertions produce the same statement. One understands an assertion and knows what statement is produced in that assertion when he comprehends that the utterance is indeed an assertion that indicates as a condition of success the applicability of a pair of fundamental criteria. Two assertions produce the same statement just in cases where they call for the same fundamental criteria. So, for the purposes of our theory, we identify a producible statement by specifying that pair of tests whose applicability would be indicated by any successful assertion of that statement.

This yields a "strong" rule of statement identity, one that readily distinguishes logically equivalent statements, e.g. *p* from $p\&(q\vee\neg q)$ (the latter but not the former is testable on occasions appropriate to the testing of *q*). This rule of statement identity assures to logic a field of investigation within which it would be possible to establish the logical equivalence of distinct statements and, incidentally, it portends a resolution of those "paradoxes of confirmation" that argue that evidence need not be equally good for all logical equivalent statements. Our rule also responds to such facts of linguistic life as the felt difference among the statements produced in (say) "each", "all", "every" and "any" assertions, slurred over by most schemes of logical and semantic analysis. I shall propose, for example, that we verify an "All ϕ's are ψ" statement by exhaustively testing for ψ objects selected from a delimited ϕ-featured class and that we would verify an "Any ϕ is ψ" statement by testing for ϕ and ψ a delimited class (details to follow in Chapt. 14, p. 419). All of this is gained without appeal to the peculiarities of the languages in which assertion is enacted.

The applications illustrate a method that is both "stronger" and "finer" than model theory. Its indifference to actual linguistic expression also separates our method from Carnap's use of the notion of "intensional isomorphism" which strengthened the model-theoretic criterion of logical equivalence by requiring that syntactically defined elements constituting the sentences used to express identical propositions could be put into semantical correspondence.

Our rule formulates a necessary and sufficient condition for statement-identity. It implicates a procedure for establishing the distinctness of statements but none for statement-identity.

Sometimes, in writing down necessary and sufficient conditions for something or other, we know from the formulation what kind of test would have to be applied in order to establish that those conditions were satisfied, but also know that there could be no such test. We are told, for example, that real numbers can be represented by non-terminating decimals. Two such representations are of the same number if they converge together. A test for that could only be an uncompletable place by place comparison of the representations. The representation does, however, implicate a test for distinctness, *viz* we find a place at which the representing decimals begin to differ otherwise than as "9" and "0". Since that condition on real number identity tells us what the identity test would be if it could be, we are also in position to prove theorems about real numbers. The situation for statements is comparable to that for real numbers. We would show that $s_1 = s_2$ by showing that the occasions of applicability and successful applicability of both the verifying and falsifying fundamental criteria for s_1 and s_2 are the same. The only effective way of doing that would be to compare terminated lists of conditions of applicability and successful applicability; that is tantamount to comparing terminated lists of conditions for doing and of success for the tests, which we have every reason to suppose cannot be done. No *procedure* for establishing the identity of statements in fact exists. However, our condition on statement identity does implicate a procedure for establishing the distinctness of statements. We would prove that $s_1 \neq s_2$ either by describing an occasion for applying one of those tests for one of the statements that was not an occasion for applying the corresponding test for the other statement, *or* by describing occasions on which a test for one of the statements could be successfully applied where the corresponding test for the other statement could not be successfully applied. The difference I have in mind as regards to conditions of applicability can in very large measure be taken from stipulations to the effect that a test could not be applied if another test were not successfully applicable, *viz* in terms of presupposition relations. Thus, application of the verification test for the statement that Smith's brother was born in Omaha requires the successful applicability of a test for being Smith's brother to a person, where that is not so for a statement to the effect that *this* man was born in Omaha. That shows that the two statements are distinct, notwithstanding the proven fact that this man is Smith's brother. Differently, we can demonstrate a difference in occasions of successful applicability by arguing that (e.g.) we could successfully falsify a statement to the effect that everyone present was someone's brother on occasions where we could not falsify a statement to the effect that everyone present was someone's sibling, *viz* on occasions where not everyone present was male. In the upshot, then, these observations suggest that we can by procedure routinely establish differences between statements where differences exist, but, where they do not, we cannot procedurally establish identity.

> The "indefiniteness" of statements confirms our readiness to draw distinctions among conditions of success indicated but not yet formulated, rather as the indefiniteness of real numbers confirms our readiness to introduce refinements of measurement in relation to exhibited magnitudes. Our definition of statement identity is analogous to Eudoxus' definitions of equality of ratio among magnitudes. One may protest that, by my formulas, it may and likely will eventuate that no two assertions ever produce the same statement. Perhaps, but then possibly and with comparable likelihood, no two magnitudes ever have the same measure.

We have now arrived at a representation of statements as ordered pairs of qualified tests, not both successfully applicable. In the following section, in "Appendix D" and throughout Part II, we shall consider various features of truth-value, modal determination, "form" and relations of consequence for statements thus represented.

I conclude this section with some recollections of earlier registered points that may usefully be adjoined as remarks on the above formulated representation of statements.

First: We have advanced, a representation of statements by test, not a "reduction" of statements to tests. The arguments against reduction are, (1) We could just as well have represented a statement by its representing pair of tests taken in reverse, (2) the representing pair of tests might represent things other than a statement, e.g. a prediction, and (3) It is obvious that a statement is not a pair of tests.

Second: The representation of a statement by a pair of tests clearly is not a paraphrase or assertion of that statement. The representation works just as well for never asserted statements known to be false as for frequently asserted true statements.

Third: This representation is designed to capture any distinction, no matter how fine, we might wish to draw among statements we are now prepared to make, but so far only those. I am not competent to legislate about how we *should* think, nor can I constructively speculate about how we might have thought had our brains been bigger or our fingertips smaller. I wish to employ these tools for characterizing how we *do* think.

The *fourth* remark is a recollection of my reply to objection #7 (p. 271), that a single statement may have several tests, e.g. that we verify height by dropping and measuring a plumb line or by optical triangulation. To this I respond that our method properly distinguishes among cases that the objection smudges over. First, different tests for length, mass, etc may give different results. We must sometimes allow that there are different concepts of length, mass, etc which enter into what must then be different statements. Second, we must also allow for distinctions among provably logically equivalent statements, e.g. that a given face is an equilateral triangle or that it is an equiangular triangle; the theorem covering this is gained from a consideration of procedures, and the use of the theorem presumes that the statements are different. As a third possibility, we certainly may sometimes describe the same tests hence the same statements in different words, hence draw distinctions which mark no differences among statements. All of this is relative to how we do think. Our description of the test for material existence, *viz* probing to touch, is noncommittal about which hand, foot or other part of the body is used. Perhaps it would be different for intellectually advanced insects whose antennae might serve for both tactual and chemical perception. Also, the concepts founded on these procedures may prove defective if the facts are other than we think, as may actually be so with our concepts of space and time. But that can be tested out too. Should it turn out to be so, presumably we would then come to the prayerful assertion of other statements in place of these which, to our surprise, turned out never to have existed at all. Our theory's readiness to accommodate such contingencies is not weakness but a strength.

Finally, recollecting objection #5 (p. 270) that a meaningful assertion may fail to raise a question over verification and falsification, my oft-repeated reply is, simply, that it is the sense of an assertion to let us know, not only under what conditions a produced statement would be true and false, but also how the world must be for an answer to be given at all. The mere meaningfulness of the assertion cannot assure the satisfaction of these various indicated conditions. It is, furthermore, altogether advantageous to separate these different ways in which assertional utterance can go wrong--in making sense, in success, in raising answerable questions and in actually being false. These different pathologies, too often confused, naturally separate along the lines of our account.

4. TESTING AND SEMIOLOGY.

Test theory may be adapted for use as a systematic language-neutral semantics.

In this section I explain how our theory of testing may be used as a systematic but limited "semiology" or theory of meaning--one which I believe comes very near to being a "fregean" theory of *Sinn*.

The meaning of an utterance is its indication of conditions of success (p. 49); the "whole meaning" of an utterance (an assertion, for example) would be defined, if it could be, by a completed listing of those indicated conditions (p. 51). The indication of any such condition is secured by the occurrence of an "expression"[20]. Any such expression may, by its occurrence, indicate a batch of conditions. A listing of those conditions defines what I call a use (p. 300, and 363, fn. 20). "Semantics", as I understand that term, affords means for resolving utterances into uses. Test-theory in its further developments provides a partial inventory of indicatable conditions presumed ultimate in relation to the human scheme of conceptualization, and also an easy explanation of the complementary concatenation of condition indicating uses (see p. 314 below). Test theory thus transfigures into a systematic but limited semantics of the uses realized in the assertional production of statements. The application carries over to the realization of those same uses in other "illocutionary contexts". Test-theoretic semantics speaks only of "conditions", without reference to particular expressions: it is "pure", "universal" and language neutral.

Among the meaning determining conditions of success for an assertion are the qualification or application conditions for the indicated pair of fundamental criteria.

Test theory is adapted to semiology through the following connections: An assertion if successful produces a statement. Since assertion is a kind of conventional action, an assertion is successful just in case all the indicated conditions of success are satisfied (p. 55). So, an assertion produces a statement just in case all the indicated conditions of success are satisfied, including the application conditions for the fundamental criteria of the statement. It follows that all the qualification conditions (of doing) for testing are included among the conditions of success indicated in the assertion. That is the principle linking our theory of statements to our theory of utterance.

While the derivation was abstract, it does I believe jibe with our common understanding of concept acquisition. I tell a child "*That figure* is a 'trapezoid', you see"; and then I show him what I mean by counting and comparing the sides of a *figure* whose existence is at once a condition of success for my utterance and a condition for applying the test I aim to inculcate.

Objection & Reply: Assertionally indicated conditions may sometimes be, not conditions for applying tests, but rather conditions for the successful applicability of those tests.

The bridge principle that carries back and forth between the analysis of the meaning of assertions and our account-to-be of testing and statements is that indicated conditions of success are also qualification conditions for testing. If *C* is some such indicated condition for applying a test, *T*, then *T* cannot be applied if *C* is not satisfied. It might now be objected that a tester who understands the assertion and who thinks the *C* is satisfied when it is not might go ahead and apply *T*. It would follow that *C* is not, as our theory demands, always a condition for applying *T*, but sometimes, at the very most, a condition of success for *T*. For example: an indicated condition might be that there is an accurate measuring instrument available to hand on the occasion of application. The tester, thinking that is so when it is not, might proceed to make an unsuccessful application of the inaccurate instrument he has.

I reply with two observations. *First*, it may be that the tester did not apply test *T* but another test *T'*: He doesn't *measure* but *estimates*. That order of reply is seldom convincing: a tester must at least be capable of knowing what test he applies. Still, *actual* self-consciousness may be absent, and what I have outlined may sometimes be the right objection-avoiding description of the case. *Second*: Taking the illustrative case of measurement mentioned above, it is clear that the availability of an accurate instrument cannot strictly be a condition for measurement, since one may sometimes make do with an instrument he *knows* to be inaccurate. All that is strictly required is that there should be some rod available, and the tester cannot be mistaken about *that*. Similarly, subject could not mistakenly think that there was enough light to see by if he were completely in the dark. At this point it would be useful to have answers, still not known, to the difficult general question about the implications of belief, *viz* a characterization, more "substantive" than that of Appendix C (p. 201) of what conditions must be satisfied to truly say that subject believes such and such, e.g. that he has light enough to see by. Failing that, I can only plead that there must always be such conditions, conspicuously including the circumstantial conditions for applying tests indicated as conditions of success for assertion.[21]

The indicated qualification conditions for fundamental criteria that define "meanings" may in considerable part be resolved in terms of the successful applicability of other tests, and it is thus that our yet-to-be-worked-out theory of testing will provide apparatus for the resolution of meanings realized in assertional utterance.

Once we have put in place an inventory of basic test-kinds-- a task undertaken in the second part of this treatise--our "bridge principle" can be given systematic semantic application along these lines: A test is applicable ("qualified") only if the sets of occasions of application are non-empty. We can often (not always) represent the qualification conditions by saying that certain other basic tests are successfully applicable, thus gaining test-theoretically formulated conditions of application for tests. These conditions of success of an assertion are indicated by the realization of "uses" ("meanings") in the assertion. We can identify and characterize those uses (no matter the expressions employed) with our test-theoretically formulated conditions of application for basic tests. So, for example, the assertional realization of *referring* uses indicates the existence of an individual as a condition for testing; otherwise put, *referring* indicates that a criterion for the statement "presupposes" (the successful applicability of) a basic individuation procedure; referring uses in their several sorts may also indicate the successful applicability of identifying, ordering, denominating and other procedures. The assertional realization of "denoting[22] uses", e.g. those conveyed by "The (each, every, any) american president", indicate the selective or the exhaustive successful applicability of a test for being a U.S. president upon the successful applicability of a delimitation procedure. Again, one use requires *complementation* by another if the indication by the one use of the successful applicability of one test requires the complementary indication of the successful applicability of some other test by the realization of some other "concatenating" use (e.g. indication of the applicability of a predicative test to "something" requires complementation with an indication of that "something", perhaps by way of the realization of a referring or a denoting use; again, a condition indicated by "The" requires the complementary indication of the successful selective applicability of a "predicative" test.) Here then a general principle of *"semantic valence"* well adapted to account for the "unity of the proposition".

A test-theoretic analysis of a use given as realized in assertion carries over to non-assertional utterances.

A test-theoretic analysis of a meaning gained in relation to assertion won't alter when that same meaning is realized in non-assertional "illocutionary context", e.g. in commanding, predicting or wondering-whether. So, for example, the test-theoretical identification of the demonstrative referring use of "this" carries over to cases where we wonder whether (ask if) *this* is blue from model cases where we assert that it is. The test-theoretic condition that we can successfully apply an individuating procedure is a condition of success indicated by the use of "this" realized in *any* illocutionary context.

The method has no application to the semantic analysis of mood--including assertion--itself or to "act-adverbial meaning".

Our method, which speaks of the conditions of success indicatable in assertion, says nothing about what assertion is itself. Obviously, conditions indicated by the realization of mood indicative uses cannot be handled in this way, and the method cannot be applied to the analysis of mood or "illocutionary force". Derivatively, the method cannot be used to say (e.g.) what is distinctive in the *addressing* or *calling* (as against *referring*) use of proper-names. It remains that if we somewhere had an account of *address* taken perhaps from an analysis of mood indicatives, we probably could use our test-theoretic method to say what is special about personal-name addresses. Similarly, test theory has no direct application as a semantics to the analysis of the uses of what contemporary linguists call "sentence adverbs" (our "act-adverbially uses", p. 60). Test-theoretic semantics is accordingly systematically limited: still, the prospect of incorporating, for example, conditionalization into statements gives hope for unsystematic borrowings, back and forward, between the indicated conditionalization of utterance and conditional statements properly taken, e.g. the lapsing of a condition that may sometimes void the issuance of an assertion may in other circumstances eventuate in a an asserted statement's being neither true nor false..

Semiology is not our chief concern in this treatise. It remains that concepts are essentially to be realized in utterance, and we shall, in what follows, regularly scout semantic applications alongside of our examination of various kinds of test.

Digression on "Sinn": A tour of semantics from Frege to Frege.

Our semiology is "fregean". I believe that the range of "meaning" accessible to the "criterial" method sketched above coincides with what is covered by Frege's *Sinn*. Those meanings which Frege disparaged as "*Farbung*" fall pretty much within the field covered by my act-adverbial uses and his "forces" are my mood indications; these *genres* of meaning, disqualified as *Sinn*, are also not directly accessible to the methods of test-theory. There is one evident but vanishing difference between Frege's approach to *Sinn* and my proposals for semiology:

Frege spoke about *Sinn* in relation to expressions, e.g. words and symbols. So, it seems, his is a doctrine of word meaning, whereas mine is a doctrine of meaning as realized in utterance or *parole*, by whatever expressions you please. I still believe that my meanings or "uses" lie very close to Frege's *Sinne*, and that my approach might have been welcomed by Frege for enabling him to improve his explanations of such "token reflexive" words as "this" (which he contemned) and proper names properly taken.

I wish in this attachment to persuade interested readers that my semiology is indeed authentically fregean, in the hope that the comparison will illuminate both doctrines and help my case. But then it turns out, for reasons that will appear, that Frege's writings have stood precedent for most of the competing semantics that have crowded contemporary philosophy, including systematic paraphrase, different kinds of extensional analysis, "model theory" and several styles of easy talk with "truth conditions".

In what now follows, I first (A) give background to Frege's theory of the determination of *Bedeutung* by *Sinn* and list what I take to be the principles of that theory. I then (B) recall the only two passages I know of from Frege's writings in which he offers an explication of *Sinn*. In both passages, the *Sinne* at question are of *Saetze*. In (C) I make a "tour" of several tendencies in contemporary semantics that take inspiration from fregean precedents. All of these save one are, in despite of their precedents, clearly non-fregean in tendency and the one (which descends through Carnap) may not tie in to fregean principles. In (D) I defend *Sinn* against extensionalists. Finally, in (E), I offer an interpretation of Frege's principles along the lines of test-theory.

A. *Background and Principles.* Frege took logic to be a corpus of laws for truth. His monumental reconstitution of the subject stands as the greatest individual contribution ever made to the discovery of laws for the determination of truth-value by truth-values and (more generally) for the determination of *Bedeutung* ("extension") by *Bedeutungen*. But now truths and falsities are conveyed by expressions and, thought Frege, there must be something about those expressions, their *Sinne*, that serves to determine their *Bedeutungen* and whose presence certifies that the laws of logic are true or false thoughts expressed in language. Little more than that was needed in order for Frege to get on with the logician's task of establishing truths having to do with the determination of truth-value by truth-values. Frege said precious little about *Sinn* and certainly had no theory of *Sinne*. Here in digest is pretty much all that I can find.

(a) Frege thought he needed *Sinn* to make sense of identity in a way that would be objective but not merely linguistic (as it was in his early *Begriffschrift*).

(b) He went on to use *Sinne* as "oblique" *Bedeutungen*, to resolve the now familiar "opacity" problem he discovered over the non-substitutability of identicals in ascriptions of belief[23].

It is evident from remarks he made in passing that Frege held *Sinne* to be (c) "objective" and non-psychological. (d) *Sinne* are themselves *Bedeutungen* that uniquely determine other *Bedeutungen*, and are cross-classifiable into objects and functions. (e) Frege regularly spoke of *Sinne* as "belonging" to the *Bedeutung* they determine; this suggests *features*, but misleadingly, I think, since this "belonging" is as propositions ("thoughts") belong to truth-values, and he conspicuously did not use "Eigenschaft" in this connection. (f) Frege once described *Sinn* as that "worin die Art des Gegebenseins enthalten ist" ("in which the manner of being given is contained"), and several times told us that expressions having different *Sinn* convey distinctions in "*Erkenntniswert*". (Much of this condenses in a sentence from "Ueber Sinn u. *Bedeutung*": "It would pertain to an encompassing knowledge of a Bedeutung that we be able to indicate of any given *Sinn* whether it attached to that *Bedeutung*" (Zu einer allseitigen Erkenntnis der Bedeutung wuerde gehoren, dass wir von jedem gegebenen Sinne sogleich angeben koennten, ob er zu ihr gehoere.")

What is always wanted is a one-many determination of *Bedeutung* by *Sinn*. So, for example, "The True" has a *Sinn* that determines The True as *Bedeutung*, but "The True" expresses, for *Sinn*, no thought, as do such sentential names for The True as "2+2=4"[24].

The following "formal" principles for the determination of *Bedeutung* by *Sinn*--all save (ix), which is, just barely, open to exegetical dispute--are fairly evident from what Frege says, often in passing[25].

(i) Every meaningful expression (hereafter, simply "expression") has a *Sinn* and should have only one *Sinn*. We may therefore stipulate the existence of a function, Σ (in the usual not in Frege's sense of "function"), from expressions to *Sinne*.

(ii) The *Sinn* of a complex expression is determined by the *Sinne* of its constituent expressions. There must be rules of semantical concatenation, and we may accordingly stipulate the existence of a "concatenation function", $\mathbf{K}_1$, according to which $\Sigma(E_1\hat{}E_2) = \mathbf{K}_1(\Sigma(E_1), \Sigma(E_2))$.

(iii) Anything may be a *Bedetungen--Gegenstaende* and functions alike, including *Sinne*. On the other hand, lots of things aren't *Sinne*: the sun isn't, nor is The True.

(iv) Expressions may or may not have *Bedeutung*. Frege thought that neither "Odysseus", nor "die am wenigsten konvergente Reihe" did; and "der von der Erde am weitesten entfernte Himmelskorper" may not.

(v) Expressions which do have a *Bedeutung* should have a unique *Bedeutung*.

(vi) The rules of an adequate *Begriffsschrift* assure us that every expression pertaining to that notation has a unique *Bedeutung*. The availability of such a *Begriffsschrift* permits us to stipulate the existence of a *Bedeutung* function, **B**, from expressions to things.

(vii) The *Bedeutung* of a complex expression is uniquely determined by the *Bedeutungen* of the constituent expressions. So here we have another "concatenation function", $\mathbf{K}_2$, such that $\mathbf{B}(E_1\hat{}E1_2) = \mathbf{K}_2(\mathbf{B}(E_1),\mathbf{B}(E_2))$. (E.g. **B**("It is not the case"^"Frege was German")= $\mathbf{K}_2$(**B**("It is not the case"), **B**("Frege was German") = $\mathbf{K}_2$(**Neg** (), The True) = The False.)

(viii) If an expression, E, lacks a *Bedeutung*, then so too does any complex expression, in which E occurs as a constituent. If $\mathbf{B}(E_j)$ is undefined, then so too are $\mathbf{B}(E_i\hat{}E_j)$ and $\mathbf{B}(E_j\hat{}E_k)$, for all expressions E_i and E_k.

(ix) Every *Sinn* determines at most a single *Bedeutung*, for any expression that has that *Sinn*. There is therefore a (partial) function, **S**, from *Sinne* to things. (The above mentioned dispute will be over whether Frege may instead hold that a *Sinn* is itself a function from *Bedeutungen* to *Bedeutung*.) In an adequate *Begriffsschrift* we try to set things up so that there is a value for every expressible *Sinn*: while **S**(Σ("die am wenigsten konvergente Reihe")) is undefined in ordinary mathematical German, we should in the *Begriffsschrift* be able to express a corresponding but different *Sinn* for which **S** is defined (Nb. The *Begriffsschrift* must sometimes put new *Sinne* in place of old, which it then leaves untended. The *Begriffsschrift* does not always provide exact translations from the German.)

(x) Since different *Sinne* may determine the same *Bedeutung*, there is no inverse function $\mathbf{S}^{-1}$ back from *Bedeutungen* to *Sinne*. Some theorists (most notably, Wittgenstein in his *Tractatus*) have supposed that some *Sinne*--those of "basic names"--are determined by their *Bedeutungen*; Frege, so far as I know, never did.

(xi) Since both the *Sinn* and the *Bedeutung* assigned to an expression might have been differently constituted, neither $\mathbf{K}_1$ nor $\mathbf{K}_2$ have inverses.[26]

B. Frege, in his various writings and correspondence, only twice, to my knowledge, offered explanations to identify what *Sinne* are. Both explanations are exceedingly "problematic". In both passages Frege was speaking of sentence-*Sinne* or "thoughts". On p. 50 at #32 of the *Grundgesetze*, Frege, after registering a claim that his primitive function-names and all allowable combinations thereof have *Bedeutung*, goes on to say that all properly assembled signs also have *Sinn*. In particular, "Every such truth-value name expresses a *Sinn*--a thought. Our stipulations determine under what conditions the sentence "bedeutet" The True. The *Sinn* of the name--the thought--is that those conditions are fulfilled." Taken literally this can't be what Frege meant: "that those conditions are fulfilled" would normally be understood to describe what it is to assert a *Sinn* and not the asserted *Sinn* itself. There is another deeper difficulty here too: Reading along, one could think that Frege was describing the determination of *Bedeutung* by *Bedeutungen*, not the determination of *Bedeutung*--in this case truth-values--by *Sinn*. We return to this below.

Frege's second explication of the sentence-*Sinne* is in a letter to Husserl dated 9 December, 1906 (pp. 70f of the Kaal translation of Frege's correspondence, Blackwell, Chicago, 1980).

> It seems to me that an objective criterion is necessary for recognizing a thought again as the same, for without it logical analysis is impossible. Now it seems to me that the only possible means of deciding whether proposition[27] A expresses

> the same thought as proposition B is the following, and here I assume that neither of the two propositions contains a logically self-evident component part in its sense. If *both* the assumption that the content of A is false and that of B true *and* the assumption that the content of A is true and that of B false lead to a logical contradiction, and if this can be established without knowing whether the content of A or B is true or false, and without requiring other than purely logical laws for this purpose, then nothing can belong to the content of A as far as it is capable of being judged true or false, which does not also belong to the content of B; for there would be no reason at all for any such surplus in the content of B, and according to the presupposition above, such a surplus would not be logically self-evident either. In the same way, given our supposition, nothing can belong to the content of B, as far as it is capable of being judged true or false, except what also belongs to the content of A. Thus what is capable of being judged true or false in the contents of A and B is identical and this alone is of concern to logic, and this is what I call the thought expressed by both A and B.

Frege here makes no appeal to logical modality but only to derivation, presumably within his own system of logic. I am a bit troubled to understand how, within that system of logic, Frege, would have distinguished between assuming that a content is true or false and simply assuming (positive or negative) contents. Clearly, however, the main thrust of the passage must be toward the "carnapian" proposal that two thoughts are the same when, derivationally viewed, they are logically equivalent; *viz* two sentences S_1 and S_2 express the same *Sinn* if the sentence S^"$\leftrightarrow$"^S_2 expresses a theorem of logic. This explanation of sentence-*Sinn* is different but no less problematic than the one from the *Grundgesetze*, and we shall return to it anon.

C. I now go on to consider a few tendencies of contemporary semantics that derive from Frege's practice but depart from his principles.

(a) *The roots of extensional semantics.* Frege's theory provides for a double determination of the *Bedeutung* of a complex expression (chiefly the truth of true sentences), both by the *Sinne* of its constituent expressions and by their *Bedeutungen*. Isn't one determination enough? Frege thought not; but others, following Frege's lead, have thought so.

The spectacular success of Frege's logic, which is a theory of the determination of *Bedeutung* by *Bedeutungen*, has encouraged several influential philosophers to repose all trust in extensional semantics. Frege's work has stood precedent for a doctrine that is patently not his own.

(b) *On to "truth-conditions".* Everyone's precedent for "truth-conditional semantics" is Frege's exemplary explanations of molecular propositions as "truth-functions" of constituent propositions: *if p then q* is false when *p* is true and *q* is false, otherwise it is true. This part of semantics, in Frege and his successors, is so far still "extensional". The term "truth conditions", which might be picked up from the passage we quoted from the *Grundgetzetze* has, however, led people off in several directions[28]. Truth conditions have been variously taken as (i) paraphrases of what the conditions are for, (ii) as actual conditions of the world in which a statement is true and (iii) as possible conditions under which a statement would be true. All of this finds precedents in Frege; none of it is satisfying, as I shall now argue.

(c) *Paraphrase.* A "canonical paraphrase" (Quine's term) or other "said-thing" is a stylized reformulation, preferably within the notation of a first-order logic of whose formulas, presumably, the truth-conditions are well-understood. Frege, for example, proposed to treat aristotelian "A"-propositions as universal hypotheticals. All S is P is true just in cases where, for every x, if that x is S then that x is P. Such proposals are plausible in degree that the paraphrase approximates to the usual formulation. We shall find occasion both to cheer and to cat-call Frege's particular proposals (in Chapter 4, #'s 3 & 4, p. 392, Chapter 14, #7, 419). Taken all by itself, the method chokes on a stutter: "The truth-conditions for the statement that the book is blue is that the book is blue" or (to stretch things out) "The truth conditions for the statement that Mary spoke is that there is an episode that is an episode of speaking and is done by Mary". If the analysis is a paraphrase of the *analyzandum*, we are but a small misstep away from taking every said-thing as its own analysis.

(d) *Conditions of the world.* Any systematic use of "truth conditions" must be within a *theory* that provides means for specifying the conditions under which that statement or other said-thing is true. The "boolean" representation of truth functions and of aristotelian "categoricals" is such a theory; again, set-theory may be useful for formulating truth conditions for the propositions of mathematics.

That's an advance, but doesn't take us far enough. Most ordinary statements, surely, must be left underdetermined by the use of any such method. The different statements that Earth is the third planet from the sun and that Mercury is the first planet from the sun are true under the same actual conditions of our solar system.[29]

(e) *Possibilities.* Some readers will feel cheated by the above objection. They will want to take the field of conditions to comprise *possibilities.* Two said-things are of different *Sinn* just in case there is a possible condition in which one is true and the other isn't. So spoke Wittgenstein in the *Tractatus*, probably prompted by the passage we quoted from the *Grundgezetze* and guided by Frege's explanations of truth-functions. The Wittgenstein formula finds a successor in the contemporary use of model theory for the analysis of intensions, a method best known from the writings of Montague. Now, in Frege, truth-functions were from actual sets of actual truth values to truth values and were the *Bedeutungen* of various so-called "logical constants". No "possibilities" here and, so far, no need for *Sinn.* Wittgenstein, however, didn't for a moment suppose that there were such things as truth-values and his "fundamental thought" was that logical constants stand for nothing (*Tractatus*, 4.0312). Wittgenstein took it that every elementary representation is true or false, possibly the one or possibly the other. In imagining a complete assignment of such valuations to elementary representations, we envisage a "truth possibility", and the indicated *Sinn* of an assertion was that it "agreed" or "disagreed" with each such possibility. A wittgensteinian *Sinn*, formally conceived, is not, as in Frege, an argument to a function from *Sinn* to *Bedeutung*, but rather itself a function from *Bedeutungen* to *Bedeutung*. This idea of *Sinne* as functions, when used as a method of semantic analysis, is hard pressed to distinguish among logically equivalent *Sinne*--a point to which we return in connection with another offshoot of Frege. The greater question is whether, in appealing to *possibilities* as *Bedeutungen*, it avoids Frege's *Sinn.* Possibilities are perilous commodities. Even Leibniz (in his letter of July 14, 1686 to Arnauld) allowed that they exist only in conception, which certainly at least suggests that they are *Sinn*-dependent. The model theoretic approach to "intensions" so far seems less frugal than its fregean inspiration.

(f) *Logical invariance.* A last semantic theory stemming from Frege and adapted by Carnap is suggested by the passage we quoted from Frege's letter to Husserl. Two sentences have the same *Sinn* or express the same thought if those thoughts are logically equivalent, that fact to be determined either by the derivation of a bi-conditional or by some extension of Wittgenstein's "semantic method" of truth-possibilities. Other expressions have the same *Sinn* when interchanging them in sentence contexts leaves the *Sinn* of those sentences unchanged or logically equivalent. Frege's own formulation of the criterion was carefully hedged, probably to avoid the awkward consequence that all the theorems of logic and arithmetic express the same thought. These "analytic" *Sinne* are conspicuously untended--fatal exemptions in view of the fact that it is sentences having just those *Sinne* we need in order to identify other *Sinne*. It is more to our point to notice that this criterion for sameness of *Sinn* seems to be at odds with those fregean principles we adduced. The two sentences, "Russell = Pope Pius X → Frege = Husserl" and "Frege ≠ Husserl → Russell ≠ Pope Pius X" are logically equivalent. (The biconditional joining them is an instance of the law of contraposition, hence logically derivable.) But surely that means only that the two propositional functions, $x=y \rightarrow u=v$ and $u \neq v \rightarrow x \neq y$ have the same values for the same arguments; but that, by Frege's principles, so far says only that the two function names have the same *Bedeutung* and nothing yet about *Sinn*.

D. *In defense of Sinn*. Extensionalists will rejoice at Frege's quandary. They will believe that Frege's appeal to *Sinn* was unavailing and unnecessary. As for being unnecessary: Some latter-day fregeans and perhaps an earlier Frege could argue that we don't need *Sinn* for identity. "The Morning Star" and "The Evening Star", it may be urged, indicate a determination of Venus from different component *Bedeutungen*. The fact that Venus is thus variously constituted is not obvious and is worth knowing. The "extensional" solution would not have satisfied the mature Frege, but for a rather subtle reason we'll return to below.

Unavailing: It might be argued that *Sinne* cannot at once be objective *Erkentnisswerte* and serviceable for the clarification of every opaque context: one may fail to know an object for what it is relative to one formulation but not to another, even when these two formulations have the same objective *Erkentnisswert*, perhaps because one of the formulations happens to be, so to speak, accidentally obscure. I think Frege would have answered (convincingly or no) that either the example establishes a difference in *Erkentnisswert* and so in *Sinn* or it reflects a merely subjective incapacity of the subject--perhaps that he doesn't even understand German.

An argument in favor of determination by Sinn. Frege's main point is always the same: something about the words we use must show us the way to Truth and other referents. That can only be their meanings, their "Art des Gegebenseins". The extensional phase of semantics, a theory of the determination of *Bedeutung* by *Bedeutungen*, ultimately drives things back to simple names--if not to "morning" and "star" then to things still "smaller". The theory would be completely undetermined unless we knew what things answered to what simple names. This "contact" is a kind of knowledge (Appendix C, pp. 188ff.) and an irreducible kind of *Sinn*; it leads to a *Bedeutung*. Couldn't we get back to the *Sinn* from the *Bedeutung*, and thereby do without the *Sinn* altogether? (I believe this was the intended conclusion of Russell's notorious argument from "Gray's Elegy" in "On Denoting".) An argument against that is that the very same *Bedeutung* could be differently named perhaps by a complex expression having a *Sinn* different from that of any simple name; if that's right, there can be no inverse determination of *Sinn* by *Bedeutung* (Principle x). Alternatively, we must hold (as perhaps did Wittgenstein in the *Tractatus*) that there are objects that can be known *only* by "contact". These *Gegenstaende* fix the whole system of representation and cannot be conceived except as simple and necessarily existing. It's a pretty theory, but not a plausible one. Consider that, if there were these simple objects, they would presumably stand in relationship to their kin, and so we should be able to refer to them "descriptively" by mention of those relationships; so, again , the same *Bedeutung* will have plural *Sinn*-determinations.

"Context" makes no escape from Sinn. Let's muster. Everyone appreciates Frege's extensional semantics--his account of the determination of *Bedeutung* by *Bedeutungen.* Since these extensions are of words or other expressions, there must, held Frege, be a determination by *Sinn*; he cited identity as a trenchant case. The opposition claims to gain the equivalent of *Sinn* by aggregations of extensions. We argued that this will not do because the same *Bedeutung* may be determined by both simple and complex *Sinne*. To obviate this difficulty, the extensionalist must find a unique path back to simple *Sinne*.

Fregeans from Wittgenstein on and perhaps an early Frege might claim to find this path in the use of Frege's discarded principle of "Context": "Man muss die Worter im Saetze betrachten, wenn man nach ihrer Bedeutung fragt." Wittgenstein seems to have had something like this in mind when he maintained that expressions are all of them sentence variables, that have *Bedeutung* only in the context of *Saetze*, that to be fixed by what they contribute to the determination of the *Sinn*--the "truth conditions"--of *Saetze* (*Tractatus* 3.3-3.317). Wittgenstein also said that he conceived the *Satz* as a function of its constituent expressions (3.318). Wittgenstein could perhaps go both ways because he also had simple names of simple objects to fall back upon, and for these there *was* a reverse determination of the "Ausdruck" by its *Bedeutung*.

Is the proposal plausible; how would Frege respond? Allow that expressions have *Bedeutung* and/or *Sinn* only because they do contribute to the determination of the truth-values of sentences. That does not enable us to fix what that *Bedeutung* and/or *Sinn* is merely by a consideration of that contribution. Actually Frege never used the principle of context save as a device for fixing *Bedeutungen* by abstracting from *equations*, and equations were the only contexts he consulted in connection with the principle. Now enter the anticipated but postponed subtlety in Frege's contention that we need *Sinn* to make sense of identity itself. Where we may perhaps extract a *Bedeutung* for "The Morning Star" from the true equation "The Morning Star = Venus", that doesn't pull out the indicated constituents of the *Bedeutung* of "The Morning Star", which Frege would have built up from functions. Formally, in Frege's system, functions don't identify, and there is no way of applying the method to them. But couldn't we abstract "extensions of functions" as surrogate *Bedeutungen* and make do with them? That, surely, can't much console the extensionalist, since the abstraction function itself takes other functions as arguments, and we cannot extract their extensions from sentential context without patent circling. Something in our expressions must determine the extensions of some simple names at least if we are ever to get rolling with our theory of the determination of extensions by extensions. We call that *Sinn*.

E. *An interpretation: Sinne are indications of tests.* Although my theory of judgment is not Frege's, I follow him in seeing a need for something like what he called "Sinn". I have an easy interpretation that conforms to his principles, namely that the *Sinn* of an assertionally employed sentence is to indicate the pair of fundamental criteria for the statement that would be produced by the assertion were it successful. The *Sinne* of constituent expressions are other tests that may "contribute" to such criteria. This gives a fairly uniform sense to

"Sinn"; it aligns with our idea that any *Sinn* is the sort of thing that can contribute to the *Sinn* of a *Satz*; finally (as I shall argue), it also meets the important constraint that the *Bedeutung* (if any) of an expression having a *Sinn* must contribute to the determination of the *Bedeutung* (if any) of any complex expression (chiefly sentences) containing the first as a constituent.

I get to this interpretation by keying on Frege's descriptions of *Sinn* as "Erkentnisswert" and as "Eine Art des Gegebenseins". I simply take the *Sinn* as an indication of how we should set about to establish the existence of a *Bedeutung*. The *Sinn* "leads us" to its *Bedeutung*, if it has one. We are to look for an object or a function that meets *these* specifications. The expression of the *Sinn* records a presumption of knowledge that there is something which can be reached in this way. When we go on to concatenate one expression having *Sinn* with another, we are supposed to be following a lead from a found *Bedeutung* to another in prospect. *Sinne* so "associate" that the indicated way of establishing the existence of a *Bedeutung* for the complex Sinn could not be applied if any of the constituent *Sinne* lacked *Bedeutung*. In this way, the existence of a *Bedeutung* for a constituent *Sinn* becomes a "presupposition" for going on to establish the existence of a *Bedeutung* for the complex *Sinn*: each "constituent *Bedeutung*" is a station we must pass through, if we are to get to our final destination along *this* way.

A full set of directions flows from the meaning or "Thought" conveyed by an assertionally employed *Satz* . Here the destination or *Bedeutung* is a truth-value, as Frege argued. So the *Sinn* of a sentence indicates a way of establishing the existence of a truth-value. But surely The True and False *must* exist; indeed, they are Frege's only primitively named objects. so it remains for the Thought only to indicate a way to get there. The way is open if all the stations along the way exist. Less figuratively, the satisfaction of all the indicated conditions encapsulated in the rule that every constituent *Sinn* has a *Bedeutung*, assures us that the Thought also does. The investigation must lead somewhere. The speaker wants it to be The True, and *that* (if I may now go beyond Frege on this one important point) is the (non-*Sinnvoll*) meaning of the judgment-sign. Speaker may be wrong of course. I take it from this that the expression of a complete thought must somehow indicate how we would establish truth-value *either* way.

This exposition of *Sinn* is figurative and vague. Still, I believe that these murmurings can be reduced to a workable system of "intensional logic" by following a line of thought that can be drawn from our theory of assertion and statements.

A sentence's *Sinn* is to be identified with that which enables that sentence to be employed in "meaningful" assertion (*Urteil* or *Bejahung*). The meaning of the assertion consists in the indication of conditions of success. The assertion may succeed or fail; if it succeeds it "gets somewhere"-- to a *Bedeutung*, if you wish. It succeeds just in case all the indicated conditions are satisfied. Chief among the indicated conditions are ones for the applicability of two criterial tests. We now confine ourselves to those features of the sentence that enable it to indicate those conditions. If either of the two criterial tests is *successfully* applicable, then the assertion gets to a truth-value as the *Bedeutung* of the sentence. (We depart from Frege in allowing assertion to get somewhere even if neither test is successfully applicable, provided they are applicable at all.) When expressions constituent of a sentence assist the indication of conditions for the applicability of criterial tests, then those expressions also have *Sinn*. We may identify that *Sinn* with the conditions they separately indicate. My theory will hold that these conditions for the applicability of criterial tests (most of them, anyway) can be given in terms of the applicability and successful applicability of other tests drawn from a small inventory of test kinds. Normally, it is the *Sinn* of a constituent expression to indicate the *successful* applicability of some such test, which then becomes a condition for the (mere) applicability of one or other of the criterial tests. If such a condition fails, then the expression having that *Sinn* lacks *Bedeutung*. But then, if that condition fails, so also must a condition for the applicability of a criterial test; in that event the assertion must fail and the sentence lack *Bedeutung*. So, to repeat, the *Sinne* of one's words or other expressions can be explained by a listing of conditions on testing made evident by the use of those expressions in assertoric context. Condition-indicating expressions have *Sinn* no matter what; however, if any of the indicated conditions are not satisfied, then the expression indicating that condition lacks a *Bedeutung*, in which event one or the other of the tests cannot be applied, and no statement, true, false or neither, is produced. If, on the other hand, all of the indicated conditions are satisfied, the assertion both produces a statement and indicates by what procedures that statement would be verified or falsified. The *Sinn* of the sentence (*Satz*) by which one enacts a successful assertion (*Urteil*) is, as "Art des

Gegebenseins", to indicate a way to find truth and falsity. The way is presented even if the presented way cannot be followed, because some of the indicated conditions are not satisfied, i.e., because one of the expressions lacks a *Bedeutung*.

This "account" is "fregean". (The reader may readily confirm that it meets the formal conditions set out on pp. 303ff.) First, expressions have unique *Sinn* and those *Sinne* should but may sometimes fail to determine unique *Bedeutungen* for those expressions.

Second, the *Sinn* of an expression somehow contributes to the determination of the truth values of *Saetze*, in which the expression occurs *viz* it is the *Sinn* of an expression to indicate conditions that must be satisfied if we are to apply procedures for establishing the truth or falsity of judgments.

Third, *Sinne* "concatenate" in that one expressed *Sinn* indicates a condition for proceeding to the next stage indicated by the expression of a complex *Sinn*.

Fourth, what corresponds to the *Bedeutung* of a *Satz* may be a truth-value.

Fifth, every *Sinn* indicates one way to find a *Bedeutung*, but the same *Bedeutung* might have been located along other routes.

Sixth, *Bedeutung*-failure is transmitted from constituents to complexes: if any expression expressing a *Sinn* lacks a *Bedeutung* because an indicated condition is not satisfied, so must any complex expression expressing a complex *Sinn* indicating that same lapsed condition. Specifically, if the complex expression is a *Satz* it must lack truth-value.

Seventh, any expression of a *Sinn* in judgment creates a presumption that speaker knows that the indicated conditions are satisfied; *Bedeutung*-failure witnesses misinformation on the speaker's part.

Eighth, a successful assertion (*Urteil*) is to the effect that such and such is so; *viz*, that one will reach Truth as the *Satz-Bedeutung* by following this route, which is to say that the verification test for the produced statement can be successfully applied. For this to be so, something in the *Sinn* of the *Satz* must show the way to the Truth as a prospective

Bedeutung. But speaker may be wrong about this too; only now we do not want to say that no statement is produced, but rather that the produced statement is false. So something in the *Sinn* of the *Satz* must show the way to Falsity as an alternative prospective Bedeutung (and *we* allow for *neither*).

5. ON THE ASCRIPTION OF TRUTH AND FALSITY TO STATEMENTS.

Statements are true or false when their verifying and falsifying fundamental criteria are respectively successfully applicable.

We have hypothesized that statements, as we have explained them, are the most fundamental vehicles for truth and falsity. Evidently, a statement is true, test-theoretically said, when its fundamental verification criterion is successfully applicable and false when its fundamental falsification criterion is successfully applicable. Formally: A Statement is true when the set of its occasions for successful verifiability is non-empty and false when the set of its occasions for successful falsifiability is non-empty. Now statements cannot be both true and false by the exclusion rule of pp. 290; they may be neither true nor false when neither of their fundamental criteria is successfully applicable.

These formulas do not carry directly carry over as explanations of the truth values of other constative products or of states of mind

I think that what I've just said sums up the whole truth about the truth and falsity of statements. It leaves much to be said, however, about the "truth-values" of other things, for which I have no general formula. Our statements are not the only bearers of truth-value, and we must give a thought in this connection to such other constative products as generalizations and conditionals. As I have just conceded, I have no general formula, nor am I ready to stake much on specific explanation of what it is for these other products of language to be true and false. Our leading hypothesis about the fundamental place of statements does indeed create a continuing presumption that we could and should explain the truth and the falsity of other products in relation to the truth and falsity of statements. Those explanation will of course vary with the product. I am confident, for example, that a generalization is true only if none of the statements it entails is false, and that a broad range of conditionals are true only if such statements as may be indicated in the protasis clause formulate what would be reasons for believing that such statements as would be formulated in the apodasis clause are true. There are also various "states of mind", chiefly belief, that stand qualified as bearers of truth-value. I lack formulas for saying what it is for such articles to be true and false. However, since beliefs are identified in relation to formulations (pp. 26, 28) and are called "true" or "false" according to the truth or falsity of those formulations, I can't doubt that formulas that may someday emerge for the truth values of beliefs would be framed in terms of constative truth and falsity.

> If we've said too little about truth-value, we've so far said almost nothing about the general sense of "true". I am persuaded that "true" is just the right English word to use for true shots, friends, and lines, as well as statements, generalizations and beliefs. Things are rightly said to be "true" in these various connections, I submit, only if they are also *true-to* something or other. I've opined that truth-value truths are characteristically true-to the facts; such truths are "fact-fitters". However, as will presently be argued, I don't think that this way of talking opens onto a definition of truth-value *truth* as correspondence to fact.

Our formulas are schematic ascriptions of truth and falsity to statements, not "analyses" of such ascriptions.

Our formulas for the truth and falsity of statements are, simply if schematically said, test-theoretically formulated ascriptions of truth and falsity to statements. We should notice and keep in mind that these formulas, which speak of the non-emptiness of sets of occasions, are, on their face, not themselves statements but something more in the order of existentials or negative generalizations. It is also to be noticed that these formulas do not provide "analyses" of what it is for statements to be true or false. They are, I repeat, schematic ascriptions of truth and falsity to statements or (if you wish) test-theoretic "paraphrases" of such ascriptions.

The desiderated "analyses" would be test-theoretic representations of these ascriptions.

These several reservations come together when we stop to consider what an "analysis" of the truth and falsity of statements would be. If our formulas were themselves (second-order) statements about the truth and falsity of statements, then, by the terms of our theory, the desiderated analyses would consist in a test-theoretic representation of any such statement as a pair of fundamental criteria, presumably trading on the fundamental criteria for their subject statements. Now the consideration that these ascriptions of truth value to statements themselves appear to be non-statemental existentials certainly depresses the prospects for completing such an "analysis". Still it's reasonable to expect that these existentials are also "governed" by criteria. It will be instructive to consider what these tests might be, chiefly in relation to the fundamental criteria for their subject statements. So, with an eye for "analysis" and with apologies for being repetitive: I assume that there are indeed these testable ascriptions of truth value to statements. We assume that these "ascriptions" are distinct from their subject statements and that they are governed by criterial analogues for verification and (more to be questioned) for falsification. With tongue only a bit in cheek, I ask what these criterial analogues might be.

Criterial analogues for ascriptions of truth values to statements presuppose the assertibility of those subject statements, but not conversely.

Now in ascribing truth or falsity to a statement *s*, we imply that *s* exists and has somehow been brought to consideration. The "bringing to consideration" might be by way of assertion itself--we shall presently suggest a stronger condition. Minimally, the criteria indicated for an ascription of truth-value to *s* presupposes the (successful) assertibility of *s*. The criteria for *s*, by contrast, presuppose nothing in regard to truth-value ascriptions. That establishes an important difference between statements generically taken and ascriptions of truth-value to those statements[30].

I digress to observe that the presupposition thesis also answers one objection to the opinion, which I share, that reference failure results in the non-existence of a statement. The argument assumes, what we deny, that an ascription of truth (s_2) to a statement s_1 is identical with s_1, and then proceeds to observe that if reference failure resulted in there being no statement s_1, true, false or otherwise, then s_2 would be false; but s_2 is s_1, so s_1 does have a truth-value after all. The short answer to this is that the non-existence of s_1 transmits to s_2; *viz*, the failure of the assertion of s_1 results in the inevitable failure of any utterance that would produce an ascription of truth-value to what is left unproduced in the originally faulted assertion.

This first half-step toward a criterial analysis of truth-value ascriptions further depresses hope for completing the task. The falsification of a truth-value ascription would presumably require us to canvass the occasions for verifying a subject-statement. That is an unlikely prospect even when the occasions in question are "at a time" (The "individuation" of occasions requires the tracking of bodies into those places. See Chapt. 12, #4, pp. 406f.) More seriously, since our general theory of testing affords no categorial analysis of anything in particular, including assertions and other "considerations" of statements, we still have no means to represent what the criterial analogue for truth-value ascriptions is supposed to "presuppose".

We hypothesize that criterial analogues for verifying truth-value ascriptions could not be applied unless the criteria for their subject statements were actually applied.

Is there then nothing we can eke out of test-theory for formulating this demand that truth-value ascriptions entail "consideration" of their subject statements? Well, the consideration would be most palpably registered in actual attempts to verify and to falsify the subject statement. Otherwise, an analogue for verifying a truth-value ascription implies the actual attempted testing of the subject statement. The stipulation proves advantageous. Some of you may find it implausible, on grounds that we have frequent occasion to asseverate that this or that untested statement is true or false. Perhaps, but I'm not so sure. Certainly someone may say, "Well, I guess it's true that Karachi numbers more than 8,000,000 people" without conveying any implication that the statement had been tested. In contrast, a declaration, perhaps hesitantly entered, that the statement that Karachi numbers more than 8,000,000 people is indeed true does seem to me to suggest that a count had been taken. It looks to me to be anyway possible that the exceptions to our experimental stipulation are ones in which "true" occurs merely as an assertion sign and not as a predicate. (See pp. 94f. and pp. 324f just ahead.) I see no principled way of establishing that that is always so. However, since it does seem to me likely true, I propose, as an hypothesis, this further stipulation, that criterial analogues for verifying truth-value ascriptions would not be qualified or applicable unless the verification and falsification procedures for the subject statement are sometime actually applied.

Criterial analogues for truth ascriptions are successfully applicable only if verification criteria for subject statements are successfully applicable and, similarly, criterial analogues for falsity ascriptions are successfully applicable only if falsification criteria for subject statements are successfully applicable.

My next (and last) provision in regard to criterial analogues for truth-value ascriptions is less "problematic" than was the first "consideration" provision. A truth ascription can't be true unless its subject statement is; similarly, a falsity ascription can't be true unless its subject statement is false. So, in our presentation, the successful applicability of the criteria for either truth-value ascription entails the successful applicability of the verification or of the falsification criterion for the subject statement.

That is a "minimal" constraint. I once thought we should strengthen it to demand that criterial analogues for truth-value ascriptions should be included in one or the other of the fundamental criteria of subject statements. I now suspect that that addition would be inconsistent with our first stipulation in regard to actual application, for we would be in danger of holding that a criterial analogue would not be qualified unless actually applied, in contravention of our "rule of inference" for testing (pp. 279-281). The weaker condition I have opted for allows that truth-value ascriptions, though fixed in reference to the statements they are "about", may still be tested "to the side of" their subject statements.

Criterial analogues for truth-value ascriptions are "restrictive upon" the criteria for their subject statements.

Criterial analogues for truth-value ascriptions must, anyway, be "with reference to" occasions of actual application of the criteria of subject statements. That enforces narrow restrictions on the occasions for testing truth-value ascriptions. Those restrictions also scotch the addition of two further clauses to our characterization of the analogues, which I once thought would hold good, namely that the always unsuccessful applicability of the verification test for the subject statement implies the sometime successful applicability of the falsification test for the truth-ascription and that the occasions for successfully falsifying the subject statement are included among the occasions for successfully falsifying the truth-ascription. It seems ever more likely that there are no analogues for falsifying truth-value ascriptions.

Another way? We have so far stipulated more than we have earned in our obstructed effort to elicit criterial analogues for truth-value ascriptions to statements. In the systematic developments of Part II we shall introduce an order of Π-tests that, among several other contributions to criteria, figure as procedures for verifying predicative statements. Π-tests are test-theoretically classifiable across several lines. In Part III we shall consider how "categories" of objects can be fixed by specifications of "criterial bases". If statements were an authentic category of objects, then we could presumably deal with their truth-values by introducing Π-tests, under appropriate test-theoretic classifications, for those determinations. But, as it turns out, statements are not qualified as objects properly taken, but are rather what I shall call "constructions" from assertion-kinds, which in their turn are constructions from persons which are, finally, material bodies. The seeming features of these several constructions are fixed by what I call "analogues" of Π-tests. Analogues of Π-tests for determinations of truth-value are in principle resolvable into tests that pertain to the categorial basis for bodies. These would include procedures for establishing the existence, individuality, separateness, identity of bodies and for delimiting sets of bodies and Π-tests for the activities of persons. However, I do not actually develop classifications of those procedures to a point where they could contribute to a theory of truth-value and I do not know how much farther we could advance this inquiry into truth-value along that route.

I have now eked out what I can in the way of stipulations on the criterial analogues for verifying truth-value ascriptions. There's precious little here, I allow, but still enough to generate two easy but satisfying "theorems". Our stipulations yield, first, a confirming if weakened version of "Tarski's Disquotation Principle" and, second, an easy resolution of the Epimenides Paradox, which so much controlled Tarski's own celebrated visitations to the topic of truth and falsity.

"Disquotation": If s-is-true is true, so must s be; if s is true, s-is-false cannot be true; if s-is-false is true, s is false; if s is false, s-is-true cannot be true; s may be true or false and the corresponding truth-value ascriptions be neither true nor false.

The familiar but inept formulation of Tarski's Disquotation Principle is: *"p" is true if and only if p* This tries to say that a truth ascription, *s*, to a statement s_1 is true if and only if s_1 is true. Something like that comes out of the account we have just given of truth-value ascriptions. Ascriptions of truth value and their subject statements are normally distinct, but they are "weakly equivalent". If *s is true* is successfully testable, so must *s* be; so, if the truth ascription is true, so must the subject statement be. Reversing it, if *s* is successfully verifiable, *s is true* may or may not be; so *s* may be true and *s is true* not be; but anyway, if *s* is true, it's falsification test is not successfully applicable, so we know that *s is false* cannot be true. Similarly, if *s is false* is true, *s* must be false; but not quite conversely for the same reason just given. We have found no criterial analogues for falsifying truth-value ascriptions; however, it is still possible that they *be* false and their subject statements be neither true nor false.

Digression on "disquotation".

The three "best and smartest" modern writers on truth-value--Frege, Ramsey and Tarski--all gave great but differing credits to disquotation. Tarski cited it as a criterion of adequacy for any acceptable *truth* definition; Frege used it as an argument against the possibility of providing a definition of *truth* and Ramsey may have used it to support his once held view that "true" is not a proper predicable of anything. Ramsey gave the disquotation principle considerable attention in his posthumously published chapters on truth and judgement, where he opined that this principle came as close as anyone could hope to encapsulating the essence of our conception of *truth* as applied to judgements and beliefs (*op. cit.*, Chapt. 1, *passim*). I don't exactly disagree, though my enthusiasm for the principle is as qualified as is my formulation of it. Tarski, while he certainly did not believe that the disquotation principle sufficed for the definition of *truth*, held that every instance of it should be generated by a proper model-theoretic definition of truth. I am pleased to agree with that, again subject to the qualifications attaching to my "weak equivalence". Frege's mention of the principle in his paper "The Thought" was obscure but interesting (*ibid*). He invoked it as evidence in support of his opinion that *truth*, strictly speaking, is undefinable. His reasoning, so far as I can make it out, seemed to be this: The principle gives an unexceptionable statement of truth conditions for a thought that another thought is true. If this rule doesn't define truth, no other would. But the condition is obviously fugitive and unavailing. More generally said, we can't suppose that just any old bi-conditional will suffice for purposes of definition. Amplifying on the point I think Frege is making in the passage under consideration, I think one might reasonably require that the *definiens* side of a definitional bi-conditional should predicate something of the free variable that also occurs on the *definiendum* side. I don't think that any of the familiar versions of the disquotation principle meets that requirement. I do think that the aristotelian formula, to which contemporary writers often repair when trying to explain the sense of disquotation, can be made to satisfy the mentioned rule of definition: we say that a judgement consisting of elements { e_i } is true when the referents of e_i's are (e.g.) "connected" (see pp. 116f.). Frege, anyway, seemed to hold that the principle, which he and Ramsey and Tarski all held to be of the essence of truth, did not suffice to define that notion in its application to propositions.

Our hypothesis that the criterial analogues for verifying truth-value ascriptions are not qualified unless the corresponding statement criteria are actually applied yields an "easy" solution to the Epimenides Paradox.

Suppose we have a truth-value ascription s_2 ascribing either truth or falsity to a truth-vehicle s_1; s_2 is itself a truth-vehicle and in principle either true, false or neither. Suppose now that s_2 is an ascription of falsity to s_1 and that $s_2 = s_1$. s_1 is to the effect that s_1 is false. If s_1 is true, then s_1 is false, which is impossible. If s_1 is false, *viz* it is false that s_1 is false, then s_1 must be true or neither true nor false; if s_1 is neither true nor false, then, *per impossible*, it would be false that s_1 is false. Now, one may readily infer from our principle that statements are not both successfully verifiable and successfully falsifiable that statements cannot be at one and the same time any two of the three options--true, false and neither. A like rule should carry over to truth-value ascriptions, whence one may conclude that there are no such items as the above defined s_1. Most readers familiar with The Epimenides and other "insolubilia" will find that conclusion too quick to stand as a "solution". The same result emerges more "positively" from our hypothesis or stipulation that truth-value ascriptions entail the actual application of the criteria for their subject statements. If the criterial analogue for a truth-value ascription $s(s_1)$ is qualified, then the criteria for s_1 are applied; so, from the assumption that a test is qualified we have inferred the actual application of two tests, hence by our "rule of inference" (p. 281), that the applied tests are distinct from those for the criterial analogues; hence that the truth-value ascription s is distinct from its subject statement s_1. No truth-value ascription can be "of" itself.

"Theories of truth, with remarks on Truth as Assessment. Graduate students of philosophy used to be put through seminars On Truth, in which a platoon of "theories" are assembled in file and reviewed. For reasons of professional civility, I am now brought to make a brief visit to the same parade ground.

Coherence Theories.

Coherence theories, recently anyway (from Joachim to Quine), have been major if ill-named players in the field. These doctrines disparage the claims of individual thoughts, beliefs, propositions or statements to be truth-bearers, and finally advocate the thesis that the only truth that can matter is a property of *systems* whose overall coherence can be scrutinized and which are externally confirmable, if at all, only by being brought globally to the world for testing. The doctrine, as it was set out by Bradley, Joachim, Bosanquet and other "idealists"--replete with "denials of the reality of relations" and other questionable theses--is remarkable for the elegant case with which Russell and Ramsey demolished it. Its attractiveness owes to the weakness it finds in the opposition. Coherence theorists aver that supposed individual truth-bearers--propositions or whatever--are comparable only to other such items, and never to the facts that make them true. If, in considering whether my current judgement that the weather is fine is true, I go on to determine whether the facts are as I think, discovering that the weather is or is not fine, I must finally be comparing my first judgement with another, that the facts are or are not as I thought. Ramsey noticed that this argument against a correspondence criterion for the truth of individual judgements depends palpably upon one's mistakenly taking his current judgement about the weather as being a judgement about truth. Judgements can be made about the truth of (other) judgements, which are facts, and the question resolved by "comparing" those facts with (e.g.) the state of the weather; thus I determine that my morning judgement that the weather would be fine this afternoon was false by observing the mismatch of what I said with the thunderstorm raging outside. ("On Truth", pp. 37 ff).

No-content theories.

Ramsey, in the essay we just cited, seemed to plump for some kind of "correspondence theory of truth, and we'll say more about that below. In his earlier "Facts and Propositions", he anyway seemed headed toward the opposed opinion that "true", when used in relation to things said or thought, was sheer redundancy and conveyed no predicative content--an opinion which Tarski following Kotarbinski was later to describe as the "nihilistic approach to the theory of truth"[31]. This doctrine, as I understand it, holds that "is true" used in reference to things said or thought always duplicates the "content" of what is said, *viz* that $s_2 = s_1\text{-}is\text{-}true = s_1$, perhaps because "is true" is itself merely an assertion sign. Now we have noticed that all assertions, and presumably lots of other utterance kinds too, do indeed "certify" or "endorse" or "register a claim to truth", which we recast to read that the assertor undertakes a "commitment" to the truth of the statement produced in his assertion (pp. 87, 94ff.). The "No-content" theorist would no doubt refuse that way of reading the matter, for he presumably believes that there is nothing to be true in the offing other than the assertion or the judgement itself. To say "true" can only be to "Yes"-say what was already said. So my way of registering the use of "true" as an assertion sign does not refute those who maintain that it has no other use in relation to things said or thought. Ramsey easily anticipated replies to other criticisms, e.g. that we can say that s_1 is false without asserting s_1, that we can say that certain statements or a range of statements are true without assertionally identifying them (see Tarski *ibid*, p. 67): To say that something is false, said Ramsey, is simply to assert its denial; again, to say that everything someone said is true is simply to say "For all p, if he asserts p, then p". I can find no counter-example fit to disprove Ramsey's sense of what is evident. But then, as it seems to me, there are three considerations at least that go against what he held. First, one must wonder whether there could be any better way of explaining *not-p* than as a truth function of *p*. Second, Ramsey's own words imply a distinction between an assertible *p* and its assertion. The assertion can be marked down on various grounds--as coarse, inopportune or (most obviously) for producing a falsehood--a possibility latent in easy talk about assertion being a "truth claim" (pp. 94). But then it looks as if there must be something other than the assertion to be said to be true or false. Finally, to scavenge one of Ramsey's own examples, we can say that a chicken believes a caterpillar to be poisonous and that belief be true; but then surely we wouldn't say that this poor chicken also believed that it's *true* that the caterpillar is poisonous, although that surely is something *we* can believe; but then, if we believe something the chicken doesn't, it must be that such and such *is true*. After all these "considerations", perhaps finally the most convincing rebuttal to the No-content theory--maybe the one that finally brought Ramsey to change his mind--is Frege's simple observation that scientists typically are concerned to find out which of several competing propositions is true.

Truth and assessment.

No-content theorists see "true" as a kind of reaffirming "Yes", useful for giving the nod of approval to someone else's assertion. And certainly there can be no doubting that "true" does function as a term of positive assessment. True is a good thing for assertions to be, by and large and for the most part. But then, I urge, truth itself must be something which does give value to assertions. Consider that "yes" itself and "nutritious" too are "evaluative terms", whose meanings differ from that of "true". "True", like "nutritious" , is an "evaluative term" with a meaning much more specific than that of "yes" or "good". The parallels are clear: Things are deemed nutritious if they are good for eating; similarly, things are deemed true if they are good for asserting or believing. Now things couldn't be nutritious unless they had a certain chemistry which interests nutritionists. So too, items could not be specifically good for asserting or believing unless they had qualities that interest "truth theorists". "True" could not have the evaluative use it does without also signifying *truth*. Here is a "fregean" (or, if you prefer, a "geachean") argument to support that conclusion: *p or q* is true only if *p* is true or *q* is true; now if the whole meaning of "true" were to express approval, then a disjunctive approval [*p or q* is true] would be equivalent to a disjunction of approvals; but a disjunction of approvals is no kind of approval at all. Since I believe that *assertion* is indeed a kind of approving and, more specifically, an approval for truth, I am buttressed in my opinion that there must be a specific non-evaluative sense of "true" appropriate, not directly to assertions or judgements, but to something more like Frege's *Gedanken* or my statements. If "true" in this specific understanding did mean something like true to the facts, then we have an immediate explanation of why truths are good for asserting, *viz* that assertion as an "institution" exists for the communication of knowledge of facts. Just as nutritious food is good for eating because it sustains life, so too truths are good for asserting because assertion serves to communicate knowledge of fact. (That's the "usual case": non-nutritious food may be dandy for dieting, and untruths mandatory for inclusion on multiple choice questions.)

"Pragmatism".

The whole plausibility of Wm. James" so-called "pragmatic theory of truth"[32] traces to this consideration that assertions, beliefs, "ideas" (James' preferred term) etc are properly praised for being true. No doubt of it, truth usually "pays", as James put it, and is "utile" and "expedient", but not always, and not everything that "pays" is true, e.g. promptness isn't. One senses that James got it backwards: something said or believed normally pays because it's true, not the reverse. In any case, this kind of "pragmatism" cannot stand as a theory of truth *per se*, as James himself all but acknowledged in various *dicta* and scattered remarks describing truth as being "verification" and as "corresponding to fact".

None of those "theories" we have so far mentioned--Coherence Theories, No-content Assertion Theories and Pragmatism--even begin to tell us what truth *is*, taken as a feature of complex representations of putative fact,. Philosophical theories of what truth is seem to me to be fairly classifiable as "mainline" and "in opposition". Aristotle's explanation (e.g. at "Theta 10 of *The Metaphysics*, see pp. 116f. above) head the mainline. He explained truth as a feature of complex representations that derives from how things otherwise are--from the "real state of the world", if you wish; truth, as Aristotle was wont to put it, is a secondary style of Being.

"Mainline" or "aristotelian" theories of truth explain truth as a feature of things said or thought that owes to how other things are. Others in opposition take truth to be a feature that attaches to thoughts or propositions without reference to how things otherwise are.

The "opposition" position was well formulated by G. E. Moore: "A proposition is constituted by any number of concepts, together with a specific relation between them; and according to the nature of this relation the proposition may be either true or false."[33] The best known version of this way of thinking is Leibniz' celebrated doctrine that true representations are ones in which the predicate is contained in the nature of the subject[34]. Moore, in his statement, went on to say, "What kind of relation makes a proposition true, what false, cannot be further defined, but must be immediately recognized." Moore, and later on Frege, argued against the mainline and, more generally, for the "undefinability of truth" on grounds that any attempted definition would be viciously regressive because a proposition would be true, by the definition, only if it *truly* had the stipulated truth conferring features[35]. Frege supported his opposition to the mainline with a plethora of other considerations, e.g. he observed that any appeal to correspondence or comparison for purposes of explaining *truth* would miscast a quality as a relation, that *being comparable* can be *more or less* as truth cannot be and that compared items must be comparable as truths and what they represent cannot be. We noticed that he invoked the "disquotation principle" to argue that there can be no proper and satisfying definition of *truth.* We long ago took notice of his observation that truths cannot represent in the manner of pictures because pictures do not of themselves and without names attached identify the subjects they picture in the way that truth determining senses must convey what they represent (p. 249). In his early writings, Frege took truth-values to be the (undefined) *Bedeutungen* of sentences which are also (contingently) determined by the *Sinne* of those sentences (see #4 above). While there is no explicit mention of any such doctrine in Frege's late papers on judgement (see pp. 123f. of Appendix A), it seems completely compatible with what he did come to hold: The *Sinn* which a sentence conveys also determines whether that sentence is true or false, and the combined doctrine is consistent with "disquotation"[36].

On "Correspondence". Mainline theories of truth are commonly styled "correspondence theories". Ramsey accepted the characterization and so did Tarski (allowing, I suppose, that the definable truth-making relation of *satisfaction* of a sentence by a model is a kind of correspondence). My own explanations of truth-value are "mainline", but have so far been offered up ungraced with talk of correspondence or even of comparison. Indeed, my doctrine is in accord with Frege's opinion that truth-value is a non-relational feature of the truth vehicle itself; I also think that Frege was by and large right in his strictures against the appeal to *correspondence* and *comparison* for purposes of *defining* truth-value. Still, while I have no particular affection for the venerable idiom of "correspondence", I would be loath to deny it cogency out of hand, in connection with matters of truth-value.

"Correspond" is of course "vague". It could and sometimes does mean *is analogous to*. That surely is not what "correspondence theorists have in mind, which must be *either* that two objects of the same kind are positively comparable *or* that for an object of the one kind *there exists a corresponding object* of some possibly other kind, e.g. for every truth there exists a corresponding fact.

Let's consider first the idea that *existence* gives the sense of "correspond", with particular attention to the existence of *facts*. Facts are controverted commodities. I trust to them myself, and I also think that things said or thought are true when "true to the facts", in a sense of "true" perhaps no different from that in which true friends are true to their friends. But then I don't think that this way of talking helps for the definition of statement-truth in particular. For one thing, a fact that verifies one statement may falsify another; so the existence of fact does not yet suffice to distinguish truth from falsity. Further, different statements may be true to the same fact, so "true to the fact" cannot distinguish the truth of one statement from the truth of another. Perhaps something better could be gained if we could somehow show that all statements are "really" statements of existence that carry on their faces indications of what "truth conditions" must exist or obtain if the statements in question are to be true. But of course not all statements are existential "really". Perhaps these observations support the idea that we may establish the existence of a correspondent for a statement by showing that statement to be true, but certainly not the converse, for the up-turning of the fact will leave still undisclosed what statement we have in mind.

> My own preferred explanation of truth appeals to tests rather than to asserted facts or conditions of the world--the position is at once in line with fregean strictures and can be used to give sense to "correspondence by comparison" (see below). I have, nonetheless, found it interesting to explore how far one can go toward connecting *truth* with *existence*. The following line of thought (which accounts for the remarks immediately foregoing) emerged in the course of a classroom discussion with Honor Students held in Urbana in the fall of 1988.
>
> (i) To establish the truth of a statement we must observe that the indicated "truth conditions" obtain.
>
> (ii) But to observe any such condition is to observe the existence of something or other.
>
> (iii) So "what makes a statement true" is the *existence* of something or other.

(iv) What about falsity? Well, to observe the conditions that make a statement false is to observe conditions that make the contradictory statement true.

(v) This seemingly entails that, for every statement, there is a matching statement of existence.

We may test the thesis by turning up appropriate equivalences. We falsify a statement that there are keys in the drawer by verifying the statement that everything in the drawer is not a set of keys, which in turn seems equivalent to the existential that here there exists a drawer empty of keys.

This "theory" makes an improvement upon Aristotle's exclusive invocation of conditions of togetherness and separation (see pp. 114f.) and also upon that theory that would hold that a statement is shown to be true if an existing object is observed to have certain features, for the "existential" definition also applies to such statements as that it's hot out, for which there may be no identifiable, featureable object to hand.

I don't so much think that this "existential" theory is false as I find it objectionable for failing to discriminate among slews of different if equivalent statements.

Talk of the existence of conditions or facts, while certainly "relevant" seems to me to be of very little service for the explanation of truth-value. Perhaps then *comparison* affords a more useful notion of correspondence. Here we must heed Frege's point that the comparison must be between comparables., *viz* between different objects of the same kind; therefore not between statements in comparison with facts, subjects, tests or any other such perceivable things.

Now, I submit, any notion of "correspondence" that applies to the discussion of the truth of statements cannot be between the statement and anything else but rather is between two *acts*, on the one flank an assertion of the statement that indicates a test for the truth of the statement and on the other flank an actual application of the test; a favorable comparison is observed with an observation of the success of the test.

I like this kind of "correspondence", *not* because it *defines* statement truth (it doesn't), but because it buttresses my account of the criterial analogues for truth-value ascriptions and my "analysis" of statement truth and falsity. That foregoing "analysis" stipulates that the criterial analogues for truth-value ascriptions presuppose both a *consideration* of the subject statement and the actual application of its criteria. The *comparison* we have now identified also requires the actual application of the criteria and the specifically assertional consideration of the statement. Moreover, I for one find it manifestly plausible to hold that the successful application of the criterial analogue for a truth-value ascription would result in the tester's observation of the successful application of one of the statement's criteria in relation to an assertional indication of what those tests are, *viz* an observation of the "comparison" we have identified.

A "correspondence theory of truth": Truth ascriptions are verified by observing that an assertion that indicates the applicability of a verification criterion for a statement is matched by the actual successful application of that criterion.

6. POSSIBILITY AND NECESSITY.

In this section I make an approach to determinations of "logical modality", especially *logical necessity*, as prospective features of statements. I take departure and inspiration from Leibniz, but come up with proposals opposed to his doctrine. Logical necessities, in Leibniz' conception, hold for all possibilities; so, to follow him, I must make some gestures in the direction of the conception of logical possibility. Readers will soon notice that I have little faith in and no taste for logical modality, and the course of the inquiry now to follow has confirmed my doubts. Still, I shall do what I can for these notions, under the caveat that my proposals are provisional, even experimental. The publisher's reader warns that my short way with *possibility* in particular is apt to give less satisfaction than offense to that contemporary tribe of enthusiasts for the topic. Only brevity not offense is intended. I don't think I'll have cut corners off the truth about this rarified matter. Let me now start with some recollections of English.

"Must" & "may" as act-adverbial modifiers.

"Necessary" and "possible" and such equivalents as "must", "may" and "might" most commonly serve as "sentence-adverbs", "parenthetically", to intensify or otherwise modify the utterance with no effect upon "what is said". "Impossible" makes for a very strong assertion of the negative. Frege thought that the German equivalents of this language had no other use, and he accordingly disparaged concern for "apodictic judgement" (see *Begriffschrift*, #4).

"Declarations of possibility", recalled.

We know that that is not the whole story about what we may want to say "must" or "may" be so. Suppose we're chatting: "He must be on his way by now," say I; "He may have missed the train," say you, undermining my assertion. Such "must"s and "may"s are everyday English and I know of no vernacular that lacks their equivalent. In the imagined exchange, "must" serves as the mark of assertion from which the second speaker refrains by saying "may". Second speaker's "declaration of possibility" is certainly to be contrasted with the first speaker's assertion, but not as the assertion of something else (See Appendix B, pp. 155ff.). Here there is a change of mood.

Some may urge that my "declarations of possibility" stand in need of evidential support by way of statements of fact that such and such is possible.

On the matter-of-factness of possibilities.

I distrust the argument. These declarations may simply register an actual lack of knowledge. Still, the question whether there are these facts, *viz* possibilities, is indeed a matter worth exploring. Possibilities, if there be such, are either *actualized* or *unactualized.* Actualized possibilities "correspond with" truths, and for convenience we may stick with true statements. But then it would seem that there are no other facts for unactualized possibilities to be. If it is riposted that not everything is possible--e.g. rational square roots of two and surfaces green and red all over are not--and that real possibilities may exist unactualized, then we must ask where and how they are to be found. We don't knock up against them with our knees. So it would seem that real, unactualized possibilities exist as a matter of fact but then certainly not as matters of actual fact. How so, then? Well, unactualized possibilities must "correspond to" untruths, e.g. statements that are false (e.g. that this sheet of paper is red) or are neither true nor false (e.g.--perhaps--that if a page of this book is green, it won't be read). But now these statements are not "of possibilities" in the way in which they may be "of facts". It would, moreover, seem, by our recipes, that there could be no such statements, for any assertion of such a statement would have to be one wherein speaker gave indications that this was something he didn't think he knew. "But there are these possibilities!" Yes! at least in the sense that there are these statements or perhaps other conventional products. Consider the cited impossibilities: There is no statement that a surface is green and red all over because there is no procedure by which a surface could concurrently be uniformly compared with two exemplars of visually distinct coloration, not, anyway, if *being the same color as* is an "equivalence relation"; again, the pythagorean proof could be interpreted as showing that there is no procedure for factoring a prime number into rational parts. Possibilities, in brief, are conceptual facts, including ones to the effect that statements exist; other constative products may be thrown in, and the notion of *possibility* further extended to comprise other kinds of "things said", even the sayings themselves. "Actual" and "unactual" when used in relation to possibilities thus conceived are implicitly, in the main instance, truth-value predicables of statements and other products of constative utterance. More generally and in the upshot, possibilities are products of our conceptualizing activities and are brought to attention in the theorist's consideration of "what we say".

Possibilities are so-called when used as assumptions.

Now, two additions. First (with thanks to Melnick): possibilities are so-called only if taken up and used or put into play as "assumptions". If someone relates to me that a colleague of ours who just left my Urbana office is visiting in New York City, I may say that that is impossible, notwithstanding my recognition that there exists that possibility. There's no place for that assumption in the business at hand. By way of contrast, a radio notice that there may be a power outage this afternoon cautions for the provision of auxiliary generators in hospitals and back-up data disks in the modern office. It remains that possibilities remain what they *are*--statements or other said things--whether or not so-called.

The second addition is that possibilities, whether or not "real", are conceived in relation to a "constituency" of presumed actualities--references, predicables or perhaps statements. So, for example, given *this man*, I conceive that he be elsewhere, e.g. in NYC; again, given the statements that this apple is green all over and that it is red all over, I conceive an "unreal possibility", that there be a statement that the apple is both green and red all over.

Domains of possibility may be variously delimited in relation to a range of actualities.

Writers of philosophy have variously held that domains of possibility are restricted to sometime truths, to statements "about" one or another collection of objects, or by reference to compatibility relationships with other constative products. Thus Aristotle, who allowed that the same "said things" could be true at one time and false at another, restricted the domain of possibility to sometime truths, actualized in the fullness of time (p. 114 and reference to Aristotle). In his exchanges with Leibniz, Arnauld remarked that it was possible that *he* have been a medical man but not unthinking. The familiar "epistemic" restriction is to statements compatible with formulations of what we know. "Physical", "medical", "human" and other such "nomic possibilities" are restricted to statements compatible with the established principles of the cited fields of inquiry. Logical possibilities, in the leibnizian understanding, are simply self-consistent. (NB: Since there may be self-contradictory statements--pp. 291--Leibniz' possibilities pass muster under the less restricted rule that possibilities exist when statements do.) The domain of possibilities taken in reference to an expanding constituency may otherwise be broadened to include fictions in relationship to their texts, moves in relationship to rules for games, and so on. So: King Lear might have had a bastard son (his fool, perhaps), but could not have been a woman. The second but not the first move of a chess game might be of The Queen. These expansions may even breach the limits of what is historically conceivable, by taking, e.g. works of fiction as constituency: Alice might have recorded an adventure in white ink on a sheet of paper that was green and red all over or H. G. Wells arrive here yesterday on a train that leaves day after tomorrow. Fastening on the thought that things called "possibilities" are taken up and used, we may even come to allow that a dog sees the possibility of food in an empty dish. (However, repairing to safer territories, I doubt that a dog, disabused of his error, would continue to be so-seen.)

Domains of possibility. Kant promoted the idea that spaces at times are "possibilities for singular existence", and I shall follow by giving preference to actual occasions for testing. I have, however, in this section, begun more broadly, allowing that possibilities may also comprise such products of the imagination as fictions and chess moves, taking these as categories of objects that span the "possible worlds" of (e.g.) fiction or chess. A chess game would be such a "world", as would any version of the ancient tale of Lear. These examples may seem to reflect credit on the fashionable mythology of possible worlds. I myself prefer to treat the alleged domains (including occasions) as "constructions" within the category of perceivable playwrights, gamesmen and other conceptualizers, happily obviating a pullulation of "worlds".

Possibilities are "conceptual facts" that reside in the imagination.

Possibilities, generally, are emergent in the imagination. Even Leibniz, who believed in possibilities if anyone ever did, allowed that they exist only in the mind--ultimately and ideally in the Divine Understanding. I shall limit my examination to possibilities we can conceive, chiefly to statements producible by ourselves, but allowing that the domain of humanly conceivable possibility may be extended to include other "conceptual facts" identified in reference to utterance.

It has been brought to me, as an *objection*, that David Lewis has argued that possibilities must exist objectively and independently of ourselves if we are to make sense of betting and other activities that involve a consideration of possibility. In betting on Dobbin, I select the possibility of his winning the race. Let it pass that it might be better said that I selected Dobbin. The possibility of Dobbin's win is objective enough, but its scarcely to be imagined unconceived by anyone. A highly competitive but unentered horse who interlopes the track may win the race but won't win any bets. His winning dash was not conceived as a wagerable possibility, and never existed as such.

In pursuit of strict logical necessity.

Holding firm, now, to this thought that possibilities, actualized or not, are "conceptual facts" of diverse sorts, we want to explore the prospects for defining a notion of "strict logical necessity"--of what couldn't have been otherwise, no matter what. A notion satisfying that description could be attained, if at all, only along a path of theoretical inquiry that takes departure from but finds no destination in our common conceptions, and would have only tenuous connections with everyday concerns. "Intuitions" about such ideas are unreliable and these ideas should be defined and justified before being imposed upon the data of everyday life. *Logical necessity* is an artifact of modern philosophy. While auguries of the notion were perhaps sensible in the cartesian philosophy, it had no explicit currency before Leibniz, whose writings sounded a keynote for a concern with the logical modalities that has become almost obsessive among academic philosophers. Mathematicians, who philosophers fancy have major interest in necessity, seldom so dignify their results: while ever alert to the demand that the truths they assert should follow necessarily from what is already established, they are normally content with simple truth. We noticed that Frege held that opinion and I have somewhere read that Gauss doubted whether there is a distinction between necessity and truth simply taken. Any explication of this notion of necessity--unlike *truth* and our everyday *must*'s and *may*'s--will be of exclusively in-house interest among philosophers. Now I, for one, cannot quarrel with Leibniz' criterion of strict logical necessity, *viz* compatibility with all self-consistent statements. But the underlying idea which the formula is meant to explicate, the idea, namely, of being so no matter what, is another kettle of fish, and Leibniz' conclusions about that notion are dubious. I shall eventually attempt to "motivate" and provide definitions for some such notion of *logical necessity*--that, of course, only for the use of philosophers; the endeavor, I own, is more apt to confirm than to dispel skepticism over whether any notion of strict logical necessity is at once respectable and serviceable for the enlightenment of our understanding.

Necessities, hold for all relevant possibilities.

What is necessary is impossibly not so and, accordingly, holds for all relevant possibilities. This is sometimes put into the formula that something said ("p") is necessary just in case its negation is not possible ("not-possibly not-p"). This rendition of *necessity* used as a predicable of statements is at odds with our earlier proposals in regard to *statement-possibility*: Statements *p* and *not-p* exist together, with their representing tests in reverse; but now to say that *not-p* is not possible could, by our earlier provisions, be understood to mean that there is no statement produced in the assertion for which "not-p" stands proxy; but then there could be no statement *p* to be said to be necessary. The argument presents no obstacle to restricted concepts of statement-necessity corresponding to restricted domains of possibilities. This way of talking about *necessity* may be extended to other "domains of possibility": Thus a statement that the first move of this chess game was not of a Queen is necessarily true because it holds of all possible chess games. Again, Hamlet was necessarily a Dane and Hera female, whatever else they may be, according to the literature that gives them their substance.

Leibniz' formula for strict logical necessity was restricted by The Principle of Contradiction. Contradictories of necessities are self-contradictions, and a truth is necessary just in case it is compatible with every non-self-contradictory proposition, *viz* it is ruled out by nothing that is "logically possible". I do not contest the integrity of the formula. I am, however, uneasy with some of the consequences he drew from it in conjunction with background assumptions about what logical necessity should be. So, accepting the leibnizian terms, let us now consider the matter a little further.

Leibniz: A truth is necessary if compatible with every self-consistent proposition. Leibniz used this with great power to draw awesome if questionable consequences

Possibilities, we said, are fixed in relation to a constituency of comparative actualities. Each elementary possibility is a (possible) combination of such things. So the simple, necessarily existing "Gegenstaende" of Wittgenstein's *Tractatus*. I believe Leibniz similarly appealed for constituency to the simple adequate ideas collectively needed to constitute any and every monad's "individual essence". In Leibniz' scheme, there may or may not be actual objects answering to essential determinations, and statements that there are such objects must accordingly be contingent. But the possibilities themselves cannot be gainsaid.

Leibniz concluded that anything further truly sayable about such possibilities is necessary: The judgement could be true only if the predicate were in the totalized possibility, and the denial that it was would be a contradiction.

One may grumble over Leibniz' plural characterization of logical necessity as (i) truth in all possible worlds, (ii) compatibility with The Principle of Contradiction and (iii) analyticity. We noticed that the possibilities in question are restricted by The Principle of Contradiction. But now, for Leibniz, all representations, necessary or otherwise, are predications of ideas to monads, themselves conceived of in relation to an ultimate inventory of basic predicables, but presumably not to include *existence*. If a predicable is among those that comprise the "complete concept" or "individual essence" of the subject, the representation is necessary but also analytic; if not, then not analytic, but also not even true, let alone necessary. So far so good; the three mentioned theses do tie into one another. However, there are troubles for Leibniz down this road, for it is not clear that the necessarily existing basic predicables can be anything else than references to actual, contingently existing monads.

Arnauld complained that, while he couldn't have not been a thinking being, he could have been a married physician instead of a cleric and a theologian. Leibniz replied that this way of conceiving of relevant possibilities in relation to contingently existing objects made possibility itself a contingency; hence necessities, true of all such possibilities, would be only contingently necessary and (in Leibniz' view) not necessary at all. Leibniz concluded that the conception of necessity required nothing less than a single, unified, utterly unconditioned, "absolute" "logical space" of possibilities. The credibility of whole systems of metaphysics is at stake in this exchange between Arnauld and Leibniz. The issue is reflected into logic as the question whether logically necessary truths may entail contingencies. Leibniz denied it. Taken altogether, Leibniz' "absolute conception of logical necessity" demands, *first* that no necessary truth entail contingencies, *second*, that there is no conception of logical necessity except in relation to a completed totality of possibilities and, *third*, that things are necessary, contingent or possible only necessarily.

Even in accepting a "leibnizian conception" of statement necessity--that to exclude contradictions as possibilities and to comprise their denials as necessities--I have found myself favoring the Arnauld side of this dispute. Take a proposition presumed necessary by leibniz' rule[37], a tautology such as *Either Leibniz was Saxon-born or he wasn't*: that statement, as it seems to me, entails the sometime existence of Leibniz, which he once cited as a contingency. That's an objection. The issue is still up for dispute, of course. I hazard that Leibniz would protest, first, that the alleged example is not yet explicitly resolved into its simple elements and, second, that it is not an authentic tautology, but rather a conditional whose protasis clause posits the (contingent) existence of Leibniz as a combination of simples.

Leibniz's absolute conception of logical necessity, as it seems to me, is unreal, unexpectedly limited and also troubled by the need for a solution to the problem of accounting for the exclusion of contraries. Let me touch on these three complaints in order.

"Unreal". Ultimate simples elude exemplification. No one has ever written out a fully articulated leibnizian judgement. The *Spielraum* of possibilities the doctrine requires is accessible only to the intellect of a necessarily existing arch-conceiver. The theological implications are daunting, though Leibniz was not displeased with them. Arnauld protested that it is a theory too arrogant for the use of us ordinary mortals who cannot but start in the middle of things (more on this at pp. 344f. below).

"Unexpectedly limited". Leibniz's way of resolving things into ultimate predicables forecloses our conceiving the possibility of time before creation, a cosmos displaced a foot to the west or in a "steady state" for even a second. Leibniz held that these "Clarke Possibilities"[38] are distinctions without a difference.

"Exclusion of contraries". Leibniz' simple determinations are invariably "positive" and so far mutually compatible. (He used this to argue for the consistency of our conception of God.) But without irreducible negative determinations, there can be no plural possibilities and, in ultimate analysis, distinctions of logical modality are effaced[39].

Kant took Leibniz' quandary as argument for descriptively unexpoundable intuition. Leibniz had found no analytic reason why an object could not be in two places at the same time. Our positioning activity, in my understanding of Kant's scheme (pp. 262f.), fixes the subjective forms of space and time and secures primitive exclusion below the level of predication by dint of the consideration that any (positive) positioning excludes every other. These positionings, not simple concepts, are constituency for the basic framework of possibility. The "real" "synthetic" necessities latent in the one conceptual scheme we know may be elicited only with reference to them.

Following from Kant, we take the occasions for testing a statement as its relevant "domain of possibility".

Testing-on-occasions answers, in my scheme, to Kant's *positioning*, and my thesis that statements be testable on occasions corresponds to his demand for intuition (pp. 257f.). The occasions for testing a statement constitute its possibilities for truth and falsity. That thought opens the way to an exposition of several "non-absolute" conceptions of the logical necessity of statements. I approach with circumspection. What follows, as I have said, is an "in house" academic exercise perhaps too much in tune with philosophy's contemporary preoccupation with trying to make Leibniz come out right in this matter of the logical modalities. My own constructions do not pose as borrowings from common understandings, but come forward as uncostumed artifacts of philosophy. They are suspect on their face. Something useful may turn up, but on this I have more hopes than expectations.

Logical necessity on the contemporary scene. Herewith a brief review of several other approaches to the subject. The theorems of logic are, by some, supposed to be logically necessary, and some simply leave it at that: logic is the one and only theory of necessity that we have. The proposal gains impetus from the "logicist" program for deriving the supposedly necessary truths of Arithmetic and Mathematical Analysis from the axioms of logic.

In order to vindicate the necessity of logic, we still must explain what it means to *say* that something is necessary. The method Frege used to certify the truth of his logical axioms leads naturally to the wittgensteinian idea that a logical truth is a tautology, an idea Leibniz would have applauded. It's unclear how to extend this to the truths of predicate logic.

In another use of logic, it is plausible to think of systems of higher-order predicate logic as incorporating theories of intensions and hence as providing (among other things) a theory of modality for lower-order logic. This would perhaps serve as a formalization of the view associated with the name of Quine, that a necessary (first-order) truth is one that remains true under every consistent replacement of so-called non-logical terms, where "not", "and", "all", etc are logical terms.

Standard logic has also been extended in a different way into systems of "modal logic" whose axioms are taken to provide "implicit" definitions of "It is necessary that", treated now as an "operator".

Others, notably Alonzo Church following another lead of Frege, attack the matter by formalizing a first order theory of intensions, including the notion of *necessity* taken as a predicable, in much the same way in which the epsilon-relation is formalized in classical presentations of set-theory. That is the method I would favor.

Recently developed "model theoretic" explanations of necessity, known from the writings of Montague, Kripke and Hintikka (among others), which adapt the method of Tarski to the ideas of Leibniz, Frege and Wittgenstein, provide a trenchant explanation of the thought that a sentence is necessary if it would remain true no matter what properties or relations were represented by its predicates (see above).

Interrelations among these different approaches to modality have been explored and used; one senses that they converge upon one another, which is mutually corroborative and evidence for the integrity of some notion of logical necessity.

An old refrain in this development has been that necessities are true "by virtue of meaning alone". Leibniz' resolution of everything into simple predicables plays in here as does Kant's explanation of the (merely) "analytic" as a judgment whose predicate is contained in its subject: by expanding the definition of "brother" in the statement that my brother is male, we engender a result proving that the statement was "true by definition". This idea has been glossed in various ways. Some simply hold that necessary truths are true by virtue of the rules of a language (Malcolm), while others have proposed that necessities are truths that follow from so-called "semantical rules" alone (Carnap). A recent linguistic version of this would try to show that denials of necessary truths are linguistically unacceptable according to the transformational rules of a language. There is less truth than unclarity in these formulas. They could apply directly only to things that have meaning, presumably either words or assertions and not to the truths that are expressed by or produced in those words or assertions. There must be something right in the idea, however. My own explanations will take departure from necessary assertions and there will remain at least some doubt over whether I succeed in escaping from that restriction.

We shall endeavor to make a transition from an explanation of the necessity of assertions to an explanation of the necessity of statements.

The thought that our knowledge of necessities arrives "a-priori" and, in a vague sense, from a "consideration of meanings" fits most examples and gives some idea of what kind of commodity our philosophers have been dealing in. Now meaning is a feature, not of statements, but of assertions. That suggests a strategy of looking first to explain a conception of logical necessity appropriate, not to statements, but to assertions. Since assertions are ephemeral where statements endure, this policy may win the applause of that band of philosophers from Gauss to Quine who have bad-mouthed the logical modalities, and I may eventually have to count myself a member of the claque. But, for now, I look ahead to the possibility that a well turned-out conception of *assertion*-necessity may stand as a platform for reaching a more abstracted conception of *statement*-necessity. The aptness of this policy, which I follow, is confirmed by a felt similarity between what is seen to be trivially and what is unexceptionally so. It will also enable us to deal in an agreeably direct way with those kinds of "intuitive truths" of which Descartes' "Cogito" is the exemplar. At end, doubts will remain whether we have indeed successfully negotiated a transition from necessary assertions to necessary statements; those doubts will, I believe, stand less to discredit my approach than to post further warnings against the very idea of statement necessity.

Something-said may seem obvious because long known. Other unobvious things-said may still be *trivial* because no intellectual effort is needed to reveal them. That is commonly so when an assertion occurs on an indicated occasion of verification: here nothing more is needed than a routine application of a criterion. An observer who understood the assertion could directly proceed to apply a test called for by the use of the expressions employed. Thus, *The first word on page 341 of this text is "thus"; 11x33=363*; *The terminating word in this sentence is the next word.* In contrast, were I now to make the statement that the first word on the page numbered "340" of this text is "follow", the reader would at least have to flip the page back before he could finally verify what was said. Generalizing on these examples, I propose the following definition: An assertion A of a statement *s* is "trivial" just in case A occurs on an occasion for the successful applicability of the verification criterion for s.

I stress three points about** assertion-triviality. **First, the same statement may be trivially or non-trivially asserted. Trivial assertions are enacted upon occasions of the successful applicability of the statements they produce; trivial statements are producible only by trivial assertions.

Second: Triviality is a definite and objective feature of assertions; assertions are trivial or they aren't, never "more or less". That is so regardless of the speaker's "intentions"; it is seldom any part of his meaning that his trivial assertions should be so. He may be the last to know.

Third: What is said in an assertion trivial by the given definition need not be immediately *obvious*. Consider: $2^{2\text{-}exp5}+1$ *is a prime number; The door is too narrow.* An assertion is obvious to an observer only if he can tell, so to speak, "by inspection" that the indicated verification test could be successfully applied. *Obviousness*, by this account, admits of degrees, and what is obvious to one person might not be to another.

The statements $6^2+8^2=10^2$ and *The second perfect square that can be represented as the sum of two squares is 100* differ in that the first but not the second can be routinely proved out by calculation. A similar difference is found between the not quite so obviously distinct statements that *The first word on this page is "Thus"* and *The first word on p. 341 of this text is "Thus".* An argument that these are distinct statements is that the latter but not the first "presupposes" a determination of page number. In the first of the arithmetic pair, the statement is such that any occasion of assertion would also be an occasion of verification; similarly, in the first of the non-arithmetic pair, employment of token reflexive expressions secures that every occasion of verification would be indicated as an occasion of assertion. In both cases, the produced statement, if it is true at all, cannot but be trivially asserted. That is a feature of the statement itself. Let us then distinguish such statements as trivial, *viz* a trivial statement is one producible only in trivial assertions. Statement-triviality cannot be expounded entirely in terms of our theory of tests, because (again) the theory affords no means for identifying assertions.

Necessary assertions assertionally imply the statements they produce.

The third example of a trivial assertion given on p. 341, that the terminal word of the sentence was "word", differed from the others. A comprehending observer of that assertion would see, not only that the produced statement was successfully verifiable on the indicated occasion, but also that the verification was "in effect" already achieved. The mere fact of assertion "makes the statement true". That was our gloss on "transcendental arguments", which purport to show that a certain fact could not be meaningfully doubted (p. 245). Doubt in such cases is defeated by the very formulation of the fact. One who understands what is said must be capable of seeing that it could not but be true. I propose to call assertions of this kind, those that convey an accomplished verification, *necessary assertions*. Bringing up terminology introduced at p. 98, we can say that an assertion A is "necessary" just in case A produces a statement *s* and assertionally implies (the truth of) *s*.

Remarks:

(i) "in effect" and "assertionally implies". The assertion is not itself an application of the test for verifying the produced statement: it is simply an assertion. However, the assertion is a fact formulated by the statement produced in the assertion--a fact which "makes the statement true" and assures the successful applicability of the indicated verification. (Our examples, by the way, are not proof against skepticism, since one may still doubt whether what was said was said.)

(ii) "assertion". The notion of necessity now in hand is clearly applicable to assertions and trivial assertions in the bargain, not to the statements they produce. The very same statements might on other occasions have been non-necessarily asserted. Saying *"There are in the sequence 1,2,3,4,5 numerals"* and *There are five numerals on the above line"* are both trivial assertions of a statement that is verifiable and falsifiable on a common occasion; but only the first assertion is necessary[40].

(iii) Assertion-necessity is an objective and definite feature of assertions; assertions are either necessary or not; never "more or less".

(iv) Necessity may be unintended by a speaker who asserts necessarily, and necessary assertions rarely if ever declare their own necessity.

(v) While our idea of assertion-necessity may initially seem unpromising because it allows for the possibility of the same one contingent statement being both necessarily and contingently assertible, it does have the advantage of accounting for our inclination to say that some *intuitive truths* of the "Cogito, ergo sum" variety are necessary. They are necessary assertions. An observation of a subject's expression of thought in language, as when he says he thinks, is proof that he does. The speaker may not have verified what he says, but on reflection will know it must be so[41].

The notion that a necessary statement is one that could be only necessarily asserted is "ineffective".

Our explanation of statement triviality--that a statement is trivial if it could not but be trivially asserted--suggests a like explanation of statement-necessity: a necessary statement is one that could not but be necessarily asserted.

That formula for statement necessity, while appealing, is, alas, unacceptable as a definition. For one thing, the occurrence of "could" intimates circularity. The weaker condition, that *all* the assertions of a necessary statement are necessary assertions, also fails as a definition, since the statement might never be asserted at all, in which case the condition is vacuously satisfied; again, a clearly contingent statement might in fact be asserted only once and then by way of a necessary assertion. There is also the difficulty already met with in connection with our explanation of statement-truth (p. 318) and of statement-triviality, that our theory of tests provides no means for identifying or otherwise talking about assertions *per se*.

The condition that all actual occasions for asserting a necessary statement should also be occasions of necessary assertion, while too weak to serve as a definition, is "suggestive". If a necessary statement were one that could not but be necessarily asserted, then certainly any actual occasion of the successful assertion of a necessary statement must also be an occasion for successfully verifying that statement. With the idea that God perhaps could assert any statement anytime, we make as if to expand the range of occasions for asserting a statement to include all the occasions for verifying and falsifying that statement. But that is pretense! Right: So, let's discount it, and deal only with statements proper, according to our representation, refusing further commerce with assertions or meaning.

Direct logical necessity: All occasions for verifying a statement are occasions for successfully verifying that statement.

Following the above "suggestions" but holding to our representation of statements as a pair of qualified tests, each in turn represented by pairs of sets of occasions of successful and unsuccessful applicability, a first notion of *logical necessity* for statements, which I call "direct", interprets the idea that a statement couldn't but be true to mean that it wouldn't but be successfully verifiable if verifiable at all. Any occasion of verification, including all those upon which the statement might be asserted but (for us, if not for God) possibly many others too, is an occasion of successful verifiability. Let's symbolize as "**N**" this notion of "direct logical necessity", *viz* a statement *s* is directly logically necessary or *Ns* is true just in case *s* has no occasions of unsuccessful verifiability.

It is evident that a directly necessary statement is true; the notion seems to satisfy the further condition definitive of a "minimal modal logic", *viz* that $N(p \rightarrow q) \rightarrow (Np \rightarrow Nq)$[42].

Direct logical necessity is applicable in relation to restricted domains of application for its subject statements and is subject to contingencies.

If we think of the actual occasions of verification of a statement as the "actual possibilities" of which it conceivably holds, then **N**, so far, is in the spirit of Leibniz. Still, what we have proposed departs radically from leibnizian principles. *First*, our "possibilities", indicated in assertion, are actualized in concrete fact as parts of the cosmos. *Second*: Leibniz' "principle of plenitude", applied to the case in hand, would require that the set of (actual) occasions for verifying any statement should be "maximal", that to include any occasion whatsoever. In our explanation, however, relevant occasions are restricted by conditions for applying the verification test. *Third*, those conditions may, in relation to their most expected formulations, be only contingently satisfied. If the statement is directly logically necessary, it is true; hence it exists; from which it follows that those contingently formulated conditions are satisfied; a necessary truth may accordingly have contingent entailments. Let us, *finally*, observe that whether every occasion for applying the verification test is also a condition of successful applicability may depend upon circumstances that would ordinarily be contingently formulated. From this we see that the direct logical necessity of a statement might itself be "contingent" (notwithstanding the likely circumstance of there being no statement of that fact, for which see p. 239). In summary, then, our **N** is "relative" not "absolute" in (at least) three ways: the range of relevant possibilities is restricted; an **N**-necessary statement might have contingent entailments; and, finally, a report that a statement is **N**-necessary may itself be contingent.

A statement is "indirectly logically necessary" when all occasions for attempting its falsification are occasions on which it could be successfully verified.

Our proposal for direct logical necessity was suggested by the idea that a necessary assertion conveys a verification. Another notion of logical necessity is suggested by the use of *reductio* arguments, chiefly in mathematics. We assume that a statement is false and then show that it couldn't be. To that end we must somehow demonstrate that the falsification test is not successfully applicable. "Intuitionists" may protest that that would not generally suffice to show that the statement was true. The *reductio* argument would carry through, however, if it showed that *any* occasion of (attempted) falsification would also be an occasion of successful verification: one might argue, for example, that any occasion on which you tried to sum the aliquot parts of an odd number with 1 to that number, you would get a number less than that number. Such a demonstration would certainly show that the statement (if it existed) that no odd number is perfect was true; and also that there was no occasion on which it could have been proven false. I call this "indirect" logical necessity, and symbolize it as "**I/I**". Our several contrasts between leibnizian necessity and **N** carry over to **I/I**, except that now we think of the "world"-constituency of the statement as (actual) occasions of possible falsification.

We define two notions of statement "logical possibility" corresponding to the foregoing definitions of necessity, according to the rule that a statement is logically possible when its contradictory is not necessary.

Our earlier discussion led us to the thought that a *possibility* exists when a statement does. We then noticed that the familiar rule that something is necessary when its contradictory is impossible holds only if the domain of possibilities ("statements") is appropriately restricted. The dual of that rule, that something is possible when its contradictory is not necessary, may now be routinely invoked to define two notions of logical possibility corresponding to our two notions of logical necessity. If a statement is represented as $\langle V_s, F_s \rangle$; then its contradictory is represented as $\langle F_s, V_s \rangle$. So, if $\langle V_s, F_s \rangle$ is possible, $\langle F_s, V_s \rangle$ is not necessary. $\langle F_s, V_s \rangle$ is not directly necessary in case there are occasions for verifying it that are not occasions for successfully verifying it. So, in brief, a statement *s* is directly logically possible when there are occasions for attempting to falsify it that are not also occasions for successfully falsifying it. Again, a statement *s* is indirectly logically possible when there are occasions for attempting its verification that are not also occasions for successfully falsifying it.

In another sense, a statement is "nearly necessary" when its truth is required by the satisfaction of a "presupposition".

Additional to the two kinds of logical necessity we have defined, there is another notion of "near necessity". The idea here is that the truth of a statement is required by the actual satisfaction of a presupposition. Test-theoretically: If the verification test, V_s, for a statement *s*, presupposing a test *t*, were sometime successfully applicable on some occasion for successfully applying *t*--if the sets of occasions of successful applicability of the two tests overlapped--then the truth of the statement is implicated in its existence. Circumstances which satisfied some of the conditions for the existence of the statement would also satisfy conditions for its truth. The presuppositions might have been otherwhere satisfied; but if they are satisfied in these circumstances, then the statement would have to be true, even though neither **N** nor **I/I**-true.

> We shall find reason to think that all true statements of identity and distinctness are "nearly necessary" and cause to speculate that there is a nearly necessary formulation of every fact. This notion of *near necessity* is so little discriminating and its metaphysical consequences so seemingly rich, we may wish to shelve it as a curiosity. So the perils of logical necessity!

Our several notions of necessity are "concrete".

A contemporary leibnizian[43] holds that the actual world is a possible world and no less "abstract" than any other. His sense of "actuality" is not that of the *cosmos*. So with Leibniz. However Leibniz may have explained the simple idea of real existence, his only "criterion" was the abstract maximization of possibility. In contrast, the "possible worlds" of our statements are actual occasions and of the cosmos we inhabit. The several logical modalities we have defined for statements in terms of those occasions are accordingly "concrete", in contrast with "abstract".

I now proceed to ring changes on the expected leibnizian *objection* to our account, that statements necessary in any of our ways, because of their contingent entailments, might not have been true. My own final opinion will be that the objection endangers my proposals in the same measure as it thwarts a rescue of logical modality for exploitation by philosophers.

It is first apposite to notice that none of my contingently necessary statements could have been false instead. If their contingent entailments were not true, then these necessities would not have existed at all as possible falsehoods. The alternative is to suppose that there is a domain of eternal truths which exist independently of and unsecured by the actualities of conceptualization. If truth is a feature of language or of its products, then surely the abstraction required by the view that necessities cannot have contingent entailments is one we should be grateful to be liberated from.

Our conceptions of necessity cover "intuitive truths", but not all similar "demonstrations".

I claim it as a merit of our account that it explains our sense that "intuitive truths" of the "Cogito"-kind couldn't have been otherwise. Descartes' *assertion* was necessary because it verified the statement it produced. But then *that* statement was verifiable only on that occasion (centering on Descartes) and so that statement is directly logically necessary. Leibniz objected that none of us need have existed at all. Agreed! The statement that

Descartes existed, is (presumptively) contingent. Still, Descartes' own simple *Sum* was a necessary formulation of that very same fact. While the fact is the same, the statements differ, and facts are not susceptible to modal determinations of the kind that may feature their various formulations.

If *I am here* is a directly necessary statement by my reckonings, what of *Something is here*? It, if true, is verifiable only on the occasion of utterance; so it too must be necessarily true; but in this case, the assertion doesn't convey a verification, so that statement *shouldn't* turn out to be necessary, even by our reckonings; surely the presence of something before one doesn't make a statement to that effect an intuitive certitude. My brief reply is this: unlike in the case where I spoke of myself, there may be occasions of testing this statement which enclose me but not the thing I ostend. So here may be occasions indicated which need not be occasions of successful verification. In the upshot, the true statement that something is here may not be necessary, and we must be gratified that we are able thus to distinguish it from those similar "intuitive truths" that couldn't be otherwise.

It is unlikely but still possible that there are "accidentally necessary" statements.

The most serious leibnizian challenge to our concrete conception of logical modality for statements is this: Apart from the contingent truths that follow from the very existence of a concretely necessary statement, it might just happen that all the occasions for verifying a particular statement were also occasions for successfully verifying that statement. In that event, the statement would be directly necessary, but only "accidentally" so. Now there probably are no such statements: occasions of verification can be divided and extended; it is unlikely that a statement whose verification was not limited to a single occasion centering on the speaker could be limited to any number; it would be most unlikely that all of these would be occasions of successful verification unless that were implicit in the conditions for applying the test. "But suppose it turned out that way!" Well, we couldn't know that except by "demonstration", and the threatening case purports to be one that couldn't be demonstrated as a "truth of reason". "That is the objection!" But then I don't see where we are going to get an example. "But there might be one anyway!". I must allow it could be. I pay that price to avoid the etherealization of truth. I really doubt that there is a middle position for logical modality, unthreatened either by excessive concreteness or by utter sublimation. Meantime, I claim this merit for my account, that it avoids the objection from illicit abstraction while leaving it unlikely that there are in fact statements that are accidentally logically necessary.

Distinctions in logical modality do not coincide with distinctions in knowledge as being either "a-priori" or "a-posteriori".

A prospect of "accidental necessity" is troublesome because threatening to our preconception that the truth of necessities can be known "a-priori" from a consideration of our conceptual operations. That thought did indeed illuminate our point of departure, and we noticed that distinctions in modality and "epistemic quality" overlap in application to judgements or assertions (p. 340 above, and fn #40). But now we must back up and take notice that the epistemic distinction between *a-priori* and *a-posteriori* knowledge cannot be the entire rationale of the sought after distinction in logical modality. Necessary truths may come to be known a-posteriori, e.g. by hearsay and even discovered by trial-and-error or by accident. School children nowadays can demonstrate by kinematics alone that a westward moving traveller keeping calendar by sunrises would, arriving back home, find himself a day behind on our orbit round the sun. I believe this fact was first discovered "a-posteriori" by the remnants of the Magellan expedition. On the other side, most readers of this book will have a-priori knowledge of the contingent truth that Saul Kripke is so-called. So we may agree with him that the traditional distinction between a-priori and a-posteriori knowledge does not exactly coincide with any distinction in the logical modality of truths.

On tautologies. One appealing account of *logical necessity* would have it comprise the truths of logic and nothing else. Authors who favor that thought are working with the notion that a truth is necessary if true for all possible truth-value assignments to some corpus of "elementary propositions". Our conceptions of statement necessity differ from that, for we delve below truth-value to the finer structure of the sets of occasions for verification and falsification for our domain of possibility. Indeed, "tautologies" of the *p or not-p* form, by our reckonings, will seldom come out as necessary, and won't in the likely circumstance that sets of occasions of either verification or falsification properly overlap the sets of occasions for the successful application of those tests. We have already noticed that tautologies may sometimes even fail to be either true or false.

Philosophers of old took mathematics as an established body of "certain knowledge". A successor opinion--one more widely held by philosophers than by mathematicians--is that the truths of mathematics, at least to comprise Number Theory and Analysis, are logically necessary, not just in the sense of necessarily following from something or as conditionals, but in and of themselves. Those "logicist" philosophers who also credited distinctions in logical modality used to think that Whitehead and Russell had demonstrated this necessity by deriving the principles of Number Theory as theorems of logic. Our analysis yields a conclusion opposed to the thesis that the truths of Number Theory are logically necessary, but also suggests a rationalization of the prejudice even to comprise Geometry and the whole territory of "natural mathematics".

While the truths of natural mathematics do not come out as being altogether necessary by our recipes, there is an adaptation of the notion of necessity according to which that may be so.

The truths of Number Theory include equations and inequations testable by arithmetic calculation. This use of calculation implicates a distinction between "basic" necessary truths of $1\_\neq\_2$ and $1<2$ sort, on which the verification system of arithmetic calculation is anchored, and other equations or inequations, such as "$2^{2\mathbf{exp}\text{-}5}+1 = 641\text{x}6{,}700{,}417$", which, though perhaps trivial, may be proven by and only by calculation. The equation, for us, is a contingent statement, since it need not be necessarily asserted. "$2^{2\mathbf{exp}\text{-}5} + 1$" is where calculations start; "$1 \neq 2$" is a place to stop. $1 < 2$ must be true if other arithmetical statements are to be proven true or false. Is it plausible to assimilate the distinction to that between necessary and contingent truth? Fermat thought that every number of the $2^{2\mathit{exp}\text{-}\mathit{prime}} + 1$ form was prime; Euler proved him wrong with the example we have cited. The question was once open in a way in which there simply could be no open question about the distinctness of 1 and 2.

Some who may welcome a distinction between $1 \neq 2$ and the Euler example may complain that the distinction is actually not sustained by our formal account of statement necessity. Suppose that the equation is true: it would be verified by carrying out the indicated calculations; but calculation, it may be argued, is conventional action, hence the rules are rules for succeeding (p. 41); hence, if the act is done, it must succeed. This seems to imply that the statement in question is indeed necessary according to the definition of "**N**". My answer is that rules for succeeding if followed assure success only if all the conditions of success for the performance are satisfied. They may not be. Recall that the measure of success for calculations taken as a test is that the agent should come to perceive something as a result of what he does. He must in particular be able to see his way from the first to the final step of the calculation. So, suppose that the calculation is very long. Night falls as the subject clicks on; by the time he writes down the final digit he can no longer see what he had written down before. Circumstantial conditions finally make the operation impossible. The occasion of testing may be one on which the calculation is not successfully applicable and the statement not be necessary.

Here now are my distressingly "vague" but I hope well directed explanations of our sense that certain theories of mathematics stand as *systems* of necessary truth. The idea of *necessity* that comes into play here is none of those we have defined, but rather an adaptation of the notion of *presupposition*. The mere existence of a statement warrants the satisfaction of the "presuppositions" for its fundamental criteria. We can say that statements of the presuppositions of a statement *s* must be true if *s* exists at all, and in that sense are necessary relative to the statement. If it turned out that a statement was necessary "relative to" *all* statements, then perhaps we could say that it was simply necessary, in this new understanding. I think that that is rather how things stand with the truths of both geometry and arithmetic in relation to other things said. Suppose I report that a carefully cut plane triangle had an area equal to half that of a rectangle with the triangle's height and base: Of course! Since it was presumably "euclidean", it had to come out that way. A euclidean geometer, reflecting on the superpositional method for determining area, shows that it is always so when his postulates are in force. If the euclidean postulates were in force, then what I reported could not have been otherwise. Or take an arithmetic example: Once we see that $(33+34)^3=33^3+3\text{x}33^2\text{x}34 + 3\text{x}34^2\text{x}33 + 34^3$ is true mustn't we also understand that there couldn't be a world in which it wasn't so? The Binomial Theorem may be invoked to show that the reported results of certain counts could not have been different, provided, as we presuppose, that the counted collection was "arithmetic". Number Theory works out the implications of this "presupposition". Mathematical theories that arise "naturally", not merely by stipulation but as a reflection upon what we otherwise say and do, prove theorems made true by the fact that we do thus think as we do, e.g. we show how-many by counting, where addition and multiplication represent possibilities for consolidating counts. Natural mathematics brings out what is bound up in the accustomed use of these methods for verifying and falsifying other usually non-mathematical statements

7. RELATIVE DEFINITENESS, PRECISION AND DETERMINATION OF STATEMENTS.

Unclarity, alas, is too often a defect of assertion and other utterance. By contrast, a certain "looseness" of *statement* is often much to be desired. Now "loose" is loose. The idea is that the statement is issued to "express reality" with a certain inspecificity, as a slack fact-fitter. People often have good reason to speak "loosely", and do so most carefully, in science as well as in diplomacy. Mathematicians, for example, may prove that a property holds of "almost all even numbers", meaning that it holds for all even numbers equal to or greater than some still unidentified particular one, and their definition of continuity, that for every *epsilon* there is a *delta*, leaves open what the precise relationship of these two magnitudes should be. I report that there were *several* students at the meeting; to have said that there were nine would have been to speak with unrequired exactness. I tell my wife that I saw her keys somewhere in the kitchen, for that's all I remember; it would have assisted her search could I have spoken more definitely and told her that I saw them on the kitchen table.

Statements may be variously "slack" as fact-fitters, in being more or less imprecise, indefinite or undetermined for truth-value.

Had there been exactly four students present, then (as it seems to me) my "several students" statement would have been neither true nor false. The terms of my assertion calculatedly left open the possibility that the produced statement was of indeterminate truth value. A statement produced in the language of "a clutch of" or "a batch of" would have been similarly indeterminate, but less so than our "several" specimen.

Suppose I had said, not "several" or "a clutch of" or a "batch of", but "six or seven": then the potential indetermination would have been forestalled, but my statement would still be *imprecise* relative to saying "six". Relative imprecision is also a feature of our mathematical examples.

The difference between our *keys in the room* and *on the table* statements is not happily assimilated to a difference in precision. Rather: to say that the keys are in the kitchen is to speak less *definitely* than to say they are on the table.

Evidently I spoke imprecisely above in saying that statements may be "slack" as fact-fitters; more precisely said, they may be imprecise, indefinite or truth-value indeterminate in varying degrees, and perhaps other such things too. The task of this section, led by an interest affiliated with the classical concern over the clarity, distinctness, truth and adequacy of our ideas, is to sort out these different notions of statement-precision, definition and determination[44]. I hope it will be agreed that this concern for the "clarity and distinctness of our ideas" has nothing to do with clarity of utterance. A formalized version of the definition of continuity with all the quantifiers properly in line is a model of clear expression. Again, I don't think there had to be anything unclear in my imagined assertion that there were several students there. There would seem to be genuine differences in the three statements, that there were seven students there, that there were six or seven and that there were several, clearly enough indicated in those terms. Statements of all orders may be more or less clearly asserted but (with apologies for repetition) *precision*, *definiteness* and *determination* are features of statements themselves and not of their assertions. In order to bring such statement features within the compass of our theory, we must turn up authentic differences among the procedures indicated for testing these statements. There do seem to be such differences: verification of the *how-many* statement requires that we count the assembled students, but not the verification of the *several* statement. However, there is a certain difficulty in the attempt to resolve this exclusively in terms of occasions of testing.

The "concreteness" of occasions throws up obstacles to the test-theoretic analysis of statement "slackness".

Suppose there were seven students present: the actual occasions for testing and for successfully testing the seven and the several statements would seem to be exactly the same. Furthermore, the several-statement in this instance would be true, in despite of the indetermination allowed for in asserting the statement. We are hampered here by the very concreteness of our representation--the source of our earlier worry over the prospect of "accidental necessity" (p. 348, also 239, 270f.). I hope to find ways of working round this obstacle, which we shall never get completely clear of in the ensuing analysis.

Two statements are "comparable" when their criteria respectively apply to the same "locations".

We have been speaking of "comparative" and "relative" preciseness, determination and truth-value determination. It is unlikely that we can always make such comparisons between statements "about" incomparables, e.g. it's not evident that we can compare a statement about keys with a statement about natural numbers for definiteness. In Chapter 5 we'll introduce apparatus for distinguishing statements in respect of what they are "about", by reference to what I shall call the "locations" to which their criteria are applied. Criteria for statements "about bodies" are applied to time-specified regions of space; "number aboutness" criteria, by way of contrast, are applied to numerical expressions. Now I shall assume that we shall be comparing only comparable statements for comparative "slackness", that is to say, statements whose criteria are applicable to the very same locations.

One statement is more definite than another if the set of occasions for applying the disjunction of the criteria for the first is more restricted than the set of occasions for applying the disjunction of the criteria for the second.

Of our three kinds of comparative "slackness", *definiteness* is the easiest for our theory, and I start with it. The statement that the keys are on the table is more definite than the statement that the keys are in the room simply because it homes in on a more definite place. Similarly, to say that she will come by 1 p.m. tomorrow instead of saying, more indefinitely, that she will come tomorrow, narrows the time of and for verifying her arrival. In both cases, indefiniteness of statement increases with the widening of the occasions of verification. That rule carries over to falsification. (Achieved falsification for the place-existential, roughly described, requires a negative result at some last place upon presumed negative results in all the others.) That suggests two options for defining relative definiteness. First, if *each* test for one statement is more widely applicable than is the corresponding test for another, the one is less definite than the other. Second, if the investigation, *one way or the other*--we pool the two criteria for each statement or take their "union"--, could be more widely carried out for the one than for the other, then the one is less definite than the other. I sense that the second idea is the better one. So, provisionally: of two comparable statements, the more narrowly testable is the more definite.

Of two comparable statements, the more precise is more narrowly successfully verifiable and more widely successfully falsifiable than the other.

I now proceed to *precision*. *6* said in reference to a given collection is more precise than *6 or 7*, *hexagonal* more precise than *polygonal* and *vermillion* more precise than *red*. Now it would first seem that tests for being vermillion and for being red are applicable on the same occasions but that vermillion-verification tests would be more seldom successfully applicable than would be tests for red; the same goes for statements that there were seven as against statements that there were six or seven. The more precise statement in both cases might be false and the other true. This must mean, for us, that of two statements compared for precision the more precise statement is successfully verifiable on a narrower range of occasions than the other and is successfully falsifiable on a broader range of occasions than the other[45]. Or so it seems, for here again the way is obstructed by excessive concreteness: were the peered-at object in fact vermillion or had there been seven in the assembly, then both statements of each pair would be true in fact and also it would seem in fact verifiable on exactly the same occasions. Now I doubt that it ever actually would turn out that way. The relevant occasions in the color case are ones which enclose the tested body from the distance of an eyeshot. It is unlikely that vermillion things could be visually distinguishable from all other reds on all these occasions; on relevant occasions, where that discrimination could not in fact be made, we may allow that the verification test for red would be successfully applicable and the test for vermillion not be. It would transpire otherwise only by the sheerest accident. My definition should cover the difference; however, I must allow that the two statements just might, unexpectedly, turn out to be equally precise. So, provisionally, my (first) definition of relative precision is this: of two comparable statements, s_1 and s_2, s_1 is more precise than s_2 when s_1 is more narrowly successfully verifiable and more widely successfully falsifiable than s_2.

Of two comparable statements, the more determinate is testable over a wider range of occasions but successfully testable over a narrower range of occasions than the other

I turn now to comparative *determinateness* of statements. A statement that there were several students present would be false if there were two students present and true if there were nine but not straightforwardly truc or falsc in casc of four. Numerical statements would be true or false in all these cases. The determination or indetermination of the "several"-statement depends upon actual circumstance. How can that be if this is a matter of meaning not fact? Here again the difficulty of excessive concreteness. I believe we can handle it as we did *imprecision*:

Because the numerical statements would be tested by counting and the "several" statement not necessarily so, it is likely that there are occasions on which several could be observed but the assemblage not actually counted. That would certainly be the case for statements in regard to *how-much*, where the exact and determinate statement could be tested only on those occasions which included a scale or balance. I suggest that counting like weighing and titrating is an assisted operation. Now, if the difficulty is overcome, our examples teach that a more determinate statement is testable one way or another over a narrower range of occasions than a less determinate one, but then successfully testable over a wider range of those occasions.

I believe that the notion of a *heap* falls under "indeterminacy" by this characterization, and that that observation contributes to the resolution of that ancient puzzle over when an aggregate of grains of sand to which other grains are added one by one becomes a heap or ceases to be a heap with the one-by-one removal of grains. There's a comparative broad band of indeterminacy between batches that clearly are and are not heaps. (We return to this topic in Chapter 23).

If these definitions hold up, then the comparisons they define may be combined. School children asked to report on the recorded shape of France, will look it up in the Atlas: Some may report that France is roughly hexagonal; others, both more precisely and more determinately, will fix the boundaries with latitude and longitude coordinates of bounding lines; again, a report that most of France lies between the Mediterranean and the Rhine is, for european comparisons, more definite but, for international comparisons, less precise than a report that over 90% of France is in Europe.

Digression on Leibniz. We remarked that our investigation into *definiteness*, *precision* and *determination* is affiliated with the early modern discussion of the *clarity*, *distinctness*, *reality*, *truth* and *adequacy* of our ideas. Many who followed in the wake of Descartes dabbled in these notions. Leibniz' explications were especially interesting, and led to some really fascinating and intriguing speculations--that relations are "unreal" and that all predicative truths are necessary, ultimately. Our definitions illuminate these speculations. Leibniz allowed that "looseness" was common and sometimes even desirable, but not for scientific purposes. Our ideas, he conceded, might be variously slack, but surely not the facts they represent. He notably included ideas that are secondary and relational in the bargain. Ideas of these kinds are all of them *inadequate* ("inaccompli") to the represented phenomena. Suppose I truly assert that one body is heavier than another: That relational statement covers a multitude of possibilities formulatable in other "more precise" statements, e.g. that one body weighed three grams and the other two or that the one weighed five grams and the other four. The relational statement is less precise than its various non-relational counterparts. Relational statements which are perhaps logically irreducible just because they cover a variety of non-relational counterparts, are slack relative to non-relational counterparts. We noted that Leibniz seems to have thought that every statement should be maximally-testable, *viz* be defined over *all* occasions. He would therefore have given little credit to our notion of definiteness. Suppose now that we adopt the "scientific ideal" of maximum determination and precision in our statemental representation of fact. Leibniz seems to have thought that every such statement, if true at all, would be necessarily true. It actually comes out that way by our definitions. A true statement is successfully verifiable; if it is also maximally determined, then every occasion of verification, will also be an occasion of successful applicability, by the first explanation of determination; in that case, the statement is also directly necessary by the definition we have given, *viz* that every occasion of verification is an occasion of successful verification. That accords with an intuition expressed in the writings of Descartes, Leibniz and Hume (among others), that existential statements are the ones most apt to be contingent: other things equal, existential statements are indeterminate in degree that they are indefinite, and that also measures the likelihood of their contingency. Counter-examples to this intuition, such as *There is a successor to (the number) 1*, and Descartes' *Cogito*, turn out to be maximally determined.

The leibnizian philosophy would be more compelling if there were some way in which we could explain and vindicate the idea of a maximally or perfectly adequate representation. I do not see either that it can be done or that it cannot. We observed that Leibniz assimilated the distinction between our ideas of primary and secondary qualities to the adequacy-inadequacy formula. I shall later suggest that secondary qualities of an object are tested for on occasions other than those upon which the object's could be shown to exist. If it could somehow be proven that everything that could be truly said about an object formulates a fact latent in its existence and ostendible on scene with the thing, then it seems to me that we would have explained and foreseen the attainment of the maximally adequate ideal. The position gained, it would come out that if a thing existed at all, then anything truly sayable of it in those preferred terms would be at least "nearly necessary". Leibniz' "rationalism" is vindicated under his assumption of "best representation".

Digression on "Degrees of Truth" and on The Probability of Statements. The publisher's reader wonders how the various kinds of statement slackness explained above line up with appeals to "degrees of truth". Reader notices that these appeals are apt to have "pragmatic" weight in matters of acting from beliefs that are "true enough", and I suspect that degrees of truth would be adaptable to the employments of those "baysians" who have followed Ramsey in his use of the notion of degrees of belief. Now Pragmatics goes beyond language, and I accordingly must think that a pragmatic conception of degrees of belief would not line up at all with our purely "semantic" conceptions of statement slackness. "Degrees of truth" might differently be thought to have something to do with those doctrines (I think of Wittgenstein and Carnap) that would assign *probability* determinations to statements. Ramsey himself, I suspect, would have argued against any such conception on grounds like those he used to criticism Keynes' notion that *probability* is a measure of logical relations between statements (see "Truth and Probability", pp. 160 *et. seq*). Here are some of my own thoughts on the matter.

Surely one may count out the number of ways in which a pair of dice may land and, of these, the number whose face-up dots add to 7 or to 3. These procedures serve to verify that the ratio of the 7-combinations to the lot is 1/6 and to falsify that the ratio of the 3-combinations to the lot is 1/36. I find no reason to doubt that there are statements for these tests. Differently, one may formulate generalizations about the proportion of 7's in sequences of tosses of pairs of dice. I see no reason not to say that these statements and generalizations about ratios and proportions are also "about probabilities". The mathematical theory of probability, assimilated to the theory of the measure of sets normalized to 1, illuminates our understanding of the relations between the successful and unsuccessful applicability of tests for these conventional products. It has been supposed that ostensible statements about the "likelihood of events" are resolvable as being to the effect that the events in question belong to "populations" covered by appropriate ratio-statements or proportion-generalizations. I doubt whether this further step can be taken while we still want for procedures to establish the "equi-probability of events".

In a different idiom, statements may themselves be spoken of as being likely or unlikely, e,g, a statement to the effect that there is a Chinese text on my study table is unlikely, to me in any event, since I know my books pretty well, and have never seen a book written out entirely in Chinese characters. Some writers, thinking perhaps of such examples, have supposed that statements themselves may be assigned probability measures ranging between 0 for logical impossibilities and 1 for necessities. The idea goes back at least to Wittgenstein's thesis that *Satz p* gives *Satz q* probability in proportion to the number of "truth possibilities" *q* agrees with among those with which *p* agrees[46]. The "analysis" is relative to truth-value assignments for fixing a domain of possibility. It implies that every "elementary sentence" has probability 1/2, a requirement which, if it can be implemented at all, carries counter-intuitive consequences, e.g. that a statement that the sun set last evening gives the statement that the sun came up this morning probability of 25%. Now, at this point, it might seem to some (as it once seemed to me) that my representation of statements in terms of sets of occasions holds out improved prospects for the definition of the logical probability of statements. Aren't sets of occasions exactly the kind of *Spielraum* that the logical conception of probability demands? Well, that hope is soon dashed. The probability of a statement should be the ratio of the measure of "positive possibilities" to some other usually larger set of possibilities. As a first shot, it might seem that the set of positive possibilities we want for the probability of a statement is exactly the set of occasions of successful verifiability of the statement. That, however, won't do, since it would yield the unwanted conclusion that the probability of every false statement is 0. Those are "actualities", one may think, and what we need are "possibilities". So it seems that the set of positive possibilities should be the *whole* set of verification occasions--successful and unsuccessful alike. What is the denominator set of occasions to be? The only reasonable candidates I can turn up are either the set of all occasions of falsification or the union of the sets of verification and falsification together. Both of these are finally disqualified by the consideration that either of those sets may sometimes coincide with the set of occasions of verification, yielding 1 as the probability of a statement that might seem unlikely. The first candidate--the set of occasions of falsification--is doubly disqualified by the consideration that it might be of smaller measure than the set of occasions of verification, in which event the statement would have a probability greater than 1.

This *denouement* of my system, let it be said, in no way jeopardizes the status of statements of probability and certainly casts no doubt upon the *use* of probability theory in statistical inference and as a guide to action. I conclude with some words on this last topic. I above instanced the case of a statement of there being a Chinese text on my study table as one that was "unlikely to me". It is natural, I allow, to speak of the probability of statements in the absence of knowledge of their truth. Some, in thinking of such examples, lay stress on "to me", whence they are brought to champion a concept of "subjective probability" as a measure of degree of assent. Confidence not statement is under assessment. I concur, with reservations. It is anyway evident that the example illustrates a concern for the perlocutionary uses of partial information, in reasoning and in action, and recalls the reader's remark about the pragmatic use of notions of "degrees of truth" in relation to further action. Those "uses" lie beyond the reach of language pure and simple. There can be no purely "semantic" or "logical conception" of such matters.

Better definitions of these three kinds of statement slackness will emerge in Part II, where we shall be able to bring in different statements of the same "kind".

The definitional efforts of this section have been obstructed by a recurring difficulty: Even though a "several" statement is true or false, we sense that it *might not have been* when a corresponding numerical statements was, or that a false "6 or 7" statement *might have been true* when the corresponding "6" statement was false. Such "possibilities" are indicated in assertions of the statements. But now every possible occasion is an actual one. In our strivings, we have managed to find only partial solution to the difficulty over excessive concreteness by exclusive appeal to the actual occasions for testing the compared statements.

Further though not complete relief from the bother of concreteness will be found through "second" definitions of definiteness, preciseness and determination for statements, that gain formal exposition in Chapt. 5 of Vol. II. The guiding idea for these second definitions is simple enough. Though a particular "several" and its corresponding "two" numerical statement are false together, we know that there are *other* "several" statements that are indeterminate when their corresponding "two"'s are false; similarly, there are *other* "6 or 7" statements that are true when the corresponding "6" is false. The wanted possibilities, here as ever, are realized in other *actualities*. We broaden the field of comparison to comprise other statements and their occasions of testing. The "other" statements must, of course, be of the "same kind". The explanation of this sense of "same kind" awaits the additional apparatus of Vol. II. It may, for now, perhaps be enough to say that we shall need the idea of a criterial-*kind* of procedure (a "proto-criterion") that yields different actual tests when applied "to" different "locations"--a notion to which we have already made advance appeal in order to secure that compared statements are "comparable" (p. 354 above). These different tests serve as criteria for different statements of the "same kind". If this actually false statement is testable by a criterial kind of procedure applied to the locations of these students, you can be pretty sure that there is another indeterminate statement testable by application of the same criterial kind of procedure to the locations of other batches of students.

I end the main text of this volume with that advertisement to the next.

NOTES

[1]With thanks to Arthur Melnick.

[2]By N. Kendrick.

[3]To be expanded at #5 of Chapter 17, p. 427. Of recent publications on perception, I have profited most from G. G. Gibson's *The Senses Considered as Perceptual Systems* and a *Scientific American* collection called *Image, Object and Illusion*.

[4]V_1 as shown in the picture is either empty or non-empty. If it is non-empty, then F_1 is empty and F_2 is non-empty and V_2 may be either; if V_2 is then non-empty (first option) then the represented statement is simply true; if it is empty (second option), then the represented statement is what we shall call "directly necessary" Now suppose that V_1 is empty: V_2 must be non-empty; F_1 may be empty or non-empty; if it is non-empty then the represented statement will be simply false if F_2 is non-empty (third option) or "impossible" if F_2 is empty (fourth option); if F_1 is empty, F_2 must be non-empty (fifth option) and the represented statement is neither true nor false.

[5]With thanks to Hugh Chandler.

[6]Due, I believe, to Hilary Putnam

[7]For which see the "Note to the Amphiboly of Concepts of Reflection", *Critique of Pure Reason*, B324-346.

[8]In seminar discussion, Nancy Kendrick properly defended Leibniz against Kant's criticism and mine on grounds that Leibniz' ideas after all are capacities for representing forms *in nature*; so they must somehow be brought to the world. Well and good; but then, if those capacities are not themselves also for getting to the world--a suggestion that cuts against Leibniz' usual way of talking about them--we need to hear more about how they are to be applied to that end.

[9]That must be the thrust of the first passage of the *Essay* in which "idea" occurs systematically, in connection with the argument that innate principles must be constituted of ideas that could be only innate (I.ii.18). It is implicit in the controlling thesis of the work, finally developed in Part IV, that knowledge is a perception of the relations of ideas, and in the thematic formula of Part III, that words stand for ideas.

[10]*An Analysis of Knowledge and Valuation*, Chapt.VII; *The Foundations of Empirical Knowledge*, Chapt. V.

[11]Now published as "Anhang B" of *Ludwig Wittgenstein u.d. Weiner Kreis*, ed. B. McGuiness, Blackwell. This work, apparently privately circulated among members of the Vienna Circle, strikes me as a kind of operationalist updating of Wittgenstein's *Tractatus*.

[12]Yes! If it were of *reality*, then metaphysics would encompass every study.

[13]Or, for readers skeptical of the interpretation I have stolen from Melnick, "possible parallels".

[14]Pp 000-000. I discuss the issue in "Hume Was Right, Almost; and Where He Wasn't Kant Was", *Midwest Studies in Philosophy, Vol. IX*, pp.135-50

[15]See his *Space, Time and Thought in Kant*, 1989, Vol 204 of this series. The constructive parts of Melnick's enquiry are currently undergoing revision.

[16]Thanks to S. Palfrey and J. Ratliff, who urged it in a seminar at Oberlin in the fall of 1978.

[17]The two-slits-open, "interference" side of the experiment has now been done in several ways, even for single electrons (See "Riding the Atomic Waves", *Science News*, Sept. 7, 1991, pp.158 f) My authorites cannot tell me whether the separate-slots-open-separately side has also been done; but I suspect it has been, for that would seem to be the easier challenge, technically speaking.

[18]These actions are all meant to be tests. Nancy Kendrick noticed that one might reach out to test for the presence of a displaceable something, which must then be distinguished from reaching out to displace something.

[19]The publisher's reader understandly wonders about the relations between my allowance for exceptions to Excluded Middle and various formalizations of truth-value-gap logics. I wish I knew! My guess is that my presentation, which abstracts from notational structures and even from assertion, affords a natural realization of the possibility of gappy valuations for the sentences of formalized languages.

[20]The expression may also be said to have a "meaning", but in a different sense from that in which an *utterance* is said to have a meaning. The meanings of most expressions give them a plurality of what I shall call *uses*, e.g. one of the meanings of "white" gives it both predicative, attributive and referential uses, where these different uses associated with the same one word meaning are realized (e.g.) in utterances to the effect that a piece of chalk *is white*, that a piece of *white* chalk is crumbly and that *white* is a neutral color. The occurence of "white" as a feature of an utterance may serve to contribute one of those or perhaps still other "uses" to the total meaning of the utterance.

[21]So our "principle" may be used to advance our understanding of the explanatory bond between knowledge-"that which" and belief.

[22]In the understanding of Russell's *Principles of Mathematics*, Chapt.V. Linguists nowadays are apt to call expressions having such "denoting uses" "determiners"; I myself fancy W. E. Johnson's "applicatives".

[23]He didn't bother over ascriptions of necessity, which, in the *Begriffschrift* anyway, he counted as non-starters: "must" affects only the "coloration" of the assertoric. I have latterly come to realize that my own doctrines, which repeatedly appeal to the principle that such "intentional objects" as *kinds*, *information* and *beliefs* are identified in relation to formulations verges close to Frege's thesis about oblique reference.

[24]Frege, to my knowledge, never cited the consideration that an object has plural meaning-determinations as evidence for *Sinn* (see p. 255 above). Nonetheless, his adherence to the principle of the plural determination of *Bedeutung* by *Sinne* is unarguable.

[25]For documentation, see my "On The Determination of Reference by Sense", *Studien zu Frege,III*, Matthis Schirn (ed), Vol III, pp 85-95. This note was adapted from that article.

[26]Documentation for this, unsupplied in Shwayder, *ibid*, is found at *Grundgezetze*, v. III, #66, at least in regard to *Bedeutungen*.

[27]I speculate that Frege used "*Satz*", meaning *sentence*.

[28]The term is appealing but sloppy. Though I occasionally fall in with talk of "truth conditions" for exposition, it is an idiom I prefer to eschew for systematic purposes since it is unclear from the literature what these conditions are supposed to be and hard to discern through this unclarity what kind of semantic theory talk of truth conditions is supposed to get us to.

[29]This is a special case of our argument against extensional semantics, see pp. 255f. I believe that Davidson's well-known theory of "Meaning and Truth", as first set-out under that title, falls under the characterizations and criticism of this paragraph.

[30]The thesis that ascription of truth to a statement, s, presupposes the existence and assertion of s, but not conversely, was apparently evident to Buridan. (See E. A. Moody, *Truth and Consequence in Medieval Logic*, pp. 68f., 106.) My "presupposition" position, buttressed by Buridan, stands opposed to the idea perhaps best if surprisingly expressed by Frege who once said that every predicative sentence has the same content as a truth-ascribing sentence (at p.20 of the reprinting of the A. M & Marcelle Quinton translation of "The Thought" [*Mind.*1956] in P. F Strawson's *Philosophical Logic*; Frege's example was in respect of the content of the sentence "I smell the scent of violets"). Not true, if the sense of that remark is that every assertion is also an assertion of truth! While every predication and other assertion is, if you wish, a claim to truth, even (in Frege's language) a "Schritt" in that direction, surely not an assertion of truth.

[31]I say "seems", for what Ramsey most depended upon in the argument of this paper was that same "Tarski disquotation principle" he retained in the later chapters on Truth, where (as I read him) Ramsey was leaning toward a "correspondence" theory. The reference to Tarski is to p. 66 of his 1969 *Scientific American* article called "Truth and Proof". Contemporary truth seminarians are most apt to know about this "Non-content" theory from Strawson's contribution to a widely read *Arist. Soc* Symp, with Austin as the other participant, held in 1950

[32]For which see his "Popular Lectures" on "Pragmatism".

[33]"The Nature of Judgement", *Mind*, 1899, p. 180.

[34]Leibniz' actual words, esp. in his *New Essays* give an unclear impression. After first, unclearly, saying that "...it would be better to assign truth to the relationships amongst objects of the ideas, by virtue of which one idea is or is not included within another (p. 397, Acad. Ed. in the Remnant-Bennett translation), he almost immediately proceeds to opt for "correspondence": "Let us be content with looking for truth in the correspondence between the propositions which are in the mind and the things which they are about" (p. 398).

[35]Moore, *ibid*. Frege, *ibid*, pp. 18 f, where the point is thrice made.

[36]For this point, see the final paragraph of A. Church's "The Need for Abstract Entities", *Proc. Amer. Acad. of Arts and Sciences*, 1951, pp. 100-113.

[37]But not by mine.

[38]I owe the name and a sense of their importance to Melnick.

[39]The doctrine of Wittgenstein's *Tractatus* shipwrecked on the same shoal. (See his 1929 *Arist. Soc.* article on "Logical Form".) To avoid such calamities, we need an explanation of the exclusion of positive determinations. Aristotle turned up an explanation in our deeper need to say what substances are as species (at *Metaphysics*, "Gamma IV").

[40]My opinion that *necessity*, is primarily a feature of assertion is "quasi-kantian". Kant wished to qualify certain *judgements* as being both *a-priori* and as *necessary*. Now certainly he was right that the *a-priori--a-posteriori* distinction has primary application to knowledge and judgement; it is of greater moment to hold that knowledge and judgement (including assertion) are also homebase for logical modality. I agree with Kant in this, despite my following efforts to move the discussion of modality off from that position.

[41]Cf. Locke's version of the "Cogito" which he uses to show that one has certain knowledge of his own existence is an "intuitive truth". *Essay*, IV.10.3.
None of these examples, *pace* Descartes, refute cartesian skepticism, for there is nothing I can see in the fact of this thought to disprove that it was all just part of a world-dream. It should be noticed that Descartes himself used the "Cogito" argument to escape from the clutches of the "Demon Hypothesis", not the "Dream Hypothesis", which he apparently thought he had already freed himself from.

[42]As will be shown on p. 384 of Appendix D, after we have provided the required representation of material conditionals.

[43]A. Plantinga, in a paper given to the Urbana Philosophy Colloquium in the spring of 1979.

[44]I daresay that recent work in "fuzzy logic", of which I remain ignorant, must lie astride the same territory, as must also those doctrines that traffick in *degrees of truth*. I'll touch on the latter notion in a digression to follow.

[45]I previously thought that two statments compared for precision had to be testable on the same occasions, hence had to be equally definite; that now seems to me to be likely wrong.

[46]*Tractatus*, 5.15. This is Keynes' relation. The probability of a *Satz* taken simply would be measured by the fraction of all truth possibilities with which the *Satz* "agrees". Wittgenstein promptly took it all back, by recasting in terms of knowledge at 5.156. To reconcile the two opposed thoughts, he went on to drag in "incomplete" representation, risking inconsistency with other things he wanted to say about every representation being complete and in good order.

APPENDIX D

INTENSIONAL LOGIC: A FRAGMENT

1. In this Appendix, I assemble the main proposals of Chapter 3 into a fragmentary system of "intensional logic". As the work progresses, I shall annex continuations. Neither this formulation nor any other eliminates uncertainty in the foundations. It secures a unified presentation which may serve at once as a pattern of organization and as an instrument of discovery; I have found it useful. This "system", save for additions having to do with logical consequence and equivalence and with the logic of molecular and modal statements, is extracted from the preceding "prosaic theory", which gives content to the formulas. I do not know how closely my presentation resembles other work in intensional and modal logic (which goes on apace) and would be gratified if specialists were to find enough in this system to provoke them to a comparative examination of its consequences.

My title is a misnomer, for this system is not a "logic"--a term I use from deference to a certain literature; it is, however, a theory of "intensions" that overlaps a field of interest classically dealt with by Alonzo Church. It is not a "semantics" in the understanding of Tarski and his followers, for we shall not be dealing with a formalized language in its relationships to a domain of objects. The ensuing development is, nonetheless, a kind of alternative to Tarski-style model theory: we do not assume a domain of objects in relationship to a language, but, among other things, shall rather try to explain the conception of a referable object (a "referent"), no matter the language, by a more discriminating appeal to a domain of occasions.

The presentation is not a "formalization" in the technical sense. Perspicuity, not "rigor", elegance or meta-mathematical convenience is my aim. Some of the formulations are marred by redundancy. Apparatus will be introduced, as needed, in step with the development of the theory. The primitive ideas needed from the very beginning are those of *occasion* (Ω, see p.241), *test* (T, see p.223) and *statement* (S, see pp. 82, 94), and I shall soon introduce the primitive predicables of *truth* and *falsity* (TR, FA, see p.317). Later the primitive ideas of *location* (Λ) and a variety of test kinds which I call *proto-criteria* will be put in place, subject to appropriate axioms (pp.393ff.). I shall introduce a succession of ideas, principles, definitions and notational conventions as it proves necessary or convenient to do so, all with whatever accompanying explanations I am able usefully to supply[1]. A continuing appeal to new axioms is needed because I am often unable to move from easily attained definitions of important statement-predicables like *truth* and *necessity* to representations of statements (if such there are) that ascribe those predicables to (other) statements. That is so for two connected reasons. First, statements, because of their amorphous character, cannot be referred to with identifying precision; *pro tanto*, things said about statements cannot be given a full test-theoretic formulation (pp. 289, 294). Second, the test-theoretic analysis of how we think about statements would have to bring in a consideration of *assertion*; but this theory of statements systematically abstracts from the fact of assertion. I should like to have, somewhere on the side, a formalizable theory of the relationship between language and its products. Failing that, my conclusions regarding things said about statements take the form of axiomatic stipulations of necessary conditions.

2. *Primitive notions and axioms.* We employ the primitive notions of

(1) *Occasion*: Ω

(2) *Test*: T

(3) *Statement*: S

subject to the following axioms:

(4) *Axiom*: Every $t \in T$, has a representation $< {}^{+}\omega_t , {}^{-}\omega_t >$ such that ${}^{+}\omega_t \cap {}^{-}\omega_t = 0$ $({}^{+}\omega_t , {}^{-}\omega_t \subset \Omega)$

(5) *Definition*: A test $t \in T$ is *qualified* just when ${}^{+}\omega_t \cup {}^{-}\omega_t \neq 0$

(6) *Axiom*: Every $s \in S$, has a representation $(V_s, F_s \in T)$ such that
(i) V_S and F_S are qualified.
(ii) ${}^{+}\omega_{V_s} \neq 0 \rightarrow {}^{+}\omega_{F_s} = 0$.

3. *Remarks.*

By "occasion" I really do mean occasions and not something else like peaches or possible occasions. Occasions are (actual) connected regions of space with their contents at times.

The theory is a "representation", not a "reduction". Tests and statements are to be identified *by* but not *with* sets of occasions (see p. 379). There may be all sorts of things other than tests and statements which satisfy these formal specifications (cf. the representation of points, vectors and pari-mutuel results by ordered triples of numbers)[2].

One novelty of this account is that statements are in a certain sense "five" not "two-valued". Every statement is represented by four sets of occasions. We gain an idea corresponding roughly to a conception of "truth value" with the determination of whether these several sets are empty or non-empty. Our axioms allow for five cases

(i) ${}^{+}\omega_{V_s} \neq 0, {}^{-}\omega_{V_s} \neq 0, {}^{+}\omega_{F_s} = 0, {}^{-}\omega_{F_s} \neq 0$ (valuation: +1);

(ii) ${}^{+}\omega_{V_s} \neq 0, {}^{-}\omega_{V_s} = 0, {}^{+}\omega_{F_s} = 0, {}^{-}\omega_{F_s} \neq 0$ (valuation: +2);

(iii) ${}^{+}\omega_{V_s} = 0, {}^{-}\omega_{V_s} \neq 0, {}^{+}\omega_{F_s} \neq 0, {}^{-}\omega_{F_s} \neq 0$ (valuation: -1);

(iv) ${}^{+}\omega_{V_s} = 0, {}^{-}\omega_{V_s} \neq 0, {}^{+}\omega_{F_s} \neq 0, {}^{-}\omega_{F_s} = 0$ (valuation: -2);

(v) ${}^{+}\omega_{V_s} = 0, {}^{-}\omega_{V_s} \neq 0, {}^{+}\omega_{F_s} = 0, {}^{-}\omega_{F_s} \neq 0)$ (valuation: 0).

These "values" are less precise than would be full specifications of the four sets of occasions. That "fine structure" affords grounds for drawing all sorts of worthwhile distinctions not available, so far as I know, to other theories of intensions.

4. We demand that every test is either applied or not applied on every occasion of application.

(7) *Postulate*[3]: For every test, t, there is a function π_t, from the set of occasions ${}^{+}\omega_t \cup {}^{-}\omega_t$ to $\{0,1\}$.

5. Some notational conventions will be useful for purposes of presentation.

(8) *Convention*: If $t \in T$ is represented by $<^{+}\omega_t, ^{-}\omega_t>$, rewrite $^{+}\omega_t \cup ^{-}\omega_t$" as "$\omega_t$".

(9) *Convention*: Rewrite "$\pi_t(o)=1$" ($o \in \omega_t$) as "$(t)_o$" ("t is applied on occasion o").

(10) *Convention*: Rewrite "$(\exists o)\cdot(t)_o$" as "(t)" ("t is sometime applied").

(11) *Convention*: Rewrite "$o \in {}^{+}\omega_t$" as "$[t]_o$" ("t is successfully applicable on occasion o").

(12) *Convention*: Rewrite "$(\exists o)\cdot[t]_o$" ($^{+}\omega_t \neq 0$) as "[t]" ("t is successfully applicable").

6. *No test requires its own application on any occasion.* One must be able to know that a test is applicable in order to know that it is applied. I have proposed that this principle be incorporated into our scheme of analysis as a rule of inference for the theory of testing.

(13) *Rule of inference.* If, from the assumption that a test t_1 is qualified ($[t_1]$) it is inferred in n-steps that a test t_2 is applied ((t_2)), then infer at the n+1st step that $t_1 \neq t_2$.

We resort to a rule of inference at this point because it is not excluded that a test be actually applied on every occasion of applicability. Were it routinely possible to move from tests to the representation of statements about tests (as it is not), we could gain formulation for the principle by stipulating that a test t_1 does not "presuppose" (the successful applicability of) a test t_2 for verifying a statement that t_1 is applied.

7. We now proceed to some definitions and consequences.

(14) *Definition*: t_1 is *commutative* iff t_1 is a test and $<^{-}\omega_t, ^{+}\omega_t>$ is the representation of a test.

(15) *Definition*: t_1 is *kinematically like* t_2 ($t_1, t_2 \in T$) iff $\omega_{t1} \subseteq \omega_{t2}$

(16) *Definition*: t_1 is *included in* ($\subseteq$) t_2 ($t_1,t_2 \in T$) iff
(i) t_1 is *kinematically like* t_2
(ii) $^+\omega_{t1} \subseteq {}^+\omega_{t2}$
(iii) $(t_1)_o \rightarrow (t_2)_o$, for every $o \in \omega_{t1}$

(17) *Definition*: t_1 *weakly presupposes* t_2 over t_3 ($t_1 \subseteq \{t_2\}t_3$) ($t_1,t_2,t_3 \in T$) iff
(i) $t_1 \subseteq t_3$
(ii) t_1 is qualified only if t_2 is successfully applicable ([t_2].

(18) *Definition*: t_1 *weakly minimally presupposes* t_2 over t_3 ($t_1 \subseteq \{t_2\}_{Min}t_3$) ($t_1,t_2,t_3 \in T$) iff
(i) $t_1 \subseteq \{t_2\}t_3$
(ii) For every test, t, $t \subseteq \{t_2\}t_3 \rightarrow t \subseteq t_1$

(19) *Definition*: t_1 *strongly presupposes* t_2 over t_3 ($t_1 \subseteq [t_2]t_3$) ($t_1,t_2,t_3 \in T$) iff
(i) $t_1 \subseteq t_3$
(ii) $\omega_{t1} \subseteq {}^+\omega_{t2}$

(20) *Definition*: t_1 *strongly minimally presupposes* t_2 *over* t_3 ($t_1 \subseteq [tf_2]_{Min}t_3$) ($t_1,t_2,t_3 \in T$) iff
(i) $t_1 \subseteq [t_2]t_3$
(ii) For every test, t, $t \subseteq [t_2]t_3 \rightarrow t \subseteq t_1$.

(21) *Definition*: t_3 is a *strong conjunction of* t_1 *on* t_2 ($t_3{=}t_2t_1$) ($t_1,t_2,t_3 \in T$) iff
(i) $t_3 \subseteq \{t_1\}_{Min}t_2$
(ii) $(t_3) \rightarrow (t_1)$.

(22) *Definition*: t_3 is a *weak conjunction of* t_1 *on* t_2 ($t_3{=}\ t_2 \wedge t_1$) ($t_1,t_2,t_3 \in T$) iff
(i) $\omega_{t3} = \omega_{t2} \cup \omega_{t1}$
(ii) $^+\omega_{t3} = {}^+\omega_{t2t1} \cup {}^+\omega_{t1t2}$
(iii) $(t_3) \rightarrow (t_1)$ and (t_2).

(23) *Definition*: t_3 is a *disjunction of* t_1 *and* t_2 ($t_3 = t_1 \vee t_2$) ($t_1, t_2, t_3 \in T$) iff

(i) ${}^{+}\omega_{t3} = {}^{+}\omega_{t1} \cup {}^{+}\omega_{t2}$
(ii) ${}^{-}\omega_{t3} = [({}^{-}\omega_{t1} - {}^{+}\omega_{t2}) \cup ({}^{-}\omega_{t2} - {}^{+}\omega_{t1})]$
(iii) $(t_3) \rightarrow [(t_1)$ and $(t_2)]$.

(24) *Definition*: t_1 is *ostensive* iff there is an occasion, o, and a test, t, such that $o \in \omega_{t1}$ & $(t)_o$.

Observe the following *consequences*.

(25) *Kinematic likeness* and *test-inclusion* are transitive, reflexive but non-symmetric relations.

(26) If $t_1 \subseteq \{t_2\}t_3$ and $t_2 \subseteq \{t_4\}t_5$, then $t_1 \subseteq \{t_4\}t_3$.

(27) $t_1 \vee t_2 = t_2 \vee t_1$ and $t_1 \wedge t_2 = t_2 \wedge t_1$ (because the definitions are symmetrical as between t_1 and t_2).

(28) (a) $t_1 \vee t_2$ is qualified (*viz* $\omega_{t1 \vee t2} \neq 0$) whenever t_1 and t_2 are.
Proof: If ${}^{+}\omega_{t1 \vee t2} = 0$, then ${}^{+}\omega_{t1} = {}^{+}\omega_{t2} = 0$; but ${}^{-}\omega_{t1 \vee t2} = 0$ only if ${}^{-}\omega_{t1} - {}^{+}\omega_{t1} = {}^{-}\omega_{t2} - {}^{+}\omega_{t2} = 0$. That would require that ${}^{-}\omega_{t1}$ and ${}^{-}\omega_{t2}$ be empty, given that both ${}^{+}\omega_{t1}$ and ${}^{+}\omega_{t2}$ are empty, all of which is incompatible with the assumption that t_1 and t_2 are qualified.

(b) $t_1 \wedge t_2$ is qualified if t_1 and t_2 are.

(c) $t_1 t_2$ is not qualified when ${}^{+}\omega_{t2} = 0$.

(29) $[t_2 t_1] \leftrightarrow [t_1 t_2]$ and $(t_2 t_1) \leftrightarrow (t_1 t_2)$ (even when $t_2 t_1 \neq t_1 t_2$).

(30) If $t_1 \subseteq \{t_2\}t_3$ and $t_1 \subseteq \{t_4\}t_3$, then $t_1 \subseteq \{t_4 t_2\}t_3$ and $t_1 \subseteq \{t_4 \wedge t_2\}t_3$.

(31) If $t_1 \subseteq [t_2]t_3$ and $t_2 \subseteq [t_4]t_5$, then $t_1 \subseteq [t_2 t_4]t_3$ and $t_1 \subseteq [t_2 \wedge t_4]t_3$.

(32) The operations of strong and weak test-conjunction and test-disjunction are associative

8. We proceed to explain a number of things commonly said about statements, beginning with truth and falsity.

(33) *Definition*: s(∈ S) is *true* iff $[V_s]$.

(34) *Definition*: s(∈ S) is *false* iff $[F_s]$.

It is doubtful whether there are statements that ascribe truth-values to statements. It does, however, seem reasonable to assume that there are truth-value ascriptions of a less structured sort that are subject to verification criteria. These, we concluded (p. 318), cannot be defined without remainder solely in terms of the criteria for the statements of which truth-values are predicated. The best we can do is to introduce *truth* and *falsity* as primitive predicables of statements subject to conditions formulated in terms of the criteria of the statements of which they are predicated.

(35) *Primitive*: *Truth*: TR.

(36) *Primitive*: *Falsity*: FA.

(37) *Axiom*: s_2 = TR(s_1) (s_1∈ S) only if (i) when V_{s2} is qualified, both (V_{s1}) and (F_{s1}) and (ii) $V_{s2} \subseteq V_{s1}$.

(38) *Axiom:* s_2 =FA(s_1) (s_1∈ S) only if (i) as in (37) and (ii) $V_{s2} \subseteq F_{s1}$

(39) *Theorem*: There are no statements, s, such that s=TR(s) or s=FA(s). (By application of Rule (13) to (6) and clauses (i) of (37) and (38).)

There are several relations of *logical consequence* and *equivalence* among statements. Speaking generally, a statement s_2 is a consequence of a statement s_1 if the truth of s_1 requires the truth of s_2. This allows for a number of different relations of *consequence*. I define three of these in order of decreasing strength.

(40) *Definition*: s_1 ENTAILS s_2 (**ENT**(s_1,s_2)) (s_1,s_2∈ S) iff $^{+}\omega_{V_{s1}} \subseteq {}^{+}\omega_{V_{s2}}$

(41) *Definition*: s_1 Entails s_2 (**Ent**(s_1,s_2)) $(s_1,s_2 \in S)$ iff $^+\omega_{V_{s1}} \neq 0 \rightarrow {}^+\omega_{V_{s1}} \cap {}^+\omega_{V_{s2}} \neq 0$. (If the first statement is true, then there are occasions on which the two statements could be verified together.)

(42) *Definition*: s_1 *materially implies* s_2 (**MI**(s_1,s_2)) $(s_1,s_2 \in S)$ iff $[Vs_1] \rightarrow [Vs_2]$.

(43) *Consequence*: Any statement is a consequence in all three senses of a false statement.

(44) *Consequence*: A true statement s_2 is materially implied by any statement, s_1, although **ENT**(s_1,s_2) and **Ent**(s_1,s_2) need not obtain.

(45) *Consequence*: If $V_{s1} \subseteq \{V_{s2}\}t_3$, then **MI**$(s_1,s_2)$. (A statement materially implies the satisfaction of its presuppositions.)

(46) *Consequence*: If $V_{s1} \subseteq [V_{s2}]t_3$, then **ENT**$(s_1,s_2)$. (A statement strongly entails its strong presuppositions.)

There are a variety of relations of logical equivalence between statements.

(47) Definition: s_1 and s_2 $(\in S)$ are Truth-EQUIVALENT iff $^+\omega_{V_{s1}} = {}^+\omega_{V_{s2}}$

(48) *Definition*: s_1 and s_2 $(\in S)$ are *Truth-Equivalent* iff **Ent**(s_1,s_2) and **Ent**(s_2,s_1). (The relation obtains if the truth of either statement requires the truth of the other.)

(49) *Definition*: s_1 and s_2 $(\in S)$ are *Truth-materially equivalent* iff **MI**(s_1,s_2) and **MI** (s_2,s_1).

(50) *Definition*: s_1 and s_2 $(\in S)$ are *False-EQUIVALENT* iff $^+\omega_{F_{s1}} = {}^+\omega_{F_{s2}}$

(51) *Definition*: s_1 and s_2 $(\in S)$ are *False-materially equivalent* iff $[F_{s1}] \leftrightarrow [F_{s2}]$.

(52) *Definition*: s_1 and s_2 are *EQUIVALENT* iff $^+\omega_{V_{s1}} = {}^+\omega_{V_{s2}}$ and $^+\omega_{F_{s1}} = {}^+\omega_{F_{s2}}$

(53) *Consequence*: If s_1 is true and s_1 and s_2 are Truth-EQUIVALENT, s_1 and s_2 are EQUIVALENT; and if s_1 is false and s_1 and s_2 are false-EQUIVALENT, s_1 and s_2 are EQUIVALENT. Pairs of statements both of which are neither true nor false are EQUIVALENT.(54) *Definition*: s_1 and s_2 ($\in S$) are *intermediate-EQUIVALENT* iff
$^{+}\omega_{V_{s1}} \subseteq {}^{+}\omega_{V_{s2}}$ or $^{+}\omega_{V_{s2}} \subseteq {}^{+}\omega_{V_{s1}}$
and
$^{+}\omega_{F_{s1}} \subseteq {}^{+}\omega_{F_{s2}}$ or $^{+}\omega_{F_{s2}} \subseteq {}^{+}\omega_{F_{s1}}$

(55) *Definition*: s_1 and s_2 ($\in S$) are *Equivalent* iff ($[V_{s1}]$ or $[F_s1]$ and $[V_{s1}] \rightarrow {}^{+}\omega_{V_{s1}} \cap {}^{+}\omega_{V_{s2}} \neq 0$ and $[F_{s1}] \rightarrow {}^{+}\omega_{F_{s1}} \cap {}^{+}\omega_{F_{s2}} \neq 0$.

(56) *Definition*: s_1 and s_2 ($\in S$) are *contraries* iff
$^{+}\omega_{V_{s1}} \neq 0 \rightarrow {}^{+}\omega_{V_{s2}} = 0$.

(57) *Definition*: s_1 and s_2 ($\in S$) are *subcontraries* iff
$^{+}\omega_{F_{s1}} \neq 0 \rightarrow {}^{+}\omega_{F_{s2}} = 0$.

10. We proposed two definitions of *logical necessity* in the text and another of *near necessity*. Looking ahead to the definition of a contradictory, gained by reversing the order of the tests $\neg s = \langle F_s, V_s \rangle$), we can define corresponding notions of *logical possibility* by appealing to the rule that a statement is logically possible when its contradictory is not logically necessary.

(58) *Definition*: s ($\in S$) is *directly logically necessary*, **N**(s), iff $^{-}\omega_{V_s}=0$

(59) *Definition*: s ($\in S$) is *directly logically possible*, **P**(s), iff $^{-}\omega_{F_s}\neq 0$.

(60) *Definition*: s($\in S$) is *indirectly logically necessary*, **I/I**(s), iff
$\omega_{F_s} \subseteq {}^{+}\omega_{V_s}$

(61) *Definition*: s($\in S$) is *indirectly logically possible*, **oI**(s), iff
$\omega_{V_s} - {}^{+}\omega_{F_s} \neq 0$.

(62) *Definition*: s ($\in S$) is *directly nearly logically necessary* iff there are tests t_1, t_2, such that $V_s \subseteq \{t_1\} t_2$ and $^{+}\omega_{V_s} \cap {}^{+}\omega_{t_1} \neq 0$.

(63) *Consequence*: A statement that could be successfully verified on some occasion upon which a condition presupposed for verification

could be shown satisfied is nearly necessary. A nearly necessary statement Entails a statement of any one of its "presuppositions". Statements ENTAILed by a statement of "presuppositions" are N-necessary

11. Definitions of relative definiteness and first definitions of relative precision and determination among "comparable" (see p.354) statements are as follows.

(64) *Definition*: s_1 is *more definite than* s_2 (s_1,s_2 being comparable statements) iff $[\omega_{V_{s1}} \cup \omega_{F_{s1}}] \subset [\omega_{V_{s2}} \cup \omega_{F_{s2}}]$

(65) *Definition*: s_1 is more precise than s_2 (s_1,s_2 being comparable statements) iff $^+\omega_{V_{s1}} \subset {}^+\omega_{V_{s2}}$ and $^+\omega_{F_{s2}} \subset {}^+\omega_{F_{s1}}$

(66) *Definition*: s_1 is more determinate than s_2 ($s_1,s_2 \in S$) iff $[^+\omega_{V_{s2}} \cup {}^+\omega_{F_{s2}}] \subseteq [^+\omega_{V_{s1}} \cup {}^+\omega_{F_{s1}}]$

Second, "better" definitions of statement precision and determination will be forthcoming in Appendix F of Part II.

(67) *Consequences*: **N**-necessary statements are less determinate than no "comparable" true statements. **N**-necessary statements are never more precise than comparable true statements that are verifiable and falsifiable on the same occasions.

12. I now define some forms of "molecular statements"[4]. We wish to gain characterizations of molecular forms embraced within "standard" or "classical" presentations of propositional logic. I shall assume a reduction of these to negation, $\neg$, and disjunction, **v**. We then define standard conjunction, $\cdot$, as $\neg[\mathbf{v}(\neg(s_1),\neg(s_2))]$ and the standard "material conditional", $\supset$, as $\mathbf{v}(\neg(s_1),s_2)$. I shall also propose independent test-theoretic characterizations of exclusive disjunction, **a**, and of a "non-material" conditional, $\Rightarrow$, in line with what was said on p.147 . In every case, we wish a characterization of the criteria of a molecular form in terms of the criteria of its "components" (arguments). These characterizations should be "truth-functional", but will have additional test-theoretic "structure".

I wish, among other *desiderata*, that these definitions should secure a regular uniform dependence of such dyadic forms, as disjunction and conjunction upon their respective "components". That is gained in every case by simply requiring that any application of either test for a dyadic molecular form should require the application of tests for both components. Otherwise, it will suffice to identify the occasions of successful and unsuccessful applicability of the molecular forms in terms of the occasions of successful and unsuccessful applicability of the criteria for components. These "interpretations" are "stronger" than those available within model theory: concentration upon occasions of testing introduces distinctions overridden by a blanketing appeal to satisfaction by models on a domain of objects.

Take note that I symbolize forms of molecular statement with notations different from what I employ in my exposition. My $\rightarrow$ in this exposition is a molecular conditional, not restricted to statements; $\supset$ is pretty much the same as $\rightarrow$ but exclusively a function from pairs of statements to statements.

13. I begin with the contradictory function.

(68) *Definition*: $s_2=\neg(s_1)$ (s_2 is the *contradictory of* s_1) ($s_1,s_2 \in S$) iff $\langle V_{s2},F_{s2}\rangle = \langle F_{s1},V_{s1}\rangle$.

(69) *Consequences*:

(i) If $s_2=\neg(s_1)$, then $s_1=\neg(s_2)$.

(ii) $s_1=\neg(\neg(s_1))$.

(iii) If s is true, $\neg$ (s) is false; if s is false, $\neg$ (s) is true. If s is **N**-necessary, $\neg$(s) is false and is not **P**-possible; s and $\neg$(s) are neither true nor false together.

Observe that when $s_2=\neg(s_1)$ and **EQUIVALENT**(s_1,s_3), s_2 need not be $\neg(s_3)$. That is because all varieties of logical equivalence are concerned exclusively with occasions of *successful* testing. Our definition of $\neg$ also brings in occasions of unsuccesful testing.

14. We consider non-exclusive disjunction and conjunction together, for I believe it is acceptable to take them as interdefinable by way of DeMorgan's Laws, as duals over negation. Take disjunction, $\mathbf{v}(s_1,s_2)$, as primitive and define conjunction, $\cdot(s_1,s_2)$, thus:

(70) *Definition*: $.\cdot\ (s_1,s_2)$ =defn= $\neg\,(\mathbf{v}(\neg\,(s_1),\neg\,(s_2)))$.

The test-theoretic situation is somewhat complicated. In the first place, we want both criteria for both disjunctive and conjunctive statement forms to be "unified" over both "components", in that both are brought in and neither given preference. That can be gained by stipulating that the application of either criterion for either form should require the "uniform" application of tests for both components. Our definitions of both test-conjunction and test-disjunction were contrived to achieve that *desideratum.* Standard non-exclusive disjunctions are true in case either component is or both are, and false if both are false; the conjunction is similarly false if either component is or both are and true if both are. Test disjunction was defined to make a disjunction of two tests successfully applicable if either is, where precautions were taken to insure that the test-disjunction should not be both succesfully and unsuccessfully applicable on any occasion. Apparently we can use test-disjunction to define the verification criterion of a disjunctive statement and to define the falsification criterion for a conjunctive statement. Strong test-conjunction, which I introduced mainly for purposes of defining conjunctive predicables, will not do for verifying conjunctive statements and falsifying disjunctive statements; the test would not be applicable if the presupposition were not satisfied, which would come over as an inadmissible requirement that one of the component statements would have to be true if the disjunctive or conjunctive statement were to exist. Weak test-conjunction, which is also symmetric over its components (see (22)), was devised to handle just that difficulty. With all those preliminaries, then, we define statement-disjunction as follows.

(71) *Definition*: s_3 is a (non-exclusive) *disjunction* of s_1 and s_2, s_3=**v**(s_1,s_2) $(s_1,s_2,s_3 \in S)$ iff
(i) $V_{s3} = V_{s1} \vee V_{s2}$
(ii) $F_{s3} = F_{s1} \wedge F_{s2}$

(72) *Consequences*: If s_3=**v**(s_1,s_2), then s_3 is true just in case s_1 is true, s_2 is true or both are true; s_3 is false just in case both s_1 and s_2 are false; if both s_1 and s_2 are **N**-necessary, so is s_3. (We cannot assume that the **N**-necessity of a single component requires the **N**-necessity of the disjunction. That would be so only if ${}^{-}\omega_{V_{s1}} \subseteq {}^{+}\omega_{V_{s2}}$ and ${}^{-}\omega_{V_{s2}} \subseteq {}^{+}\omega_{V_{s1}}$.) s_3 is not **P**=possible just in case both s_1 and s_2 are not **P**-possible; s_3 lacks truth value just in case both s_1 and s_2 lack truth-value or one of them does and the other is false.

(73) *Consequence*: $V_{\cdot(s1,s2)} = V_{s1} \wedge V_{s2}$ and $F_{\cdot(s1,s2)} = F_{s1} \vee F_{s2}$.

15. Although the familiar "material" interpretation of conditional-statements as being equivalent to $v(\neg(s_1),s_2)$ has been notoriously unsatisfying, nothing better has hitherto come forward, and this "truth-function" is undeniably a useful instrument for the formulation of theories. We have substantially adopted it to our own devices throughout this work, and would do well to give it a noticed place in our table of molecular statements[5].

(74) *Definition*: $s_3 = \supset(s_1,s_2)$, s_3 is the *material conditional* of s_2 on s_1 ($s_1,s_2,s_3 \in S$) iff $s_3 = v(\neg(s_1),s_2)$.

16. Under the two assumptions that every statement is either true or false and that truth-functions are defined for all arguments, it can be shown that all truth-functions can be introduced in terms of $\neg$ and **v**. Statements do not satisfy either assumption, by our characterizations. First, our statements are, in a manner of speaking, "five valued": a statement may be either (i) contingently true (=+1) or (ii) contingently false (=-1) or (iii) **N**-necessary (=+2) or (iv) **P**-impossible (=-2) or neither true nor false (=0). We can, moreover, introduce functions not defined for all pairs of statement arguments. Those dispensations allow us to introduce functions, not reducible to $\neg$ and v, which stand closer (I believe) to our everyday understanding of "if-then" and exclusive disjunction than what are available within orthodox presentations of propositional logic.

17. Let's now consider the prospects for a "non-material" but "truth-functional" conditional. I suggested that the *general* force of utterances of the form "*if p then q*" is that, while what would be formulated in the protasis clause *p* is pertinent to, usually as a reason for, asserting or otherwise saying what would be formulated in the apodasis clause *q*, the truth of what would be formulated in *q* is not contingent upon the truth of what would be formulated in *p*. That formula holds equally for examples like "There are some biscuits on the sideboard, if you want some" (Geach), for conditional assertions, for conditionals and for material conditionals (pp.146f.). The question arises, after we have discounted these differences, whether truth-functional conditional statements other than material conditionals are ever asserted in everyday life. I believe so. A rabbit hunter who, while forcing smoke into one end of a hollow log, says, "If he's gone into here, he'll come out the other end now." receives no credit for the creature's getting clean away; "If that's a jonathan, it's not ripe" speaks neither truly nor falsely of a pippin; the mathematician who proves that, if The Axiom

of Choice is true, then every set can be well-ordered, has presumably made a connection other than what would be established by the proven falsity of the Axiom of Choice. I believe that any analysis of such statements is subject to at least these constraints. First, if s_1 is true and the conditional s_1 *then* s_2 is true, so too must be s_2; s_1 *then* s_2 would be false in case s_1 were true and s_2 false. Second, s_1 *then* s_2 is not merely a conjunction of s_1 and s_2. Third, the falsity of s_1 leaves the question whether s_1 *then* s_2 essentially undecidable. However, fourth, we do not want to say that s_1 *then* s_2 would not exist if s_1 were false, since we usually assert conditional statements when we do not suppose we know that s_1 is true or false. We can accommodate these demands by stipulating as one condition on s_1 *then* s_2 [which I'll now write as "$\Rightarrow(s_1,s_2)$"], that occasions upon which s_1 is falsified are occasions upon which $\Rightarrow(s_1,s_2)$ has no truth value [${}^{+}\omega_{Fs1} \subseteq {}^{-}\omega_{V\Rightarrow(s1,s2)}$ and ${}^{+}\omega_{Fs1} \subseteq {}^{-}\omega_{F\Rightarrow(s1,s2)}$]. We may additionally stipulate that $\Rightarrow(s_1,s_2)$ is successfully verifiable on occasions upon which s_2 could be successfully verified, on condition that s_1 were successfully verifiable; using our notation for strong conjunctive testing, ${}^{+}\omega_{V\Rightarrow(s1,s2)} = {}^{+}\omega_{Vs2Vs1}$; similarly, $\Rightarrow(s_1,s_2)$ would be falsifiable on occasions upon which s_2 were falsifiable on condition that s_1 was successfully verifiable , *viz* ${}^{+}\omega_{F\Rightarrow(s1,s2)} = {}^{+}\omega_{Fs2Vs1}$. We gain a full specification of the criteria for $\Rightarrow(s_1,s_2)$ by making the additional plausible stipulation that no truth values could be assigned to the conditional if a truth value could not be assigned to either or both components.

(75) *Definition*: $s_3 = \Rightarrow(s_1,s_2)$ $(s_1,s_2,s_3 \in S)$ iff

(i) ${}^{+}\omega_{Vs3} = {}^{+}\omega_{Vs2Vs1}$

(ii) ${}^{-}\omega_{Vs3} = {}^{-}\omega_{Fs3} = {}^{+}\omega_{Fs1} \cup {}^{-}\omega_{Vs1} \cup {}^{-}\omega_{Fs1} \cup {}^{-}\omega_{Vs2} \cup {}^{-}\omega_{Fs2}$

(iii) ${}^{+}\omega_{Fs3} = {}^{+}\omega_{Fs2Vs1}$

(76) *Consequences*:

(a) $\Rightarrow(s_1,s_2)$ ENTAILS s_2

(b) $\Rightarrow(s_1,s_2)$ materially implies s_1

(c) $\neg(\Rightarrow(s_1,s_2)$ has no truth value if s_1 is false. [(c) is relevant to certain of the "paradoxes of material implication" that involve counter-position.)

(d)$\Rightarrow(_{s1},s_2)$ is never either **N**-necessary or **P**-impossible.

18. If a Roman said "aut" rather than "vel", intending that exactly one of the posited alternatives was so "but I don't know which" ("Either he's not coming or he's already on his way"), then, if both alternatives obtained, what he said would be less false than a mishap. The statement would be verified disjunctively and falsified by showing that one of the alternatives did not obtain on condition that the other didn't either. If other alternatives obtained, the falsification test indicated by "aut" couldn't be applied, and the statement wouldn't exist. Symbolize an *exclusive disjunction* as "**a**"; then, in formulas:

(77) *Definition*: $s_3 = \mathbf{a}(s_1,s_2)$ $(s_1,s_2,s_3 \in S)$ iff
(i) $V_{s3} = V_{s1} \vee V_{s2}$
(ii) $F_{s3} = F_{s2}F_{s1} \vee F_{s1}F_{s2}$.

(78) *Consequences*:
(a) $\mathbf{a}(s_1,s_2)$ is true just in case either s_1 is true or s_2 is, but not both are;
(b) $\mathbf{a}(s_1,s_2)$ is false in case both s_1 and s_2 are false;
(c) $\mathbf{a}(s_1,s_2)$ statements are never **N**-necessary;
(d) $\mathbf{a}(s_1,s_2)$ statements are **P**-impossible when both s_1 and s_2 are **P**-impossible;
(e) $\mathbf{a}(s_1,s_2)$ lacks truth value (=0) just in cases where either s_1 or s_2 is false and the other lacks truth value;
(f) $\mathbf{a}(s_1,s_2)$ is undefined when neither s_1 nor s_2 is false.

19. The following tables summarize much of what has been presented above.

+1 = contingent truth; +2 = **N**-necessary truth; -1 = contingent falsehood; -2 = **P**-impossibility; 0 = neither true nor false; X = undefined.

(79)

s	¬(s)
+2	-2
+1	-1
0	0
-1	+1
-2	+2

s_1	s_2	$\mathbf{v}(s_1,s_2)$	$\cdot(s_1,s_2)$	$\supset(s_1,s_2)$	$\Rightarrow(s_1,s_2)$	$\mathbf{a}(s_1,s_2)$
+2	+2	+2	+2	+1	+1	X
+2	+1	+1	+1	+1	+1	X
+2	0	+1	0	0	0	X
+2	-1	+1	-1	-1	-1	+1
+2	-2	+1	-1	-2	-1	+1
+1	+2	+1	+1	+1	+1	X
+1	+1	+1	+1	+1	+1	X
+1	0	+1	0	0	0	X
+1	-1	+1	-1	-1	-1	+1
+1	-2	+1	-1	-1	-1	+1
0	+2	+1	-1	+1	0	X
0	+1	+1	0	+1	0	X
0	0	0	0	0	0	X
0	-1	0	-1	0	0	0
0	-2	0	-1	0	0	0
-1	+2	+1	-1	+1	0	+1
-1	+1	+1	-1	+1	0	+1
-1	0	0	-1	0	0	0
-1	-1	-1	-1	+1	0	-1
-1	-2	-1	-1	+1	0	-1
-2	+2	+1	-1	+2	0	+1
-2	+1	+1	-1	+1	0	+1
-2	0	0	-1	+1	0	0
-2	-1	-1	-1	+1	0	-1
-2	-2	-2	-2	+1	0	-2

20. Test-theory for propositional logic, whether presented by truth-table, through derivations from formal axioms or by reference to rules of Natural Deduction, is pretty much "by inspection". So-called "tautologies", *viz* formulas expressible in terms of sentential variables in any number, with "connectives" $\neg$ and v and that take on value True for all True or False valuations for the incorporated sentential variables, may become Neither-True-Nor-False (=0) but never False if any of the valuations are switched to 0.

Again, familiar rules of natural deduction are satisfied for the "molecular" forms we have introduced. Formally:

(81) : If s_1 and s_2 are statements

(a) *Modus Ponens*: if s_1 is true and either $\Rightarrow(s_1,s_2)$ is true or $\supset(s_1,s_2)$ is true, then s_2 is true.

(b) *Addition*: If s_1 is true, then $v(s_1,s_2)$ is true.

(c) *Deduction*: If s_2 is a consequence of s_1 in any of the senses of "consequence" defined at formulas (40)-(42), then $\supset(s_1,s_2)$ is true. No comparable rule holds for $\Rightarrow(s_1,s_2)$.

(d) *Conjunction*: If s_1 is true and s_2 is true, $\cdot(s_1,s_2)$ is true.

21. I wish now to consider to what extent our conception of a statement satisfies several alleged rules of modal logic under the definitions of *necessity* and *possibility* we have advanced. The restrictedness of our conception of a statement foretells limited prospects here, scant cause for dissatisfaction in view of the disputed status of those modal rules. In this presentation I draw heavily on one of Saul Kripke's classical presentations[6]. Kripke stipulates two rules of inference and five axiom schemata from which different selections can be made in order to achieve differing systems of modal logic. We use "■" to represent *necessity* and "♦" to represent *possibility*.

R1. *Modus Ponens*: If True(s_1) and True($\supset(s_1,s_2)$), True(s_2)

R2. If s is derivable, then ■s is derivable.

A1. $\supset$(■s, s)

A2. $\supset$(s, ■♦s) (The "Broweresche Axiom")[7]

A3. $\supset$[(■$\supset(s_1,s_2)$, $\supset$(■s_1, ■s_2)]

A4. $\supset$(■s, ■■s)

A5. $\supset$($\neg$■s, ■$\neg$■s)

R1 holds for our statements R2 has an indeterminate meaning within our scheme of analysis; however, if we suppose that derivations are made within a standard system of propositional logic, then the rule is not valid, for derivable *s*'s may have no truth value and hence be necessary in neither of our senses.

The inapplicability of derivation as a test for necessity excuses us from the ban against modal *predicables* which Montague claimed to have established by showing, via an appeal to incompleteness considerations (hence *derivation*), that $\supset$(■s,s) is not generally valid. ("Syntactical Treatments of Modality", *Acta Philosophica Fennica*, 1963, pp. 153-66).

We have shown that both **N**- and **I/I**- satisfy Axiom A1.

A3 is satisfied by **N**-. Proof: $\vee(\neg(s_1),s_2)$ is N-necessary iff $^{-}\omega_{V_{\vee(\neg(s1),s2)}} = 0$ and that is so just in case $(o \in {}^{-}\omega_{F_{s1}}) \rightarrow o \in {}^{+}\omega_{V_{s2}}$ and $(o \in {}^{-}\omega_{V_{s2}}) \rightarrow (o \in {}^{+}\omega_{F_{s1}})$, for every occasion, o. But if s_1 is **N**-necessary, then ${}^{+}\omega_{F_{s1}} = 0$, hence ${}^{-}\omega_{V_{s2}} = 0$. Q.E.D.

Similarly, **I/I**- satisfies A3. Proof: If **I/I**$\mathbf{v}(\neg(s_1),s_2))$, then $({}^{+}\omega_{V_{s1}} \cup \omega_{F_{s2}}) \subseteq ({}^{+}\omega_{F_{s1}} \cup {}^{+}\omega_{V_{s2}})$. But if **I/I**$s_1$, ${}^{+}\omega_{F_{s1}} = 0$; so $\omega_{F_{s2}} \subseteq {}^{+}\omega_{V_{s2}}$ Q.E.D.

$\Rightarrow(s_1,s_2)$ is never **N**-necessary; it is **I/I**-necessary only if s_1 is true and s_2 is both **N**- and **I/I** -necessary, in which case A3 is also satisfied by $\Rightarrow$.

ENT(s_1,s_2) and **N**(s_1) does not require **N**(s_2): $\omega_{V_{s1}}$ may be included in both ${}^{+}\omega_{V_{s1}}$ and in ${}^{+}\omega_{V_{s2}}$, and $\omega_{V_{s2}}$ not be included in ${}^{+}\omega_{V_{s2}}$, as when s_1 is N-necessary and s_2 formulates a contingent strong presupposition for s_1. *This husband is male*, which is **N**-necessary, ENTAILS *This is a husband*, which is contingent.

22. Further examination is obstructed. Our analysis rules out any general appeal to the closure principle, that if s is a statement so too is $\blacksquare$s, and indeed there is a question whether that is ever so. Without that, there is no obvious way of providing a criterial analysis of iterated modality. The situation shapes up as follows: Any statement to the effect that a statement, *s*, is **N**-necessary would be paraphrastically equivalent to the universal conditional $(\forall o)\ (o \in \omega_{V_s} \rightarrow o \in {}^{+}\omega_{V_{s1}})$; similarly, a **I/I**s would be equivalent to $(\forall o)\ (o \in \omega_{F_s} \rightarrow o \in {}^{+}\omega_{V_s})$. If we could give criterial analyses of those universal conditionals, then we could immediately proceed to examine whether our statements satisfy A2, A4, and A5. For that, we need procedures for delimiting the total sets of occasions of verification and falsification for statements. We do not generally have these, for two reasons. First, the occasions for testing most statements are spread over time, and I see no way of delimiting those occasions on an occasion specified to a time. Second, and more fundamentally, the operative determination of a set of occasions for testing a statement would require that we establish that all relevant conditions of application be satisfied; furthermore, since we are also concerned with occasions of successful applicability, we must establish that all the conditions of success for the test application are satisfied. But the inspecificity of these conditions, which has been a constant theme running through our whole account, rules out any such determination, and that would remain so even if there were but a single occasion of verification or falsification appropriate to a statement. In the upshot, then, ascriptions of logical modalities cannot be criterially defined and there are no *assertions* of logical possibility and necessity.

23. The situation is not utterly desperate. The mentioned universal conditionals certainly make good sense, though not as assertions. (Cf. "All terrestrial climates support some form of life.") They may be true or false, and there should be ways by which we can bring out their consequences. We can argue heuristically in at least two ways, of which I favor the second:

First, some (not I!) may wish to urge that any occasion for verifying a statement s_2 about all the occasions for verifying a statement s_1 must at least be an occasion for verifying s_1. They might argue that any other occasion for verifying s_2 would be extraneous to the identity of s_1. It might seem, then, that occasions for verifying s_2 are included in the occasions for verifying s_1. But certainly to verify s_2 we must consider all the occasions for verifying s_1. The conclusion would be that $\omega_{V_{s1}} = \omega_{V_{s2}}$. Now if we suppose that there were some way of circumscribing $\omega_{V_{s1}}$ on an occasion belonging to that set and that $\mathbf{N}(s_1)$

were true--so that the verification test for s_1 was successfully applicable on all occasions belonging to $\omega_{V_{s1}}$--we could see that that was so on any occasion belonging to $\omega_{V_{s1}}$, hence on all occasions appropriate to the verification of **N**(s_1) (=s_2), hence **N**(**N**(s_1)) would be true. This manner of thinking, which leads to the conditional EQUIVALENCE of s_1 with **N**(s_1), requires the conclusion that the iteration of modalities does not change the domain of verification, provided the "origin" statement is true. The domain of verification for different such nests of statements defined on different origins go in packages whose members are related by some appropriate equivalence relation-- {s_1, **N**(s_1), **N**(**N**(s_1))...} is one such nest and, if $\omega_{V_{s2}} \neq \omega_{V_{s1}}$, then {$s_2$, **N**($s_2$), **N**(**N**($s_2$)...} is another. This line of thought seems unconvincing on reflection: either it rigidly segregates occasions for testing different statements not in the same nest of necessities or it requires *a-la*-Leibniz that every statement is testable on every possible occasion. I find both options unrealistic.

The *second* line of thought appeals to the desired "sense" of logical necessity. Most statements have contingent conditions of test-application and -success. So, whether a statement, s_1, is in fact **N**-necessary depends upon whether all the occasions of verification ($\omega_{V_{s1}}$) meet conditions for successful application. Even if that is in fact so, it might not have been so. When we ask whether a statement which is necessary must have been necessary, we must take that possibility under review. So, even though any statement is defined on actual occasions, we can allow that the world could be different in respect of the occasions for verifying that statement, and we express that supposition in other formulations, statemental or otherwise. In the concrete terms we favor, we suppose the statement testable on (actual) occasions other than those upon which it is in fact testable. That gives sense to what we say we think, that any occasion in its physical fullness might have been other than it is; we simply posit another occasion differing from this one in the requisite ways. A consideration of iterated modalities, which we pretend yield statements true or false, has the effect of broadening the relevant range of (actual) occasions of testing: imagined occasions for testing **N***s* are possible occasions for testing *s*. Formulations **P**(*s*) or **N**(*s*) ask us to consider those other occasions. Modal predicables have a liberating tendency opposite from that of truth-value predicables. If there were occasions for testing those further possibilities, we would, in formulating **N**(**N**(*s*)), go on to consider a still broader range of possibilities. Everything has now become exceedingly abstract. I do not for myself see how to take more than a single step into modality, for possibilities on possibilities elude my comprehension, and there must be a place for anyone, however more ingeniously grasping than I, beyond which his imagination too cannot go.

24. Axioms 1 and 3 (p.384 above) are satisfied by our statements, where ■ is interpreted either as **N** or as **I/I**; Axioms 2, 4 and 5 were left hanging. Following upon the second line of thought described just above, let's now reconsider how they may fare. These axioms are reformulated as shown below, where "N" indicates my direct logical necessity as a paraphrase for "■" and "/" indicates a paraphrase in terms of my indirect logical necessity.

AN2: $^{+}\omega_{V_s} \neq 0 \rightarrow$ ■$(^{-}\omega_{F_s} \neq 0)$

A/2: $^{+}\omega_{V_s} \neq 0 \rightarrow$ ■$(\omega_{V_s}\text{-}^{+}\omega_{F_s} \neq 0)$

AN4: $^{-}\omega_{V_s} = 0 \rightarrow$ ■$(^{-}\omega_{V_s} = 0)$

A/4: $[(\omega_{F_s} \cap {}^{+}\omega_{V_s}) = \omega_{F_s}] \rightarrow$ ■$[(\omega_{F_s} \cap {}^{+}\omega_{V_s}) = \omega_{F_s}]$

AN5: $^{-}\omega_{V_s} \neq 0 \rightarrow$ ■$(^{-}\omega_{V_s} \neq 0)$

A/5: $[(\omega_{F_s}\text{-}^{+}\omega_{V_s}) \neq 0] \rightarrow$ ■$(\omega_{F_s}\text{-}^{+}\omega_{V_s}).\neq 0$

I don't think that any of these propositions need be true. Consider AN4: If $^{+}\omega_{V_s}$ had any single member that did not meet the condition of success for V_s, then $^{-}\omega_{V_s}$ would not have been empty, although in fact it is. A similar pattern of thought can be fitted to A/4, AN5 and A/5. AN2 and A/2 are more interesting, for they may seem to have "greater chance" of being true than do the others. Assume $^{+}\omega_{V_s}$ had a single member; then, if we forced that out of $^{+}\omega_{V_s}$ and into $^{-}\omega_{V_s}$, we would, by doing that, also have to force every member out of $^{-}\omega_{F_s}$ and into $^{+}\omega_{F_s}$ in order to find an example contrary to the apodasis. That would happen if ω_{V_s} had a single member and F_s were $<^{-}\omega_{V_s}, {}^{+}\omega_{V_s}>$, the commutation of V_s.

The above kind of argument is confirmed by Kripke's model-theoretic analysis of modality[8]. Kripke introduces the idea of a model structure <G,K,R>, where R is a reflexive relation on the non-empty set of "possible worlds" K to which G belongs as a designated "real world". The sense of R is that H_1RH_2 ($H_1,H_2 \in K$) means that H_1 is possible relative to H_2, where that would be glossed to read that any statement true in H_2 is possible in H_1. He then argues that the reflexivity of R assures the satisfaction of Axiom A1. Additionally, R is symmetric if and only if A2 is satisfied. A4 is a consequence of the transitivity of R, and, if R is a full equivalence relation, we also get A5 and the complete system of S5. I propose to interpret every K as the set of occasions of verification appropriate to some statement, *s*, and G as the set of occasions of successful applicability for *s*, reading "o_1Ro_2" to mean that, if o_2 is an occasion of successful application for a statement *s*, then o_1 is an occasion of application for s: $o_1 R o_2$ iff $(o_2 \in {}^{+}\omega_{V_s}) \rightarrow (o_1 \in \omega_{V_s})$. Kripke explains "$\blacksquare s$" to mean that s is true in every model in which it is possible relative to models in which it is true, which (in my interpretation) means that, if $o_1, o_2 \in \omega_{V_s}$, then, if $o_1 R o_2$, $o_1 \in {}^{+}\omega_{V_s}$, *viz* every occasion of possible verification is an occasion of successful verification or **N**s. Since ${}^{+}\omega_{V_s} \subseteq \omega_{V_s}$ for all s, the R-relation in our explanation is reflexive as required. **N** does indeed satisfy Axioms A1 and A3, as now seems to have been expected. R, under our interpretation, is not generally either symmetric or transitive, and it should be no surprize that A2, A4, and A5 are not satisfied by **N**

I do not have a confident grasp on a sense in which $o_1 R o_2$ might be significantly symmetric and hence insufficient comprehension of what the bearing of this would be on the truth of ${}^{+}\omega_{V_s} \neq 0 \rightarrow \mathbf{N}({}^{-}\omega_{F_s} \neq 0)$ (=AN2). We might handle it by bringing in a second statement: R is symmetric iff for every s_1 there is a s_2 such that $(o_2 \in {}^{+}\omega_{V_{s1}} \rightarrow o_1 \in \omega_{V_{s1}}) \rightarrow (o_1 \in {}^{+}\omega_{V_{s2}} \rightarrow o_2 \in \omega_{V_{s2}})$. Something like this would hold if we supposed that $\omega_{V_s} = \omega_{F_s}$ in all cases and that ${}^{+}\omega_{V_s} = {}^{-}\omega_{F_s}$. That is implausible. (I shall later argue that there are conditions where we can successfully apply a test for proving distinctness upon which a corresponding (falsifying) test for identity is inapplicable.)

It is fairly clear that R's being an equivalence relation, would yield the parcelling effect on occasions alluded to above: any two statements which share a verification occasion share them all, with the result that $\omega_{V_s} = \omega_{V_{Ns}}$, for all s. This follows from the appealing but implausible leibnizian assumption that all statements have the same total set of occasions of verification. A tendency to move toward that assumption is manifest in logic treatise stipulations by which it is secured that statements are defined under all conditions. If that tendency were followed to the limit, it would have the further consequence that statements would themselves be necessary existents, *viz* they could be successfully shown to exist in all circumstances for testing any statement at all.

NOTES

[1]Universes corresponding to the primitive ideas will normally be symbolized with capital greek or latin letters (as above), subclasses schematically by subscripted small greek latters ((e.g. "ω_t" is to be taken as a schematic name for a class of occasions), and individuals by subscripted small latin letters.

[2]Equations that seem to say that a test equals a pair of sets of occasions should be taken rather in the geometricians' way of saying that a line equals a pair of determining points.

[3]Dignifying this as a "postulate" is a "safety play". The principle seems a tautology; further, the existence of the stipulated function may follow from the axioms of set-theory. However, I am not altogether confident of such derivations; so...

[4]A term I'm told that "real logicians" no longer much like. The same goes for the once popular "truth-function". I don't much like them either, but shall use both from a professional deference to teachers of philosophy.

[5]The utility of the material conditional for logic is, I think, fairly easily accounted for. Inference at its most fundamental is simply the reduction of alternatives. That may always be made to take the form of the excision of disjuncts. Material conditionals are disjunctions with one member expressed in the negative. Modus Ponens is a sufficent rule for the excision of the negative disjunct, that by virtue of our principle (6ii) that a false statement is not true. A weaker, more broadly applying generalization of this rule for the excision of $\neg p$ would require that we have to establish, not the truth of p, but simply that p is not false. The interesting question is why we should wish to express such disjunctions as conditionals. I explain that as follows: Under one formula, a conditional is true when a statement or other constative product indicated in the protasis clause gives reason for believing that a constative product indicated by the apodasis clause is true (see pp. 147f.). And of course the rejection of an alternative, on grounds that its contradictory is true, does indeed give reason for thinking that what remains after the excision of the appropriate disjunct is true. The conditional describes or gives expression to the way we think about the reduction of alternatives--our most fundamental rule of inference.

[6]"Semantical Considerations on Modal Logic", *Acta Philosophica Fennica*, 1963, pp 83-94. Our present line of investigation gives no opening onto the field of quantified modality, but it favors the doubters.

[7]Atwell Turquette once told me that this is a weakened version of what C. I. Lewis (following O. Becker) called by the same name.

[8]This approach differs in two ways from mine. First, for Kripke, every sentence is either true or false, whereas my statements may be neither. Kripke, for this reason, is able to apply a fairly straightforward extension of the familiar matrix analysis of truth-functional connectives. Second, Kripke assumes a designated "real world" which has no counterpart in my analysis, since every occasion in my scheme is part of the one and only real world. This militates against my introducing an idea corresponding to what Kripke called a "connected model", and indeed what would correspond to that idea in my scheme would normally be empty, hence not generally equivalent (as in Kripke) to ordinary models as originally defined. These differences do not disallow the adaptation of his results to my analysis.

SYNOPSIS OF PARTS II AND III

The foregoing chapters and appendices comprise the first part of three. In the last complete revision of the entire treatise, summarizing captions, corresponding to the marginal summaries of this volume, were annexed throughout. This appendix brings those captions together in order as a summary of the expected contents of Parts II and III. What follows will have to suffice as text for the advance references scattered across this volume, and may incidentally give interested readers some idea of where I am going. Since I cannot anticipate all of the revisions I may have to make in preparing Parts II and III for publication, these "contents" are only provisional.

PART II: STATEMENT-FORM AND *SYNCATEGOREMATA*.

CHAPTER 4. BACKGROUND AND PROGRAM.

1. *Two hypotheses concerning criteria.* To advance our theory of statements, we must fashion apparatus for classifying and selecting tests and for resolving their fine structure. The undertaking is directed toward our actual conceptualization of the cosmos and deals with a range of possibilities that are "real" and contingently existing, because dependent upon how things are with us. The investigation looks back to our theory of language. It also looks ahead to metaphysics. We distinguish between tests that are and are not applied TO something or other. Objection: The account is circular. Objection. You must resolve YOUR conception of an *applicandum*. Statements of all "forms" may be made "about" objects of different kinds. Our statement forms comprise all "logical" forms. Two statements about different things share a form only if their respective criteria are "formally" alike. Hypothesis: The only tests needed for the representation of statements are "location-applicable". Hypothesis: Every criterion can be resolved into basic components classifiable under a few general "proto-criterial" orders. Regular relationships among proto-criterially classified basic components are uniform dependencies.

2. *Statement-form stands in need of theoretical exposition.* A statement's forms are completely determined by conditions indicated in any assertion of that statement. The systematic challenge becomes one of finding classifications of indicated conditions that may contribute to the determination of form. Grammatical theory is an expected adjuvant to our inquiry, but notoriously unreliable. It may be thought that connectives may be eliminated in favor of

direct connection. One may thence improperly conclude that a proper logical syntax would eliminate form markers and abolish concern with form. Syntax is no less significant than morphology. Assumption: qualified singular statements are of the singular form, and so on. The assumption throws our inquiry into a course parallel to that followed by certain developments in First Philosophy. We must scan some of this metaphysics. The traditional literature of logic and philosophy has provided theories of denotation and quantification for the analysis of statement forms.

3. *The formulary of logic*. On the uses of logic. A logical notation is a well-defined set of formulas employed in accordance with definite rules. In order to assign a formula to a statement one must be able, first, to identify the elements of the statement separately marked by various features of the representing formula and, second, to show that the marking is correct. Any given statement may be assigned several different formulas from the same notation. We do not secure an analysis of form by "translating" the utterance into formulas. Criticism of "Regimented Paraphrase". Fregean precedents. Boolean foundations for categoricals. In Defense of Theory.

4. *Two theories of form: Quantification and Denotation.* Theories of quantification and denotation ("suppositio")--the first installed as foundation for contemporary propositional-cum-predicate logic and the second for traditional "term" logic--are instrumentalities for the assignment of statement-form. Russell on form. Quantification theories are constructed in stages. (i) They begin with an assumed basic class of singular predicative statements which are form-classified according to the number of distinct references they incorporate. (ii) Most quantification theories annex to the singulars a class of identities. (iii) Molecular statements are form-classified according to how their truth-values depend upon the truth-values of a finite number of other "component" statements. (iv) We define classes of statements that differ from each other only in respect of a single incorporated reference. A statement is quantificationally form-classified as a universal with respect to such a class just in case it is true whenever all members of the class are true, and otherwise false. (v) A statement is universal for a predicable if it is true just in case all members of a class of statements that differ only in respect of a single incorporated predicable are true, otherwise false. These definitions are recursive, with a resulting proliferation of forms. Frege's Theory of Quantification. Denotational theories of statement-form start with the assumption that every statement can be resolved into at least one subject term and a predication. Statements are form-classified according to the number and the nature of these terms. The terms may be simply referential or they may be denotative. They are referential if they refer to particular objects by name, number or other "given" appellation. They are denotative if they indicate a specific selection from the extension of a particular predicable. Denotative terms are distinguished and the statements into which they are

incorporated are form-classified by what these terms indicate about the extension and by the selections they prescribe. Russell's Theory of Denotation. Model-theory and Montague Grammar. Quine. Denotational and quantificational doctrines differ both in their principles and in their coverage.

5. *A program for statement-form.* The forms of a statement are fixed by the pattern of proto-criterial kinds of test that constitute its criteria. We claim a method for distinguishing any difference in the forms of humanly producible statements. Our account, as an organon of inference, has broader coverage than the other two. We shall additionally provide a "formal" explanation of the idea of singular reference.

6. *The logical insufficiency of form.* "Logical" forms are statement-forms useful for charting relationships of entailment among statements. Our method promises a larger inventory of logical forms than other theories. Are all logical entailments among statements exclusively assignable to the forms of those statements? Some inferences appear to involve considerations of "content". Our theory gives hope that a formal distinction can be drawn between body and beam placement predicables and the appearance of non-formal entailment dispelled. The general question over whether all relationships of logical entailment among statements can be formally expounded is still open.

7. *Metaphysical aspects.* This inquiry has a metaphysical obverse. First Philosophy and "Ontology". Proto-criterial classifications of tests are also components of our conception of a referable object as something that is at once existing, individual, separate and such as to have an identity and various features. Consideration of our conceptions of existence, individuality et. al. as testable facts is a species of First Philosophy. Various procedures of the same basic proto-criterial kinds determine different "categories of objects". "Absolute Idealism" is an alternative.

CHAPTER 5. PROTO-CRITERIA.

1. *Basic test-kinds and the conception of a category.* We now begin our attack on the problem of formalizing the notion of a location-applicable test. Patent similarities in testing and evident relationships among the myriad of forms anticipate a reduction of different criteria to a few orders of basic components. We seek a manageably short inventory of basic kinds of test adequate for the resolution of all others. These basic test-kinds are related among themselves in regular ways, and, in train, relations among resolvable criteria can be standardized. Form-invariant features of different statements, including entailment potentials, are fixed by the supposed invariant relationships among basic orders of testing. A full complement of basic test-kinds constitutes a "basis" for a category of referents. Any assignable point of difference among corresponding entries in a pair of categorial "bases" establishes a "categorial distinction" among referents. The hypothesized reduction of test-kinds predicts

that common forms of statements can be made "about" objects in all categories. Selections of resolvable tests may stand among themselves in relationships formally analogous to those which obtain among the basics. A clutch of such resolvable procedures that are analogous in their interrelationships to the basic complement constitutes a "basis analogue" for an internally determined range of things "constructed on" the referents of the category. Classes are a kind of "construction". My identification of basic kinds will be by trial and error and hopeful conjecture, guided by a consideration of how we actually think and talk about the world. Our inventory of basic test kinds and our exposition of the relationships among them will be guided by the thought that a full complement of basic procedures defines a "category" of referents. Hypotheses will be confirmed by the ease with which they permit us to identify forms of statement available to use in the everyday. Basic tests need not themselves ever serve as (fundamental) criteria properly taken. The accreditation of "proto-assertional" uses of language affords soft evidence for dependency relations among basic test kinds. The notion of an identifiable object of reference is "problematic". "Absolute Idealism" remains an option. There are minimally six basic orders of testing which I call proto-criteria of existence, individuation, distinctness, identity, completeness and featuredness. It is convenient to introduce a seventh order of "primitive stems". A proto-criterion, basic or defined, is itself a kind of test. We do not actually identify a test within a particular proto-criterion until we have specified a location or a set or a pair, triple, etc of locations, to which the procedure is applied. A test results only when a proto-criterion is applied to appropriate locations. A full complement of proto-criteria is assembled by specifying a single E, a single T, a single Δ, a single I, a single A and Π's in any number. I hypothesize that nothing more is required to define a category of referents about which any form of statement can be asserted. Complements of proto-criteria may be variously less than full, commonly in ways that reflect fixed dependency relationships among our several basic proto-criterial orders. The assembly of basic complements is governed firstly by the consideration that a dependent proto-criterial kind is applicable only to locations to which a less dependent kind is applicable. I hypothesize that there is a fixed relation of "local dependence" among our listed proto-criterial orders of tests. This pattern of dependence joins complementary proto-criteria into a categorial basis, all applicable to "locations" comprised within a range of locations brought in as *applicanda* for a most basic E. The basic proto-criteria of a categorial basis stand, secondly, in various systematic relationships of presupposition. Relatively dependent basic proto-criteria are not definable in terms of less dependent ones. Different forms of statement, related though they are through their proto-criterial representations, do not stand in one-way dependency relationships among themselves. There are also dependency relationships between basic proto-criteria of different categorial bases. Preliminary Illustration for Bodies and Numbers.

2. *Evidence for dependence: Categorial truncations and branchings.* Cases of proto-assertional utterance are soft evidence in support of our dependency hypothesis. Two confirming developments would be found in the existence of categories with underfulfilled complements of proto-criteria or "truncations" and in "branchings" where two or more proto-criteria of a given order are associated with a single proto-criterion of another "less dependent" order. Truncations and branchings are only presumptions. Truncations: The category of real numbers is a "standard" Δ-I truncation. The real numbers stand also as an ostensible example of a Δ-A truncation. Bodily sensations are an Δ-I truncation that is not also a Δ-A truncation. Categories of physical particles may be I-A truncations. Sounds are a possible E-T truncation. Winds are a possible E-truncation. Branchings: Bodies and material surfaces are an ostensible instance of branching at the T-node. Bodies and flows of liquid and gas are a likely branching at the Δ-I node. The category of flows is dependent upon the category of bodies. Bodies and the quantities of stuff, were they authentically distinct categories, would stand as an example of a branching at the Δ-I node, and the distinction illustrates that possibility. Luminous Branchings: The use of differing methods of occlusion and spectroscopy for individuation define a branching at the E-T node. *Visibilia* individuated in the first way may then be separated differently and "branchingly" at the T-Δ-node either from a position, by differential occlusion or, by "parallax", across an alteration of perspective. The contrast between "trackable" and perspectively identified *visibilia* illustrates a putative branching at the Δ-I node.

APPENDIX E. EXTENSION OF THE FORMAL REPRESENTATION TO PROTO-CRITERIA.

CHAPTER 6. EXISTENCE

1. *Principles of existence.* Everything exists. Facts of material existence are formulable in utterances that something exists somewhere. An expressible conception of material existence is founded on one's knowing-how to apply tests for material existence. One would seek out experience of what exists by applying a procedure of this kind. Our expressible conception of existence is founded upon our knowing-how to apply "E-tests", and the applicability of such a test is indicted in any formulation of existential fact. We shall explain existence with an account of E-testing. E-tests may be involved in the representation of many forms of statement. Are there different kinds of existence? Categorial pluralism is consistent with physical unity. Some specimens. There are puzzles about the conception and the fact of non-existence. The conception of non-existence implicates the possibility of

existence: An E-test is applicable. We prove non-existence by putting an object into location.

2. *E-tests are monadic location applicable tests.* E-tests are location-applicable. E-tests are applied to locations take one at a time.

3. *Among proto-criteria, E's are maximally independent.* Every proto-criterion is locally dependent upon some E, and E's are locally dependent only upon themselves. E's may be otherwise dependent upon other proto-criteria. Proto-criteria may or may not depend upon E's otherwise than locally. Arguments for local dependence: E-based proto-assertions, branchings and truncations above E's. Objection and reply: E-testing is for features of constitution or of placement. Remarks on Berkeley: Materiality is a fact but not a quality. Existence is primitive fact.

4. *Realms of being and the material principle.* Different E's define different "realms of being" each possibly comprising many different categories. The E-procedure fixes a "material principle" shared by the objects of any category falling within a realm of being. The maximal independence of E-testing implies, with reference to traditional debates, that "esse" is a more fundamental factor in the constitution of substances than is "hypostasis".

5. *Why existence is not a predicable.* The rule that proto-criteria of featuredness both are locally dependent upon and presuppose E's endows the maxim that existence is not a predicate with the right sense, which is that existence is not a predicative feature of things truly said to exist. The maxim has nothing to do with the contrast between "real" material existence and other "unreal" kinds of existence. The maxim has little if anything to do with the predicative ascription of extantness. We are interested in a rule found in writings from Plato to Russell, perhaps best known to modern readers from Hume's declaration that whatever we conceive we conceive to be existent. Hume's statement of the rule raises the question of how we do in fact conceive non-existence. The Kant-Russell argument that supposed predications of existence are not existentially quantifiable. We must provide for unindividuated existence and for non-existence. We cannot conceive an object except as existing because individuation presupposes E-testing. This yields a generalization upon Plato's solution to the problem of non-Being. We conceive "blank" non-existence by imagining the unsuccessful applicability of an E. To conceive blank non-existence we must imagine the use of an actual object as a place-filler. Existence could be a predicable only if the applicability of an E locally presupposed itself. That would be the case only for dubious categories of necessary existents. Can existence be replaced by an occupancy-predicable of locations?

6. *Existential import.* Any statement either of whose criteria presupposes (successful) E-testing has "existential import".

7. *Further features of E-testing.* E-proto-criteria are non-individuative. E-tests do not commute into E-tests and probably do not commute into any other proto-criterial kind of test.
8. *Co-occasionality of locations.* A generalization of the idea of synchronous body locations, which we call "co-occasionality", can be defined entirely in terms of E-testing: Two locations are co-occasional if an E-proto-criterion can be applied to both on a common occasion.
9. *Formalities.*

CHAPTER 7. INDIVIDUALITY

1. *Principles of individuality.* Every identical or distinguishable material object is an individual. There appear to be material individuals which are not identifiable and even some that are not distinguishable from others. Material individuality is the lowest grade of non-accidental "unity" in a sense of "grade of unity" familiar to readers of Aristotle's *Metaphysics*; it also corresponds to Locke's idea of solidity. Comparison with Plato. Our notion of material individuality is included in that sort of particularity which Kant maintained was immediately apprehendible within the forms of space and time by intuition. Any fact of material individuality could be somewhere, sometime formulated in an utterance that a thing is here. Such an utterance may occur in response to the speaker's having succeeded in "individuating" something on the spot. Chief among the conditions of success for such an utterance are conditions for securing an individuation. Our expressible conception of individuality is founded on our knowing-how to perform individuations. We shall explain our conception of individuality by providing an account of individuation. A consideration of utterance affords little guidance to such an account. Individuations are location-applicable tests that pertain to proto-criterial kinds, applicable always to locations taken one at a time, that locally presuppose a particular E. We designate these proto-criteria as T's. T's which locally presuppose different E's are distinct. Distinct T's may locally presuppose the same E. Examples: bodies; surfaces; geometrically bounded *visibilia*; household colors; natural numbers.

2. *Features of* T*-testing.* T's locally presuppose E's. The E-presupposition is "strong" and T's are locally homogeneous with the E's on which they depend. T's are weakly but not strongly individuative. Some T's do and others do not commute into other kinds of tests.
3. *The dependence of distinctness and identity testing on individuation.* Among the various proto-criterial orders, individuation seems to depend upon existence testing only; distinctness and identity testing, on the other hand, seem to depend upon individuation. Objection and reply: There is no individuation of bodies without separations, for one could not know what he had individuated unless he separated it from something else. As evidence for dependency, we exhibit relevant branchings and truncations.
4. *Categorizing realms of being.* T's which locally presuppose different E's are distinct. Different T's may locally presuppose the same E, resulting in there being different categories of individuals within the realm of being. Further categorial distinctions, chiefly in relation to identification, explain why individuals of some categories are "constituted of" individuals of another category.
5. *Feature-testing depends upon individuation.* Tests for features of individual things locally presuppose individuation. the hypothesis that feature-tests presuppose individuation seems needed to support our sense that predicables may be opposed as contraries. Comparisons with Plato and Aristotle. The distinction we have drawn between "primitive stems," and feature-tests lends support to our intuitive sense of a difference between stuffs and predicables of composition.
6. *On the rule that individuality is not a predicable feature of individuals.* Questions of whether individuality is a feature of things run parallel to questions of whether existence is a feature of things. The "tradition" is mostly against treating individuality ("oneness", "thisness", "haecceitas") as a predicable; but there are notable holdouts and the arguments are inconclusive. Individuation does not test for the feature of having that location at which individuation is secured. Feature-testing locally presuppose individuation, and individuality could be a feature only if there was no occasion of unsuccessful individuation. Individuality could perhaps be such a "necessary feature" of things like natural numbers, but not of material bodies.
7. *Individualized existential import.* A statement has individualized existential import if either or both of its criteria locally presupposes a (successful) T. Individualized existential import is common and various.
8. *Reference.* Our theory of individuation has its most important semantic application to the recently controverted issue of reference. We seek a semantic characterization of reference in terms of assertionally indicated conditions of success. Reference is a kind of individualized existential import that requires uniqueness. Uniqueness implicates individuation. Uniqueness presuppositions

may be inspecific about locutions. We stipulate that denotationally indicated presuppositions of uniquely individualized existential import are non-referential. Referring uses indicate as a condition of success for assertion that an individuating procedure is successfully applicable at a particular location. Objection and reply: All locations are equally "particular". Demonstrative reference secures the indication of a particular location as one within or in relation to a proximate occasion of assertion. Objection and reply: This characterization of demonstrative reference is, inadmissibly, with respect to assertion: please put it solely in terms of conditions for the application of criteria. Objection: This obliterates any distinction between the statements that This is my watch and The body in P at t is Shwayder's watch. Answer: The realization of an "identifying" referring use in assertion indicates as a presupposition for the application of a criterion that an identification procedure is successfully applicable "from" a particular location. Identifying references may be complicated by further conditions. Not all reference is demonstrative. Demonstratives are "basic". For every statement containing a reference there sometime exists another truth-Equivalent statement containing a demonstrative reference. A rule of singular paraphrase: Of this x, ϕx. "Individual Concepts" vs. individuation. "Intuition" and the persistence of substance.
9. *Formalities.*

CHAPTER 8. INHERENCE AND PREDICATION
1. *Principles of inherence.* Individual material things may turn up in different places at different times and take part in various happenings at or between those places, are usually of several sorts and have qualities and magnitudes of several kinds and stand in various relationships to other things. The Doctrine of Inherence: Sorts are to be observed with things of those sorts; magnitudes, qualities, relationships and happenings are always of, in or among things, and there is reason to think that places and durations are given only "in relation to" things. Celebrated authors have, in different ways and degrees, denied the doctrine of inherence. Facts of inherence are formulable in "predications" about material individuals, and difficulties in regard to inherence are to be cleared away by gaining an adequate understanding of the expressible conceptions of sort, quality, relationship, etc which are realized in these formulations. Aristotle's Theory of Predication. Predications are utterances, commonly but not necessarily fully assertional in force, "about" individuals, that produce formulations of facts of inherence. Different predications may issue in formulations of the same inherence. Predications may occur in response to successful investigations of individuals. We hold that a subject has an expressible inherence-concept just in case he knows how to apply such a "Π-test". Π-tests are of proto-criterial kinds. Testing and Nominalism. Π-tests apply to individuals and presuppose (successful) individuations. Difficulties

about inherence overcome. Successfully applicable Π-tests which presuppose the same individuation define concurrent or congregating inherences. The inherences in a thing are as variously classifiable as are the kinds of Π-tests which apply under the same presupposition of individuality. Tests applicable but never successfully applicable together under the same presupposition of individuality define excluding inherences. Different Π's that are successfully applicable on the same occasions define the same inherence. Successful predications combine the realizations of predicative and referring uses. The predicative uses thus realized indicate the applicability of a proto-criterial kind of Π. If a predication is an assertion, then we say that the utterance contributes a predicable to the resulting statement. Predicables may also be contributed to statements by non-predicative assertions. Predicables, which are "in" statements, are to be contrasted with corresponding inherences and stuffs which are "in the world". It will, nonetheless, be sometimes useful to speak of individuals as "having predicables", when the appropriate predications are true. Locke and Leibniz on "ideas" and qualities. Utterance indications that a Π-test is successfully applicable to something or other realize "attributive uses". A corresponding "attribute" occurs in a statement when either criterion for that statement presupposes (the successful applicability) of a particular proto-criterial kind of Π-test. A necessary condition for the existence of an attribute is the existence of a corresponding inherence in something or other. Referable "universals", if they exist, are to be identified, not with inherences but with predicables. Π-proto-criteria and their associated predicables are said to "consociate" categorically if they presuppose the same procedure of individuation. Predicables can be test-theoretically classified into many kinds.
2. *Π-testing.* Π-tests are instances of proto-criteria that uniformly, locally presuppose a unique T applicable to a fixed, finite number of separate and distinct locations. The rule of uniformity is theoretically well-motivated, pictorially appealing and confirmed by applications. A counter-example countered.
3. *Predicability and distinctness.* A simple extension of the rule of uniform individuation for Π-testing, to require that relational Π-tests locally depend upon and non-locally presuppose distinctness testing, affords elucidation of the idea that distinctness is not a relation.
4. *Must individuals have predicables or be conceived to have predicables?* Individuals must have but need not be conceived to have predicables of location; individuals conceived of as identifiable are also conceived to have those predicables.
5. *What predicables are not.* Some "things said" of individuals do not seem to be predicative. Some predicables are "non-extensional" and do not directly identify inherences. Ascriptions of mental states and statements of explanation are disqualified as predications by our rule of uniform individuation. Numerical

statements, such as "The Apostles were twelve" are quantificational, and therefore not predicative by our rule. "Greater Forms" are not predicables. Schematic predicates may be used to predicate various meant but lexically unspecified predicables of mentioned individuals. We conjecture that some of our non-predicative sayings may be replaced with schematic predications. We argue that dispositional predicates are schematic and that no classifiable kind of inherence is a disposition.
6. *Formalities.*

CHAPTER 9. IMPRESSIONS OF DISTINCTNESS AND IDENTITY
1. *On the conceptions of "pure numerical difference" and "identity".*
2. *Problems of identity.* The fact of change at once occasions identifications and baffles our understanding of it. The fact that one body may be a part of another creates apparent contradictions in regard to body-identity. Since bodies are identified from one time to another and since concurrent observations at those two times are impossible, the mere passage of time may seem to make the observation of identity an impossible task and leave our conception of identity unsupported in experience. The rule that things are identical only with themselves throws doubt on the existence of facts of identity. The rule of self-identity poses a problem of explaining how true statements of identity can ever be anything other than trivial. Are true identities necessary?
3. *Reductions of identity.* Predication is the alternative to identity. While identity may be treated in logic as an equivalence relation, it is introduced as a primitive and needed for the definition of other relations. Objects of different categories may be identified and distinguished. Self-identity could only be a "necessary" feature of things. Is "being what it is" a predicative feature of objects? Objects may have bona-fide predicables of location which cannot but identify them for what they are. Cannot statements of identity be construed as relational statements about the occupancy of pairs of locations? Identification and the presumption of indiscernibility. Proposals for reducing identity to predication would replace statements of identity with their analyses.
4. *Identity in the conceptual order.* Identity and particulars. Identity and the discrimination of categories. Identity and objectivization of time. Identity and the assignment of location. Identity and reference. The assertion of unchanging truths implicates a capacity for identification. We speculate that only subjects who know-how to make identifications can have our conception of truth. Identity and rounding out the conceptual order.
5. *Testing and asserting identity, and the way around paradox.* Testing for identity is one thing; asserting identity is another. The distinction assists the resolution of paradoxes of fusion and fission.

6. *Identification depends upon distinctness.* Truncation: Real numbers are separable but not identifiable. Branching: perspectival and trackable seen-things may be a case. An aristotelian explanation: The priorities of perception.
7. *Identification and sortal determination are independent.* Identification and Π-testing for sort are prima facie distinct and independent. Substantial identification in Aristotle. Two contemporary "aristotelians": Geach and Wiggins. Arguments opposed: Counting. Cross-sortal identifications are meaningless. Bodies are identified "as bodies". "Several-sort" examples are evidence that not all identifications are sort-specified. Model-theory for sorts presumes "pure identity" in the background. Identification may be sort-specified. We should not confuse questions about essence with questions of identity. Sortal specification solves fewer problems than some have supposed.

CHAPTER 10. SEPARATION AND DISTINCTNESS
1. *Principles of distinctness.* Every individual body is distinct from every other. An individual body is distinct from any other from which it stands separate on a given occasion; any individual body is distinct from any of its individual body parts, and every such part is distinct from the whole. An individual body is distinct from any individual body from which "it" previously stood separate. Facts of distinctness are formulable in utterances which we call "distinctions". We say "These are different". Distinctions may occur in response to successful investigations or tests of distinctness, which we may generally denominate as Δ-tests. Body-Δ's separate one individuatable body from another by displacing the one while holding some part of the other fast. A "part" is separable from a "whole", but not conversely. Body-separations are applicable only to co-occasional locations at a time. Separations and other Δ's establish "immediate distinctness". A body currently separated from another would have been shown to be (non-immediately) distinct from something else by having been tracked from a previous location at which it was separated from the something else in question. It could also be distinguished from something else once in an earlier location by being separated from an object tracked from that earlier location. Such extensions upon distinctness testing presuppose both a Δ test and an identification. One who knows how to make body Δ-tests and who is capable of responding in utterance to his successful application of such a test has an expressible conception of body-distinctness. There appear to be different "kinds" of distinctnesses based on different kinds of Δ's. Body-surfaces. Natural numbers. Real numbers. A Δ may be always unsuccessfully applicable by being applicable only upon the occasions of unsuccessful applicability of the T it presupposes. True distinctions, while not always strictly N-necessary, are "nearly necessary" in that some occasion of successful applicability for a presupposed individuation is also an occasion for successfully applying the verifying Δ-test in at least one direction. Identification locally presupposes Δ-

testing. Δ-tests may (or may not) be "specified" by presuppositional constraints of (successful) Π-testing at both locations.
2. *Formalities.*

CHAPTER 11. IDENTITY
1. *Principles of identity.* A distinct body is identical with itself at all the locations it occupies during a term of its continued existence. Specifically, a body is identical with itself in all of its previous locations. A fact of body-identity of this latter kind may be reportorially formulated in an identification. Such a report may occur in response to a successful investigation which we also, eponymously, call an "identification". Such an investigation is a kind of dyadic location-applicable test that would consist in immediately distinguishing a body tactually followed or "tracked" from some previous location at which it was separable from something else. Tracking may be "in place". Body identifications accommodate possibilities of fusion and fission. Body-identification is "transitive". Body-identifications may not be "totally connected". Body-identifications are never successfully applicable, if applicable at all, to locations "at the same time". Body-identifications are "a-synchronous". If a body-identification is successfully applicable from one location to another, then there is no applicable body-identification from the other to the one: Body-identification is "temporal". One who knows-how to apply body-identifications has an expressible conception of body-identity. We explain that conception by detailing the features of body-identification. The applicability and successful applicability of such tests is variously indicated in different kinds of utterance. There appear to be different categories of objects with their own styles of identity defined by different proto-criterial procedures which we may generically denominate as I-proto-criteria. Identification and sortal Π-testing are independent. Identification and delimitation are independent.
2. *I-testing and D-testing: A formal statement.* I-proto-criteria are dyadic; they are included in and locally presuppose Δ's; they are "transitive"; they may or may not be "totally connected", "a-synchronous" or "temporal". Δ-testing may be extended from immediate separation parallel to I-testing to yield what I call "D-proto-criteria". Formalities.
3. *Statements of identity.* Statements of identity may be represented as verifiable by application of an I to a pair of locations and as falsifiable by application of the corresponding D to the same pair of locations. Objection: Assertions of statements of identity effected by use of proper-names may give no indication of locations. Reply: Proper-names may schematically indicate particular locations of application or, differently, identifying presuppositions. Objection and reply: Some statements of identity contain references fixed to a single location. Equations are verifiable by disjunctions of I's and falsifiable by

disjunctions of D's. Is identity an "equivalence relation"? Specified statements of identity are verifiable by specified I's and are falsifiable by similarly specified corresponding D's. True statements of body-identity are nearly-necessary. Statements of identity made in regard to co-occasionally located bodies, if they exist at all, are neither true nor false. Formalities.

4. *Predicables of location and other "essential" features of things.* Statements of identity entail predications of location. Locating Π-tests, by which predications of location are verified, are successfully applicable on the same occasions as identifications are. Locating Π's may be more widely applicable than their corresponding identifications, and the identification is accordingly included in the corresponding locating Π. Every locating Π is included in some or several identifications. *In-Lm* locating Π's are (monadic) "*lm-projections* of identifications from *L*m. *L*m-projection-"complements" are applicable whenever some *L*n-projection, with *L*n co-occasional with *L*m, is applicable or successfully applicable. A principle of indiscernibility. A "transcendental argument" for the Principle of the Identity of Indiscernibles. Essential features. True predications of location are nearly-necessary. On the relation between classical and contemporary "essentialisms". Formalities.

5. *Temporal existence and bilateral identification.* Bodies may be truly said to exist at times. The Π-tests by which statements of temporal existence are verified must segregate a "temporal factor" in the location of bodies. The desired Π's may be described as bilateral identifications. Π's for temporal existence are included in L_m-projections of an I which presuppose another projection from an L_n co-occasional with L_m. Bilateral identifications of bodies verify statements of facts that must obtain if questions about identity are to be successfully answered.

6. *Particularized existential import and identifying reference.* Statements either of whose criteria presupposes an identification are said to have particularized existential import. When the presupposed identification is from a particular location, the statement is additionally said to contain an identifying reference. The employment of grammatically designated "singular terms" may or may not, in assertional occurrence, convey an identifying reference. "Identifying reference" has been a wasteful topic in contemporary philosophy. Speculations on the grammar of time: Subject-terms indicate temporal presuppositions; predicates indicate fallible identifications.

7. *Temporal-order and the space-around.* I-procedures may or may not be "a-synchronous" in the formal test-theoretic sense of never being successfully applicable to co-occasional locations, and "temporal" in the formal test-theoretic sense that the successful applicability of an I to a pair of locations may require that the I be inapplicable to the locations paired in reverse. Both notions, that of a-synchronicity and temporality of I-testing, may be generalized and strengthened. Pairs of locations belonging to the domain of application of an I

are "immediately synchronous" if co-occasional and if applicable Δ's are successfully applicable and applicable I's unsuccessfully applicable. Locations are "synchronous" if they can be "linked" by relations of co-occasionality to a pair of immediately synchronous locations. An I that is never successfully applicable to a pair of synchronous locations is said to be "strongly a-synchronous". I's that, when successfully applicable to a pair L_1,L_2, are not applicable at all from locations synchronous with L_2 to locations synchronous with L_1, are "strongly temporal". A location belonging to a domain of application of an I is "directly prior" to another if and only if the I is successfully applicable to the pair in that order and not applicable at all to the locations in the other order. A location pertaining to the domain of applicability of an I is "earlier than" another if the one is synchronous with a location that can be linked through a sequence of identifications and switches between synchronous locations to a location synchronous with the other. A domain of application of an I is a "temporal order" if and only if every pair of locations in the domain are either synchronous or exactly one is earlier than the other. A "space around" is a total set of synchronous locations within a temporal order. What is space? Conditions for the existence of a temporal order. Bilateral body identifications based on a single comparison object effect a partial "temporal" ordering of locations. Formalities. Generalizing time?

8. *Primary location.* Some schemes of testing provide for "necessarily occupied locations" to which the E is successfully applicable on all relevant occasions. Objects sometime-individuated in necessarily occupied locations are "weak necessary existents". Necessarily occupied locations of a set into which every object of a category is uniquely identifiable are called "primary locations". Only very "abstract" theories have criterial bases that provide primary locations. The specification of a set of primary locations defines a complete category of maximally abstract, minimally temporal or weak necessary existents. Though true statements about primarily located objects may be sometimes contingent, the corpus of such truths may be taken as necessary RELATIVE TO other conceptual methods of which they constitute the theory and as formulating knowledge gained "a-priori" merely by a consideration of these other conceptual methods. Mathematics is "abstract" in proportion to its distance from its "natural" foundations. Formalities.

CHAPTER 12. DELIMITATION AND GENERALITY

1. *Principles of bunching.* Bunches of bodies exist with their memberships in delimited regions of space at particular times. A fact of bunching may be formulated in an utterance that "These are all". Such a report may be issued in response to a successful investigation or "A-test". A body-A-test exhaustively surveys for material existence the whole of a delimited region at a time ("as quickly as you wish"). Delimited body regions are determined relative to a

probe. A condition of application of a body A-test is that a tester be on the scene. Any location within a delimited region may be "occupied" or "unoccupied". Body-A-tests may fail in being "unthorough". We capture the idea of an A being unsuccessfully applicable by stipulating that occasions of successful applicability are ones on which the underlying E is applicable to every location in the delimited region. Knowing how to apply body A-tests is a condition necessary and sufficient for having an expressible conception of a bunch. The verification of universals. The falsification of existentials. Conceptualizing locations and classes. Aggregative predicables. Further operations. There appear to be available methods appropriate for delimiting sets of locations for categories of things other than bodies. A-procedures do not generally depend upon I's. We stipulate that A-procedures require the sometime applicability of a unique Δ to delimitable pairs of locations. A-testing is a consequential addition to the conceptual order.

2. *A-testing.* Every A-test is of a proto-criterial kind applicable to sets of locations, any pair of which locations are co-occasional in respect of the same E, and of which one is "terminal"; every such test is homogeneous with the application of the E to the terminal location. The E on which the A-proto-criterion locally depends is also one on which a unique Δ also locally depends; that Δ is successfully applicable to any pair of locations in an A-delimited region co-occasionally occupied by individual objects. An additional condition: Every region of A-testing is matched by a set of objects equally numerous to the locations in the region. Objections confirm the need for a matching set of objects. An A-test is successfully applicable on occasions of application upon which the underlying E is applicable to all of the locations belonging to the region of application.

3. *Reference-analogues and constructions: A preview.* A-procedures provide means for the conceptualization of locations and classes as "constructions from" their occupants or memberships. Constructions are appearances in thought and language of referents for reference-analogues. The assertional realization of referring-analogues indicate as conditions of success the applicability of tests which stand in relationships analogous to those that obtain among the tests indicated by the realization of genuine referring uses. These tests are definable from the proto-criterial basis for the category of objects of which those constructions are constituted. In this chapter we seek to show that locations and classes are constructions.

4. *Region- and location-reference analogues.* Our capacity to conceive that locations may be unoccupied argues for the conclusion that delimitation is necessary and sufficient for location-conceptualization. The analogues for both E and T-testing for regions is A-testing. The analogues of Δ and I-procedures for regions are applicable on occasions upon which a (weak) conjunction of two A's with different terminal locations is applicable; the Δ-analogue is

successfully applicable and the I-analogue unsuccessfully applicable on occasions upon which only one of the two A's is successfully applicable. Reduction of location-analogues. Locations in their analogous sense of existence, exist necessarily and necessarily as and where they are. Formalities.

5. *Class-reference analogues.* The membership of a class occupy synchronous locations within a delimited region. Discussion of the question whether regular set-theoretic practices are consistent with the narrow idea of a class on a region. The E-analogue underlying class-reference-analogues, which we call "EC-testing", is simply A-testing. An individuation-analogue for classes (TC), is applicable on all occasions upon which an A is successfully applicable and then unsuccessfully applicable on all occasions upon which an underlying T is unsuccessfully applicable. We say that a successfully applicable TC "bunches" a class on a region. We call that set of locations, in a delimited region, at which individuation can be secured the "class on the region". This notion of "class" may be recursively extended to indefinitely wide sets of synchronous locations. A separation analogue for classes or (ΔC), is applicable whenever TC's are concurrently successfully applicable with respect to distinct regions and is successfully applicable if either region is empty and the other not or if the location of a "selected" individual in one region is "exhaustively" successfully Δ-testable with (all) the locations of objects in the other region. An identification analogue for classes (IC) is applicable on occasions when TC's are successfully applicable to distinct regions and then successfully applicable if (i) both regions are empty or (ii) the regions are concurrently delimited and any occasion of the successful applicability of the underlying T to a location of one region is also an occasion for the successful applicability of that T to a location of the other region or (iii) every occupied location of the one region is successfully I-testable into an occupied location of the other region and there is no occupied location of the other region not successfully I-testable from an occupied location of the first region. Statements analogically ascribing features and relations to classes are tested by Π-analogues for classes. Tests for class-emptiness and non-emptiness do not presuppose the successful applicability of class-individuation-analogue and stand as analogues to "primitive stems". A class is "included in" another if the bunching of the first class individuates only members of the second. Formalities.

6. *Featured and identified classes, exhaustive surveys, selections and denumerations.* We propose definitions of test-kinds affiliated to delimitation, such as exhaustive surveys of and selections from possibly featured or identified classes on regions; these are needed for the representation of different statement-forms. The set of locations in a class on a region occupied by individuals that have a common feature, ϕ, determines the ϕ-featured class on the region. A ϕ-featured class is bunched by exhaustively applying an appropriate Π-test to the occupied locations of a bunched and non-empty class

and is successfully applicable on occasions upon which the disjunction of the Π with a "contrary" test is collectively successfully applicable to all of the occupied locations of the region. We "select" or we "identify" from a location in a bunched non-empty class by applying appropriate individuation and identification procedures to the location. We bunch an identified class by exhaustively applying an identification procedure to locations in a bunched and non-empty class. Exhaustive application of a test to a region or to a class on a region is secured and successfully secured on the intersection of occasions for applying or successfully applying the test to the eligible locations. A φ-examination is the exhaustive application of a Π for φ to the class on a region. Selecting from a region is definable as a test applicable and successfully applicable on unions of occasions for individuating and successfully individuating at locations in regions. Selections may also be identifications and may be coupled with other tests. Some operations on two delimited regions: Identifying from-to. Relating, R-pairing, Denominating and Denumerating are cases of operations on pairs of classes of possibly categorially distinct memberships. We handle these as exhaustive applications of conjunctions of (possible proto-criterially different) individuations.

CHAPTER 13. ON THE CHARACTERIZATION OF PREDICABLES

1. *General introduction.* Individuals are indefinitely featurable. Π-procedures may, accordingly, be indefinitely annexed to a categorial basis. The principles of Π-testing set forth in chapter 8 are meant to cover the lot. To resolve detail, we append "little test theories" of features. These "little theories" afford a finer resolution of "logical form" than that provided by the use of predicate logic. Our interest is less in applications to logic than in applications to several traditionally controverted issues in "metaphysics". Our first concern is with predicables or (in Locke's usage) with "ideas", and not with the various qualities, "inherences", "modifications", "magnitudes", stuffs, sorts, et al. differently "represented" by these ideas. We aim to find test theoretically exponible characterizations of Π-procedures. The characterizations are more or less broad, and may be variously coupled. The apparatus routinely results in an expansion of the semantics which stands companion to our theory of proto-criteria. In this work, such applications are only occasionally made and, in every event, are only illustrative. The results are general and "a-categorial". The selection is unsystematic and the results sometimes incomplete.

2. *Division by proto-criterial involvements.* Π's and their associated predicables may be broadly characterized by reference to the relations they do and do not bear to other orders of proto-criteria.

3. *Primitive Π's.* Testing an unparceled ambience. I have argued that there is a significant difference, both sensed and vernacularly marked, between these procedures and testing for the hardness, temperature, or material composition of

an individual stone. The Π is "included in" the other test which we have dubbed a "primitive stem". The stem is itself a location-applicable proto-criterion. The stem does not locally presuppose individuality. Primitive Π's are included in primitive stems. Formalities.

4. *Ινδιϖιδυατιϖε Π's.* No predicable "individuates" in a strict sense; in a deviant sense, they all do. Some predicables, however, "individuate" in the still different sense that they "attach to objects only as wholes". If only wholes can instance the predicable, the Π is "strongly individuative" in the sense that every occasion of application is restricted to a single location. "Weakly individuative" Π's are never actually applied to more than a single location. Formalities.

5. *Separative Π's.* Some Π's do and others do not incorporate a presupposition of distinctness among the objects occupying the locations to which they are applied. If the presupposition is that the Π is never applied except to ordered sets of mutually distinct objects we call it "strongly separative"; if the presupposition is only that the application could be to distinct objects, we call it "weakly separative". Formalities.

6. *Π's with identity.* Predicables may or may not "involve" identification in a number of different ways. In the first place, the ascription of a predicable may (or may not) merely imply the possibility of identifying objects to which the Π is applied. Non-I-dependent Π's normally apply to a wider domain of locations than do I-dependent counterparts. I-dependent Π's may or may not, strongly or weakly, locally presuppose the I on which they depend. We call those which do (weakly or strongly) "temporal" Π's. such I-dependent Π's as are actually included in I's we call "identifying". Formalities.

7. *Basic predicables.* We have the idea that some monadic predicables are "most basic" because "directly observable" in the sense that their Π's in no way "involve" applications to "other" locations. These supposed "most basic" predicables are shareable features in respect of which an object may change. Supposed "most basic" predicables are "projectable" in the sense of Goodman. We provisionally define "basic predicables" as ones whose Π's are non-temporal and non-I-dependent. Formalities.

8. *Aggregative predicables.* Aggregative predicables presuppose bunch delimitations. Strongly aggregative Π's additionally presuppose that the subject of predication be a member of the bunch. Strongly aggregative Π's locally presuppose individuation at the terminal location of the presupposed delimitations. Impredicative predicables. Formalities.

9. *Π's in opposition: Contradictory Π's, contraries and negatives.* The selection of Π's as "contradictory" is merely formal and presupposes prior determinations. It is sometimes necessary to assume that we do have a Π procedure which is contradictorily paired with another for a range of statements in actual production. Here we may formally dub one such Π the contradictory of another "relative to a statement". Some Π's are naturally opposed as

"contraries". Paired tests for such predicables seldom if ever serve to define statements in actual production. Analysis of contrariety. In quest of "negatives". Examples and analysis: Negative Π's are applied on occasions appropriate to their positives, to locations appropriate to their positives, where every occasion of successful applicability for the negative is an occasion of unsuccessful applicability for its positive and occasions for successfully applying the positive which are also occasions for applying the negative are occasions for unsuccessfully applying the negative. There may be gaps between predicables paired as positive and negative. Formalities.

10. *Π's with Π's: An introduction to "complexity".* Predicables "involve" other predicables in a bewildering variety of ways. These cases are all of them problems for analysis. We consider a number of these "involvements" of Π's within Π's in greater detail, with an eye out to find test-theoretic analyses.

11. *Composite predicables and simples.* Perhaps the most natural kind of theory of anything at all is one that provides a uniform scheme for resolving phenomena into simple elements. Locke's "New Way with Ideas", set forth in Books II and III of his *Essay*, is a chief case in point. Simples are "relative". As a restriction on the lexical selection of composite ideas, we require that the indicated components could occur both predicatively and attributively. Specimens of compositeness. If there are contrasting simples, then they are perceptual irreducibles take in at a glance, sniff, or reach relative to our natural state, and to lack structure exponible in text theoretical terms. Tests for simples are unstaged relative to the subject's capacities. Analysis of constituency. Simple Π's have no constituents. Concurrently testable predicables needn't be constituents. The simple-composite distinction cuts across others. Classifying compositeness. Formalities.

12. *Subordinate predicables.* Some predicables stand in need of complements. These predicates are not elliptical for what would in full formulation be complex. Complementation is not compositeness but presupposition. The presupposition is "local". We say that a Π that locally presupposes another Π is "subordinated". "Application" fixes the presupposition. Subordination may or may not be to a "sortal". Formalities.

13. *Conglomerative predicables.* Objects are "recognized" to have certain features from other features they may have. That recognizability condition is required for certain predicables. Features by which an object is recognized to be what it is may be various. The recognizing features need none of them be absolutely essential to the sort or the stuff. One may not notice or even be prepared to ascribe a feature from which he recognized the other. For sake of a name, I call the "recognized features" "conglomerative predicables" and, following Locke, I call features by which the recognized feature is recognized "leading properties" of the conglomerative predicable. Conglomerative predicables include what have been called "Gestalt" properties. Conglomerative

predicables pose a severe challenge to the methods of test-theory. Conglomerative predicates are "schematic" for leading properties. As an opening thought, we take note of the fact that testing goes beyond recognition. In testing for the conglomerative predicable we must also test for some subsidiary leading feature, possibly unnoticed in recognition, where that would also be one among many other recognizing features. In testing for a conglomerative predicable we must also test for some leading property or other, but no one in particular. A testers knowledge that there be this leading property is grounded in actual experience with the conglomerative feature. Test-theory approaches to this demand for actual experience by making reference to actual application in its characterizations of the test kind. We stipulate that a conglomerative Π is actually applied to a set of locations on a definite occasion only if another consociating Π is also applied to those locations on that occasion. Conglomerative predicables may or may not be "ostensive". Formalities. It is understood that successfully applicability of a leading Π does not assure the successful applicability of the conglomerative Π. We have not said enough about what leading properties are and that leaves the definition incomplete and provisional.

14. *Relational predicables.* Relations are "problematic". We distinguish between relational predicables "said of" things together and the relationships that obtain as inherences among things. "Denials of relations" may be to the effect either that there are no irreducible relational predicables ("ideas") or that there are no relationship-facts. Theorists who conceive assertion as a concatenation of references implicitly disallow irreducible predicables of any kind, relational or otherwise. The name-concatenation theory of assertion, which denies relational predicables, represents assertions themselves as a kind of relationship. Derogating relational predicables. Things are undistinguished by their relations. Relational predicables give imperfect representations of facts. The charge that relational predicables are imperfect representations would be more telling if we could be sure that some "monadic" representations are more perfect than any relation is. The argument that relata need not change over changes of relation is an irrelevant conclusion: ordered sets of relata change over changes of relation is an irrelevant conclusion: ordered sets of relata change over changes in relation. The use of one-place relational predicates argues for, not against, the need for irreducible relational predicables. Π's for relational monadic predicables are included in relational Π's. Relational Π's may be included in primitive stems or in monadic Π's; but even then the presumption of distinct relata argues for irreducibility. Not everything is "relative". Every relational predicable is uniformly relational: Any Π is always applicable to the same number of distinct locations. "Among" is not a relation. Prospects for the logical classification of relations increases with their order of relationality. "Basic" relations are non-I-dependent and non-temporal. "Basic"

n-adic relations hold of n-distinct objects. We stipulate that every n-adic relational Π must be "separative" in the sense of being sometime applicable to n locations occupied by n mutually distinct objects. Relational predicables of all relational orders can be subclassified. Some questions about "basic relations". Formalities.

15. *Token-reflexive predicables: A conjecture.* Token-reflexive predicables are things said in relation to ourselves. Analysis: Token-reflexive Π's are applicable only on occasions occupied by a testing subject. The first conclusion is that token-reflexive Π's are "ostensive". Not all ostensive Π's are token-reflexive. Occasions for applying a token-reflexive Π must also contain a testing subject. The test is applied "from" the subject "to" the location. The suggested conclusion is that the location to which a token-reflexive Π is applicable determines only a single occasion of application for the test. Token-reflexive predications are strongly "non-extensional". True token-reflexive predications are, by this account, N-necessary. Formalities.

16. *Determinant predicables.* A "determinable" is a range of mutually contrary predicables some one of which must be truly ascribable to any appropriate subject or set of subjects; any one of the predicables comprised in the range is a "determinant" of the determinable. Determinants are special. No determinable range of determinants is finite. The provision of measurable determinants typically waits upon and witnesses to a scientific advance. Conditions of failure for Π's of non-determinant predicables make succeeding determinant Π's inapplicable altogether. A first test-theoretic condition on determinants is that they are tested for by consociating Π's of the same relational order. They are, in a phrase, applicable to the same objects. Second, the Π's in question are all mutually "homogeneous". Third, the Π's are mutually contrary. Fourth, some one or other of the Π's is successfully applicable to any appropriate set of locations on some occasion. Finally, every determinant has a "commutation". Our stipulation that determinates have commutations raises the question whether the determinable is itself testable and, if so, testable by a Π. Formalities.

17. *Primary and secondary predicables.* The controverted distinction between "our ideas of primary and secondary qualities". Locke's examination and use of what he invoked as the distinction between "ideas" of primary and secondary qualities, for all of its notorious lack of precision, is useful for starting and for centering discussions. Locke's Listings. Locke's Explanations. With reference to physics. An explanation can come only from an accurate investigation into our "conceptual order". Concentrating on Existence. Primary predicables are Π-tested by touch. This condition on primary predicables may be immediately generalized to all categories and read off as a general formula that Π-procedures for primary predicables are locally homogeneous with their presupposed E's. Existence-co-incident predicables and E-co-incident Π's. Objects needn't have certain sorts of E-co-incident predicables. Not all primary Π's, as we have

characterized them, are E-co-incident. Explaining the sense in which any primary predicable is connected with the existence of a thing: their Π's coincide with "E-contractions". The argument does not carry over to secondary predicables (e.g., color), for here we could not have maintained that the corresponding union of Π's was indeed included in the underlying E. Our definition excludes certain forms of heat and certain other percutaneous-perceivables. Formalities. Jibing with Locke and others.

18. *Perfected predicables, with remarks on "reality" and on the identification of inherences.* Predicables of every kind are equally "representative" and equally "real". It may be that some predicables are "better", "more resembling" representations of inherences than are others. This notion of "resemblance" is a figure of speech which requires elucidation. An elucidation by "perception". Two assumptions for a "leibnizian" analysis. For Leibniz, an idea is "resembling" in degree that it is "determinate". A problem over the identification of inherences. Predicables are "of the same inherence" just in cases where their respective Π's are consociating and successfully applicable to the same location. The inherence is "fixed" in its "best" "most resembling" "most determinate" predicable. "Perfected predicables", as we shall now call them, have Π's that are uniformly successfully or unsuccessfully applicable to appropriate sets of locations. Primary predicables needn't be perfected and perfected predicables needn't be primary. Formalities.

19. *Homeomeric predicables.* What things are made of: Stuffs and their predicables. We are concerned with "predicables of composition", and it's not quite obvious what they are. Mass terms and their adjectives. The paucity of different kinds of sortal predicable stands in telling contrast with the varieties of stuff predicables. We need an "account". To distinguish our objective I shall technically dub these predicables of being-composed-of-stuff as "homeomeric". Homeomeric predicables have Π's of a certain kind. It is clear, first of all, that these Π's are monadic. Homeomeric-predicable Π's are conglomerative. Some of the leading properties of an homeomeric predicable have secondary Π's. Homeomeric-predicables have primitive Π's. Homeomeric-predicables are "homeomeric" in the sense that instances have parts that are similarly featured. Test-theoretically: The Π (or its stem) is always applicable on a "wider" range of occasions and, if successfully applicable to a location, then sometimes co-occasionally applicable to another location. Homeomeric predicables could not be homeomeric in the prescribed way unless they were also primitive. Formalities.

20. *Sortals.*

A. <u>General:</u> We assume that some predicables are distinctively "sortal". Sorts in Aristotle. Paradoxes of confirmation are an isolated problem for whose solution the distinction of sortals may be useful. We do not give so much importance to substantial species as did Aristotle. We neither subscribe to nor deny Aristotle's important thesis that every "basic predicable" also "introduces" and maybe even "is" a species-sortal within a subordinate category. I neither subscribe to nor deny Aristotle's extremely important thesis that every substantial individual belongs to a single species. I disbelieve the aristotelian doctrine that the identification of particulars is sort-subordinated. I less strongly disbelieve that sortal predication presupposes identification. I also disbelieve that sortal predicables are "aggregative" or presuppose delimitation. Sortal predicables have monadic Π's. They are not "relative" *a-la*-Geach. Sortal predicables are weakly individuative. Sortals are not "primitive". Sortal predicables are "conglomerative". Sortal predicables do not "lead" others. I am convinced that we need at least this much for the definition of sortals. I am not confident that we do not need something more for the general definition of a sortal. I conjecture that body-sortals are also secondary and have secondary leading properties and that this further stipulation is enough to secure us a definition for that most important case.

B. <u>Species, modes and the quest for essence</u>: Species in the classical sense are sorts thought of as "definable" *per genus et differentiam*. A "definition" comes "after" the sort. If the object "satisfies" the definition it must be of the sort in question. On the other side, an object may be of the sort in question and yet fail to meet the specifications given by the definition, perhaps because of some defect or other. The definition is meant to give us the "essence" of the species in the sense of telling us what a perfected specimen would be. In what sense an amputated dog is a quadruped. Specific differences are not what they seem. One could check the specifications given by the definition of a species only if the object were known to be of that sort. In test-theoretic summary. The main traditional problem about species has been to work up a plausible account of species definition. Aristotle's theory of definition. Traditionally, species-definitions are discovered by an investigation of instances of the species. The classical concern for species definition has been both neglected and misunderstood in "modern philosophy". The question of what a species definition should be "logically speaking" remains open and interesting. Urmson's contribution. A suggestion: permissible differentiae are "individuative". A species definition Π is a strong conjunction of an individuative but non-sortal differentiating-Π into a sortal generic Π, both components applicable to the same locations, where it is additionally required that the locations to which the generic Π is applicable are included among those to which the differentiating Π is applicable. We leave open the question whether a species Π also presupposes the generic Π's of its definitions. We may further distinguish "modal sortals" as ones that are composite.

C. Partitives: Certain sorts of individuals are "essentially" "constituted" of parts. We seek a test-theoretic characterization of essentially part-constituted sortal Π's. We assume that there are such sorts. The parts of a partitive-sort instance are themselves sorted. A partitive sortal Π is "locally homogeneous" with the Π's for each of its parts. Having a (relevant) part is a leading property for the partitive sortal: The leading Π presupposes the part-sortal Π applicable to the "smaller" included location. That a basic relation obtains between a part and "the rest" is sufficient for an individual to be an instance of a partitive sortal: occasions for successfully applying a basic relational Π are included among those for successfully applying a sortal Π. If a constituting relational Π is always applicable to two part-specified sub-locations, we say that the sortal is "totally partitive". Instances of partitive sorts may or may not be identified through their parts. We call those partitive sorts "strongly partitive" for which there is at least one part that is co-occasionally identified with the identification of any instance. Instances of partitive sortals may have features not assignable to any particular combination of parts. They may even be partially constituted of stuffs not exclusively assignable to any part.

D. Family resemblance: What Wittgenstein called "family resemblance concepts" are sortal predicables. There may be no feature common to all members of the family except indeed that they are members of that family. Wittgenstein thought it important to say that certain sorts were of this kind. The elucidation of family resemblance sortals is a mildly interesting task for test-theory. The general idea is that of a sort held together by a chain of different leading properties. Test-theory for family-resemblances.

E. *Formalities for sortals.*

21. *Time-specified predicables, origins and changes.* Major cases are predicables of location and of temporal existence, schematized, respectively, as being in *L* and existing at t. We consider "essential" predicables schematized as "being ϕ at t" and "being sometime ϕ". A Π for being ϕ at t is included in a bilateral identification and presupposes a Π for ϕ at the location from which the application is made. A test for being sometime ϕ is applicable and successfully applicable on unions of occasions for "being ϕ at t" tests under the additional presuppositions of (successful) identifications from some "earliest" locations to all of the t-locations in question. Some will object to time-specified predicables on grounds that temporal specifications cannot pertain to the proposition properly taken and others on grounds that a predicable must be time-specified if it may be. Time specified predicables obviate opacity difficulties endemic to the representation of tense-operators. We need a distinction between un-time-specified ϕ and time-specified ϕ-at-t predicables in order to explain the sense of saying that an object changed from being ϕ to something else in such a way as also to preserve our sense of continuance between being ϕ at t_1 and being ϕ at t_2.

What explains ϕ's being more fundamental than ϕ-at-t is that the latter requires, as the former needn't, a full-running conception of identity. The distinction matters "metaphysically"; time-specified predicables are "essential" and cannot be "universals". In defense of a modest "essentialism". Origins: An object originates in a location if it can be identified from but not into that location. There seem to be no specifications available for a procedure to prove that result. Objects are shown to originate as sorts in locations from which a location-specified sortal is applicable, where that sortal Π is never successfully applicable to locations from which identification into the location of origination could be successfully made. Such "sort originations" are essential predicables. Our sense of simple spontaneous creation and destruction is apparently unfounded. We test for an object changing from ϕ to ψ by applying a procedure included in a non-identifying Π_1 for ψ under the presupposition of an identification from a location L_m at which the object was ϕ-testable by application of another non-identifying Π contrary to Π_1. Π-tests for changes need not specify the location in which the change occurred. The test posits an interim and it leaves it at that. Π-tests for changes presuppose without having to be included in identifications: they are "temporal" without having to be "essential". Π-tests, for changes may (but need not) be further specified, as being in respect of a certain determinable. Formalities.
22. *Properties.* "Property" is a somewhat technical term for our commonest representations of the perceivable shareable qualities in respect of which objects may change. Property Π's are not token-reflexive. To accommodate our sense that properties are "extensional" we stipulate that property Π's are I-dependent, hence, "non-basic". Property Π's are non-identifying and "non-essential". Properties are not sortals. Properties are not negative. Properties are not relations.

Chapter 14. STATEMENT FORM
1. *The "forms" of a statement are determined by the patterns of proto-criterial orders exhibited by the criteria for verifying and falsifying that statement.* We shall consider several "non-molecular" forms of statement. A statement's forms are determined by the patterns of proto-criterial involvement exemplified by the criteria for that statement. There may be different statements of the same forms about objects of possibly different categories. Statement-forms determine neither "object" nor "attribute". Every statement has a form. Statements are "aristotelian" in respect of form. Superficial grammar may cover over distinctions in form. Something in the assertion of a statement must give indication of the forms of that statement. Applicatives and copulas. The method is "experimental". We build a stock of form-patterns in conformity with the theory of proto-criteria worked out in foregoing chapters. The inventory is open-ended. The stock of definitions may be expanded by the adjunction of

presuppositions, by defining additional operations on classes, by different pairings of verification and falsification criteria and by the formal discrimination of Π-tests. Relationships between verification and falsification criteria are various and often "problematic" because of the difficulty of providing assurances that not both procedures are successfully applicable. The problem is especially acute for singular predicative forms of statement. Non-molecular forms of statements have different degrees of complexity and some may be more "basic" than others. Our theory is merely analytic and is initially uncommitted on several controversial issues, to whose resolution it may come to make some contribution. Our statement-forms properly comprise "logical forms" in the traditional understanding. Forms as statement predicables: a better way?

2. *Operations.*

3. *Singular statement-form.* A singular statement is "about" an individual something, standardly determined by the referential employment of a "singular term". The singularity of any term--and there are different candidates for the distinction--is open to dispute. One or both of the criteria for a singular statement presupposes an individuation at a particular location. This gives singular pride of place to demonstrative pronouns. Other candidate singular terms would indicate other conditions as well. Singularity goes deep. Our stipulation covers a variety of cases. It leaves some disputed cases among the non-singulars.

4. *Predicative statement-form.* Predicative statements predicate something of something. Not all statements are predications. We must find a test-theoretic characterization of "predication". At least one of the criteria for a predicative statement must "contain" a Π-test component. We further stipulate that one of the criteria of a predication should terminate in a Π. Is one enough? We stipulate that both criteria for a predicative statement should terminate in Π's. Predicates.

5. *Singular-predicative statement-form.* Statements may be both singular and predicative. We designate as singular-predicative only statements in which an indicated ordered set of individual referents is the "subject" of predication. Terminating Π's for verifying such statements are applicable to that ordered set of locations over which a uniform individuation presupposition is indicated. What of falsification? The terminating Π's by which a singular-predicative statement is verified and by which it is falsified are applicable to the same set of locations. Our general definition of a statement entails that, in the matter of falsification, one or other of the terminating Π's for a singular-predication should not be successfully applicable to the locations for which individuation is presupposed. A further question: Should one or the other of the terminating Π's of a singular predicative statement be "homogeneous" with the other? On the varieties of predication. Complications of reference.

6. *Identifying statement-form.* That there is an identifying form of statement. The sense of an identification is that of two references to the same thing. Our examples are all "singular", and we shall consider no other kind. A statement of identifying form--a "statement of identity"--is one for which the verifying criterion is an I-procedure applied to a distinct pair of locations and for which the falsifying criterion is a D-test applied to the same pair of locations. Statements of identity may be more narrowly form-classified according to what their criteria presuppose. Most statements of identity in actual production are sortally "specified". The criteria for such statements uniformly presupposes a sortal Π at both locations of application and at every "intermediate" location. Equations are something "more than" statements of identity.

7. *Universal and existential forms of statement.* We contemplate a test-theoretic exploration of several universal and existential forms of statement.

A. Some linguistic data, schematically assembled: Universal and existential forms of statements are, in standard formulation, assertionally produced by use of "applicatives". I conjecture that applicatives in assertional context indicate delimitation or A-testing. Not all statements formulated with applicatives are universal or existential in form. A guiding list of assertional schemata in English and Peano notation. Any "extensional reduction" of cases proceeds from a recognition of their *prima-facie* diversity, and cannot be achieved independently of theory. Our choice of specimens cannot be vindicated except by appeal to a theory empowered by its other applications.

B. General definitions: We seek generic characterizations of universal and existential statement forms. Showing what we mean. We show what we mean in asserting a universal by displaying a relevant "totality", and prove it out by showing that we get the "right" result across that range of cases. Universal statements are verified by the exhaustive application of a procedure to a delimited region. We propose, then, that a universal statement is verified by the exhaustive application of some one or other proto-criterion to an A-delimited region. Universal statements are falsified by exhibiting a "counter-example" selected from the same totality. Generically, the falsification of a universal consists in applying a proto-criterion to an object locally selected from the indicated region. Another class: Universals falsified by application of existence or individuation procedures within regions? The generic definition of universal form leaves open myriad possibilities. The delimited region may be variously qualified. The procedures exhaustively or selectively applied may also be of any kind definable from proto-criteria. Existential forms, corresponding to first class universals, are verified by selectively testing regions and falsified by exhaustively testing regions. Existentials corresponding to the second class of universals may be similarly gained by simply reversing criteria. "The ϕ-

ψ is χ" schematized statements are not falsified in the style of existentials. "A ϕ is χ" statements are not verified in the style of existentials. Generic statements "about sorts" (if there be such) are neither verified nor falsified by exhaustive tests and do not qualify as universals. Universal statements and generalizations.

C. Null-predicate second-class cases; null-predicate first-class cases.

D. An illustrative helter-skelter of one-predicate one-applicative cases--some universal, some existential.

E. Some one-applicative two predicate universal "categorical" forms of statement.

i. "Categorical" A-statements, in traditional denomination, fall within the enormous class of one-applicative two predicable universals, which also comprises the universal quantification ("$(x)\cdot\phi x \rightarrow \psi x$").

ii. Data

iii. Proposals for "any". Vendler; Davison; *Principles of Mathematics; Principia Mathematica*; Quine and Geach.

iv. A quantificational representation of the sense of "any". We seek existential quantifications equivalent to universals. A rule: A universal quantifier represents "any" just where there is an equivalent formula containing an unnegated existential quantifier. Jiggling some questionable data. An extended rule of scope can be made to handle data otherwise disallowed. The rule does not yet distinguish "any" from "each". Test theory must explain the success of the quantificational representation.

v. Test theory for ALL, EVERY, EACH and ANY. We assume that there are these universal forms of statement. Impressions of criteria. Proposed criteria. A thought on "basicness". The account proves out. Digression on questions. Application to specimens incorporating two applicatives. Heuristic Extensions: A Semantics of Encouragement. Class-theory for the differences: A first-order formulation.

F. Test-theory for one-applicative two-predicate existentials. Schematic examples of such forms in actual production are Some ϕ is (are) ψ, A ϕ is ψ, There is (are) a $\phi\psi$, Certain ϕ's are ψ. A survey of possible criteria.

G. Two-predicate quantificational forms. The methods of first-order predicate logic gain these as substitutions into "$(x)\cdot\phi x$" and "$(\exists x)\cdot\phi x$". To make sense of this, we must, with enormous presumption, take the "$\phi x\cdot\psi x$" of the existential conjunction to signify a predicable defined by a Π-procedure that is also a conjunction of Π's for ϕ and ψ. Similarly, the "$\phi x \supset \psi x$" of the universal material conditional presumably signifies a "molecular" predicable. Both pairs of criteria (suspectly) involve presumed Π-tests for falsifying corresponding singulars. The requirement that we are able to substitute truth-functionally composed predicates is "problematic".

To take matters any further we have no recourse but to assume that the molecular stencils substituted under the quantifiers do indeed signify complex predicables for which Π-procedures are available. Test-theory for $(x)\cdot\phi x \supset \psi x$ and $(\exists x)\cdot \phi x\cdot\psi x$.

H. Ruminations on multiply quantified forms. We speculate that criteria for forms that do not contain irreducible relational elements may be gained, if questionably, by repeated application of simpler procedures. The test-theoretic resolution of forms containing irreducible relational components may require exhaustive test applications with respect to two classes. We assume that the variables of quantifications are associated with the regions of testing or "domains of discourse", which may or may not be distinct. We also assume that relations have negatives. These formulas may be haltingly extended by "iterative" methods to "prenex" formulas under which no variable of quantification occurs more than once. Test-theory for formulas containing duplicated variables of quantification would require that we allow selecting at different locations from the same region, simplified perhaps by judicious appeal to "arbitrary selection".

I. Rough thoughts about some "mixed" cases: "how-many", "most", "lots".

APPENDIX F: LOGICAL EQUIVALENCE OF FORMS AND THE VALIDATION OF PREDICATE LOGIC

1. On the logical equivalence of several universal forms.

A. While insisting upon the distinctness of the several two-predicable one-applicative universal forms we have distinguished, their sometime logical equivalence is intuitively evident. We wish to establish the conditions under which those equivalences obtain. The theoretic advantage of quantified universal material conditionals is that they will exist and have truth-values when A-categoricals may not.

B. On the Equivalence of *All* (*every, each*) *of the* ϕ's is ψ statements with $\forall\cdot\phi x \rightarrow \psi x$ statements. Truth-equivalence. False-equivalence.

C. The equivalence of 'Any'-categoricals with $(x)\cdot\phi x$ statements.

D. Class-theoretic reduction. While the Algebra of Classes permits us to give a common "boolean" representation to these several universal forms of statement, these representations are not themselves equivalent to universal statements.

2. A test-theoretic validation of first-order predicate logic.

A. We extend our earlier test-theoretic examination of propositional logic to quantificational first-order predicate logic.

B. We work with the Church formulation of a *Principia Mathematica* and *Grundgesetzse* style system.

C. Seven principles of interpretation. First principle. Second Principle: Every object variable, whether it occurs bound or free or as part of a quantifier, shall be taken to indicate a non-empty region of testing. Third Principle: Functional variables represent predicables appropriate to whatever objects might be individuated within the supposed delimited regions of testing. Fourth Principle: "$(x_1)Fx_1$" will be verified by exhaustively applying a Π for F to objects selected from a non-empty class on region ρ_1 and be falsified by applying a presumed Π for F to an object selected from a class on region ρ_1, where "F" under the quantifier must signify a predicable whose proto-criterion can be applied to objects in regions associated with the quantifier variables. Truth-functionally composed expressions occurring under quantifiers must be thought of as if they expressed predicables. Fifth Principle: Truth functional signs, adjoined to "closed" sentences will be taken according to the explanations of Appendix D. "Open sentences" are taken "schematically" and so is "p v Fx_1"

D. Rules of substitution.
E. Rule of generalization.
F. Axiom to restrict quantification.
G. Schematic axiom of instantiation.
H. Definition of existential quantification.

3. *A test-theoretic validation of the traditional rules of immediate inference for categorical statements.*

B. Rules of contradiction.
C. Rules of conversion.
D. Rules of subalternation.
E. Rules of obversion.
F. Contraries and subcontraries

PART III: CATEGORIES, REFERENTS AND CONSTRUCTIONS, WITH SPECIAL ATTENTION TO THINGS MET WITH IN SPACE AND TIME

CHAPTER 15. METAPHYSICAL CATEGORIES AND DEPARTMENTS OF LANGUAGE.

1. *Toward a metaphysics of referents and categories.* In the foregoing account of proto-criteria we appealed throughout, for purposes of introduction and illustration, to the ideas of an object and of a category of objects. We are concerned with objects conceived of as referents, and the data for the investigation is to be drawn from our actual conceptualization of things. Referents are metaphysically "categorized" according to the criterial basis that underlies reference. The technical idea of a "metaphysical category" corresponds to the everyday idea of a subject-matter. We call those "conceptual

schemes" of which a category of referents is the "subject-matter" a "department of language." The burdens of categorization are among the implications of theorizing. Such "theoretical notions" normally have "substantial counterparts" from which they must be held distinct as concepts. The two concepts are systematically related and may be joined by the '"is" of essence'. A mixed mode "corresponds to" a "substantial counterpart" only if it can be applied for purposes of illuminating and explaining what the counterpart is. The use of a mixed mode for purposes of illuminating a substantial counterpart can be secured only by disregarding "irrelevancies". Our understanding of a substantial counterpart is adjusted to instances and in such a way that we must come to understand that the sort can never be finally codified in a list of specifications. The "idea", in Locke's terms, is "inadequate"; the corresponding mode is, however, precisely something for which we seek specifications and these, once achieved, are not subject to adjustment to instances; rather, we say that an instance fails to meet specifications in such and such respects or degrees; mixed modes are "adequate". Modes are used as standards in ways in which substantial counterparts are not; they are modes. A fully-formed category is defined by specification of an identification procedure. Less fully formed categories are defined by specification of separation procedures or procedures for individuating objects. Evidence for the definition of categories: Authoritative precedents; sense of form; ease of illustration; predictive pay-offs. All of this assumes that categories are real. The criterial basis of a category may also enable reference-analogues to "constructions", which are not referents proper.

2. *Contra categories.* A category distinction is a distinction among things founded on a uniform distinction in thought or "intension". First objection against categorization: Distinctions in thought do not establish distinctions in fact. Second Objection: Categorization is classification. Third Objection: Category-grounding distinctions in I-testing may themselves be only distinctions in thought and not in fact. Fourth Objection: Allegedly distinct categories may be consolidated; categorization is arbitrary. Fifth Objection: Categories are ineffable.

3. *Alternatives to categorization.* Some putative categorizations are classifications of basic reality. Some putative categories are kinds of or "aspects of" or "constructions from" basic reality. Abstract things like numbers are mere "notions", non-objectual syncategorematic features of our thought about anything, hence not objects of any category at all.

4. *Categorization not classification.* We seek arguments that categorization can be distinguished from classification. Questionable Data: "That's no body; it's a mirage"; "There is no natural number such that". First argument: Bodies and quantities of stuff cannot be taken as contrasting classifications. Second

argument: Category ascriptions implicate, not determinate leading properties, as do sortal predications, but determinables, which cannot be leading properties.
5. *Ineffability spoken: hypothesis and hope.* Are categories ineffable at least in the denial or the mistake? Category contrasts are sometimes negative existentials, with the categorizing term occurring as an index to the quantifier. Affirmations of category may be categorially specific statements about sorts or references. One or the other of our interpretations handles false assignments to and denials of category.
6. *Categorization and predication.* A category consists of uniformly comparable referents. Our proposal is superior for purposes of categorization to the use of contrasts in predication. Our proposal predicts and explains the drawbacks of categorizing by predication. Predicable sharing from primitive stems. Our theory implies that only predicables which presuppose identification will do for purposes of categorization.
7. *What objects are essentially.* The members of a category may be identified by reference to the (various) locations from which they may be successfully I-tested. An identifiable object is essentially in its locations. An object needn't of necessity be what it is essentially. Features of an object entailed by its predicables of location are, by our explanation, also "essential" to the object. A richer "aristotelian" essentialism would be entailed just in case objects could be identified only within their species.
8. *Matter.* A notion corresponding to the classical idea of "material principle" is readily and plausibly explained as given with the specification of the E-procedure on a categorial basis. Aristotelian applications. Making the connection from aristotelian assumptions. Questions about degeneration. Prime substantial matter is also prime matter for subordinate categories. Extensions by analogy of the material principle outside the ten categories are less doubtful for our explanation than for Aristotle's. Assimilation of stuffs to matter is "by analogy". What's matter? An alternative? Is prime matter the condition of individuality? Matter and Existence. This conception of matter is also "physical" and "modern". Proximate matters and primitive stems. Matters for other categories. Speculations on the "reduction" of chemistry to physics.
9. *Classifying categories.* Categories are classifiable according to the features of their criterial bases. Temporality. It appears that categories of objects are "temporal" only by association with bodies. Categories of objects for which primary locations are provided are maximally "detemporalized" and are, in a sense, "necessary existents". Different categories may be related through shared proto-criteria. One category may "presuppositionally" depend upon another.
10. *On the relative abstractness of referents.* We seek a rule defining the relative abstractness (or concreteness) of categories of objects. This is an "in shop" issue among philosophers, and we take as data their agreements and disagreements about what things are more abstract than or equally abstract with

other things. I assume (i) that things are more or less abstract or less or more concrete by category; (ii) that one category cannot be both more and less abstract than another one. Some rough formulas for comparative abstractness. One category of objects is "directly more abstract" than another when occasions for individuating objects of the first kind are always properly included among occasions for individuating the other kind, but not conversely. Comparison of abstractness may prospectively be generalized to categories that are not directly comparable. Speculations about why temporal things seem to be maximally concrete. Categories of "mathematical objects" are always abstract and those for which primary locations are provided nearly maximally so. Because these objects are also "intensional" elements of conceptual skills, they may be actualized on virtually unlimited occasions, sometimes by making more fundamental use of those expressions in which the "mathematical objects" are primarily located.

CHAPTER 16. CONSTRUCTIONS

1. *A theory of reference analogues.* The appearance of reference in utterance may give rise to the appearance of a referent in fact. The paraphrastic elimination of referring expressions is not enough to disprove reference. We denominate as "referring analogues" those uses which give the appearance of being referential without actually being so, and use "construction" to cover their apparent referents. Referring-analogues indicate the applicability of tests systematically definable from the elements of a criterial basis; these defined tests stand among themselves in relations analogous to those that hold among the basis proto-criteria. Reference-analogues and construction-kinds are classifiable by what they require and incorporate from a criterial basis. Constructions may be provisionally distinguished by "category arguments". Construction-kinds are not categories apart, for constructions are always secured within categories. Constructions are "nothing new". Constructions are never referents.

2. *Non-substantials in Aristotle.* Aristotle's declaration that everything is either substance, sayable, or inherence poses questions about what inherences and sayables are and of how he would have dealt with species, abstracta, and with such other things as surfaces and *visibilia.* Aristotle's non-substantial categories of being are very nearly (our) constructions-kinds, even in his presentation. Aristotle's predicative "sayables" are among what he called "universals". Though he held that there are no separately existing individual universals, he acknowledged a need to talk of universals nonetheless. An explanation of this talk is that predicative meaning is secured with an indication that one or another subject could be used as a standard of comparison to test whether another subject was of that kind or not. Talk of species and other forms may, in the aristotelian analysis, be represented as by use of a kind of reference analogue

that indicates that an individual substance may be taken as a representative object of investigation. Species are "intentional objects". Aristotle held that talk about geometrical objects was non-temporal talk about bodies or body-talk restricted to certain qualifications; differently, he seems to have held that number-talk was an a-categorially standardizable way of talking of how-many and how-much, but always in regard to some sort of thing; geometric things and numbers both come out as constructions by this analysis. Aristotle would have analyzed surfaces as body-boundaries and as constructions. The light by which things are seen is defined by Aristotle as an activity, and would presumably fit under one of the last two categories as a kind of construction.

3. *Qualities.* Qualities "exist" just in case one or another of a restricted class of predications is true. Our program for the "analysis" of qualities is to define for them criterial analogues of existence, individuation, etc. Talk about qualities is "of" something observable in the world. Qualities are shown to "exist" and are "individuated" in and "identified" relative to identified "proto-bearers". We prove the "non-existence" of a quality by disproving the presence of any quality of the family in the proto-bearer. Qualities may be shared by objects which "match" the designated proto-bearer. Bearers may change in quality. Qualities are not generally "identifiable" over bearer identification. We could introduce a notion of quality retention. On the question of whether the same quality may be represented by predicables defined by Π's belonging to different families: "Theoretical identity" is a definable possibility. Roughly speaking, qualities are themselves featureless. Formalities.

4. *Sorts.* Sorts "exist' when an instance does. Forms and meanings in Aristotle. There is much to be said about sorts; this distinguishes them from qualities. Our wish to provide an analogue for the identification of sorts corresponds to the traditional demand for essences. Sorts are "located in" individuals of a category. A sort is concurrently shown to "exist" and "individuated" by applying a sortal Π to an individual. An alternative suggested by the rule of the exclusion of species. The test for sort-existence seems to hold for every kind of sort. Sorts are "recognized" from subsidiary "leading features". Sorts are not generally to be "distinguished" or "identified" by mention of any one or several of the leading properties from which instances are recognized. Locke on essence. The requirement that an examined specimen be "typical" or "representative" is "something new" that may make the provision of a criterial analogue for the identification of sorts impossible. To say what a sort "is" is to provide an explanation. We say what a sort "is" by giving a definition with a theory; we provide explanations by fitting sortal specimens into those definitions. The definition or "essential specifications" are "mixed modes" in Locke's sense. An elaboration of Locke. The use of definitions for purposes of explanation requires that an object may be both an instance of the sort and of the definition. A sort may have several definitions of several kinds. The determinants of the s

definition may attach to specimens of the sort as "leading properties". Sorts are analogously "distinguished" and "identified" by testing pairs of specimens for such features. Representative specimens may be inspected for "features" of a sort. Sorts may also be classified and otherwise spoken of "metaphysically" by reference to the features of the sortal Π-procedures by which they are individuated. These features do not carry over to instances, and in speaking thus of sorts we may not be treating them as "constructions". Products and other "particularized forms". Formalities.

5. *A potpourri of constructions.* "Logical constructions", in Russell's sense, are constructions. The construction of rational numbers from integers is a case. Configurations, teams, processes, happenings, careers, quantities of stuff, etc are constructions from bodies or other perceptible things. We briefly and informally consider "presences" as an illustrative case in point. Facts, which are "intentional objects" of knowledge are, problematically, constructions from "things said".

CHAPTER 17. BODIES

1. *Categorizing bodies.* We seek explanations of and justifications for a technical usage of "body" to cover and to categorize the referents most familiar to everyday thought. Explanations are needed because of disagreements about whether certain classifications of things are of bodies. The categorization of everyday classifications of referents is open to question. What gives substance to the idea that there may be a category of bodies is the thought that bodies are basic. Bodies are basic in the sense that our conceptualization of other categories of referents requires some capacity to conceptualize bodies, but not conversely. Some honored ancestors and distinguished cousins. We shall be chiefly concerned with forms of conceptualization accessible to the common thought of human beings.

2. *On saying what bodies are.* We need a characterization of "body", unarbitrary because responsible to agreed-upon examples. We explain "body" as a categorizing term, by describing a criterial basis for bodies; we test this against examples of body-kinds and against characteristic determinables. Examples. Preliminary Characterization. We now proceed to describe a criterial basis for bodies.

3. *Body-location:*$_B$ Λ . Body-locations are closed connected regions of space at times. Body-locations are not objects proper, but constructions analogously individuated with the application of body tests. Locations are introduced in relation to a proto-criterion, by specifying a function from locations to which the proto-criterion is applicable to sets of occasions upon which it is applicable and successfully applicable. Occasions, like body-locations, are connected regions of space at times, tactually accessible throughout, with their contents included. A body-location is the overlap of occasions for applying an E-test for bodies.

4. *E-testing bodies: $_BE$.* A body E-test is an act of probing at a place with a tactually sensitive part of one's own body; the measure of success is the agent's awareness of tactual perception. A circumstantial condition of application for an E-test, affecting our preferred representation, is that nothing should make it impossible to enter the tested place from the position of testing at the time. A general condition affecting the determination of the occasions of successful applicability is that there should be tangible matter in the location of application. A general condition of application for body-E-testing having to do with the state of the agent is that he be "kinesthetically aware" of the movements by which he enters the region. A second condition for doing such an act is that the tester should actually enter the region with a part of its own body. Among the agent-centered conditions of success for body E-testing is that agent should actually make contact with something in the region with a tactually sensitive part of its own body. The "measure of success" of body-E-testing is that the agent be self-consciously aware of touching something in the location of application.
5. *Conditions for perception and sensation, with special attention to touch.* Perception and sensation are processes by which information available from an environment affects the behavior of organisms. Failing formulations of tests for these, we shall adduce necessary conditions for the occurrence of these happenings. A subject (S) perceives an object (O) only when a phenomenon induces in S a state of readiness for becoming directly aware of O. On the sense of "directly aware". The phenomenon which induces perception is a relative change in a "medium" such as matter or light. We distinguish the object of perception from the phenomenon. There are difficulties in explaining tactation. Some of these suggest the need for a general distinction between perception and sensation. A subject has a sensation of a certain kind only if subject is directly aware of its own body. Digression on "feeling". A sensation's "quality", as a tickle, itch or a pain, is puzzling. Perception and sensation may occur independently. Kinesthesis is awareness in sensation of movement. Kinesthesis is present in all cases of voluntary action, including testing. Remarks on tactation as a form of perception in which the phenomenon induces a change of resistance along contacting surfaces. Non-contactual perception through the skin, e.g. of radiant heat, seems to be always sensation mediated.
6. *Individuating bodies:$_BT$.* Bodies are individuated in the largest connected regions of space they maximally occupy at times. A procedure for drawing in the complete boundary around contactable matter would suffice for purposes of individuating bodies if it could be generally and effectively applied. However, there seems to be no such procedure available that does not presuppose accomplished body-individuation. Smoothing over at a place of contact or drawing a closed curve on a body would not suffice for individuating bodies, since those procedures fail to distinguish the body from its accessible surface. Locke's discussion of the origin of our idea of solidity is a useful lead. Bodies

are individuated by displacement. We need additional constraints to distinguish the individuation of bodies from that of liquid flows. A condition for applying a body-individuation is that tester should not perceive a flow, and that there should be no flow in the tested location is a presupposition of body individuation. A flow-test is a "primitive-stem" of which the measure of success is the self-conscious perception of motion. Body-individuation presupposes the commutation of such a test, which is also a primitive stem. Occasions of body-individuation are constrained by the condition that everything in the location to which the test is applied should be at rest relative to the center of the occasion of testing. A $_BT$ is a test in which an agent, who has secured contact with non-moving matter, seeks to displace what it touches by tug, pull or relaxation of pressure; the measure of success of such an act is the animal's kinesthetic awareness of its movement of pushing, tugging against resistance or of yielding to pressure. Observations: Self-movement of agent is not by itself a body individuation. Individual bodies may contain liquids. Bodies are of the same order of size as testers. "At a time". Individuation presupposes existence and tester individuates what exists in the tested locations. The displaced stuff is tactually connected. Body-individuation is not identification. Body-individuation is "individuative", but not "strongly" so. Two counter-examples countered. First counterexample: Displacing mud underfoot. Second counterexample: A man sitting in a car with the motor on releases the clutch and therewith causes the displacement of the car with himself enclosed.

7. *Testing for separation or immediate distinctness of bodies:* $_BD$. Individual bodies and other things individuatable by $_BT$ that occupy unconnected regions of space at a time are distinct. Distinct bodies may be connected as part or whole or as different parts of a single whole or merely by lying adjacent. A test for distinctness applicable for all these cases would be to hold one individual body fast in part while displacing the other. A test of this kind requires that the agent self-consciously secure tactual perception at both locations and the measure of success is that the agent then kinesthetically registers firmness at one position but not at the other. A $_B\Delta$ is a strong conjunction of holding fast into an individuation at the other co-occasional location. On the question of whether a $_B\Delta$ is ever applicable at all. How a $_B\Delta$ can be successfully applicable. How a $_B\Delta$ can be unsuccessfully applicable. Our procedure is superior to that of jiggling both bodies concurrently. Part and whole and other "relations of distinctness".

8. *Testing for the identity of bodies:* $_BI$. Bodies exist "for a time" and may be verbally identified by reference to the locations they occupy at different times. Such a verbal identification of a body would be verified by applying an identifying procedure "from one location to another". The burden of this section is to elicit conditions for applying and successfully applying such tests. Bodies

are identified from and to locations in which they might be individuated. $_{B}$I's locally presuppose two $_{B}$T's. Bodies come to exist and cease to exist. Bodies cannot be individuated hence cannot be identified to or from locations at times when they do not exist. Bodies cannot be identified to or from locations at times before they come to exist or after they ceased to exist. Self-conscious awareness of continuous tactation is an agent seated condition of application for a $_{B}$I. Testers awareness of continuous tactation requires that he be kinesthetically aware of keeping with and of being at rest relative to what he touches, hence that he be kinesthetically aware of himself moving or holding still. Corresponding to the agent-seated condition of continuous tactation is the already cited circumstantial condition that something tangible should continue to exist over an interval of time. This puts constraints on the occasions of application for $_{B}$I. What exists continuously at the intervening locations may fuse, split, change in material conditions and be unindividualized. These explanations provide warrant for saying that $_{B}$I's are applied to pairs of distinct locations. I now assume that $_{B}$I is different from both $_{B}$T and $_{B}\Delta$. The two locations to which a $_{B}$I is applied are distinct and non-co-occasional. The locations cannot be "at a time". The temporal separation of the locations to which a $_{B}$I is applied is "definite". Since, in our understanding, tests are not applied if not completely applied, we may assume that the occasions for applying a $_{B}$I are only those associated with the location "to" which it is applied. $_{B}$I's presuppose a (successful) $_{B}\Delta$-separation on occasions associated with the location from which the $_{B}$I is applied. A $_{B}$I is successfully applicable only if the contacted matter in the location to which the test is applied is separable from something else. Formally, the occasions for applying a $_{B}$I to a pair of occasions L_1,L_2 are included in the occasions of applying $_{B}$T to L_2 and the occasions for successfully applying the test are included in the union of occasions for successfully applying $_{B}\Delta$ to L_2 with some other co-occasional location. I believe that the specifications on $_{B}$I-testing we have given are "enough" and that we have defined the category of bodies in accordance with recipes of Chapter 15. $_{B}I(L_1,L_2)$ and $_{B}I(L_2,L_1)$, if they both exist, are different tests. A demonstration of the evident "temporality" of $_{B}$I-testing will be attempted later. What we have already observed entails that $_{B}$I is "weakly temporal" in the sense that not both $_{B}I(L_1,L_2)$ and $_{B}I(L_2,L_1)$ are successfully applicable. Our characterization of $_{B}$I leaves open possibilities in regard to fusion, fission and specification of sort which can be closed down by stipulations. A body in location L_2 may be distinguished from another in a location L_1 at another time either by separating the body in L_2 from a body identified from L_1 or by identifying the body in L_2 from a location in which it was separated from L_1. $_{B}$D: Whenever a $_{B}$I is successfully applicable so too is a $_{B}$D and conversely; both may be successfully applicable to the same pair of locations on the same occasions. The "paradoxes" of body-identity. Body-fusion, fission and reassembly of parts give rise to apparent violations of the rule that two things equal to the same thing are equal to each other. Sortal-specification doesn't solve the problems. Sameness of matter does not solve the problems. Preservation of parts does not solve the problems. Our solution is that there may be no statements of identity and distinctness defined over fission and fusion. This is theoretically well-founded and seems to be in accord with common understandings. Fusion: Is temporality the solution? Evidence for "no statement". Fission: Evidence for "no statement". Solutions to the Delian Ship, in two versions.

9. *Delimiting body-regions: $_{B}A$.* Bunches of bodies are collectable within closed and connected regions of space at a time. The collection of a bunch of bodies is within a delimited region of body-locations. Such a region is defined by the successful application of a body A-procedure or $_{B}A$. Delimitation is agent-dependent. Regions are delimited by use of individual body probes. $_{B}A$ is locally dependent upon and homogeneous with $_{B}E$. $_{B}A$-testing, strongly presupposes but does not locally presuppose $_{B}T$. Body-region-delimitation is probe-relative. $_{B}A$-testing is the foundation for a relational geometry of space. The demand for an individual probe body is "something new". The probe used for $_{B}A$-testing is always part of the testers body, and the tester must have a conception of his own bodily parts as individual bodies. The occupied locations of a delimited region need not be occupied by individual bodies. Nevertheless, in applying a $_{B}A$ the tester has the idea that the delimited region might be occupied by a number of separate bodies. $_{B}A$-testing requires the existence of a "comparison set" of bodies with which the tester is familiar. $_{B}A$-testing presupposes $_{B}\Delta$. A circumstantial condition of success for the application of $_{B}A$ to a set of locations is that every member of the set include the occasion of testing: $_{B}A$ is successfully applicable only on occasions upon which $_{B}E$ is applicable to every location in the set.

10. *Some body-Π's (for later reference).* We consider a few body Π's (or $_{B}\Pi$'s), important for geometric characterization, of which the measure of success is self-conscious perception by sight or touch. E-co-incident token-reflexive $_{B}\Pi$'s provide a foundation for features of demonstrative position. These may be "schematized" or "objectified" hence replaced by procedures of translating and rotating identifiable reference bodies. These tests are the most basic subject-matter for geometric theory. Tests "by sight", always "on light", may be body-applicable $_{B}\Pi$'s when the light emanates or is reflected from or is transmitted through bodies. Color Π's applicable to bodies differ from those applicable to visual things, because they presuppose different bearers. $_{B}\Pi$-tests of body co-incidence may be either tactual or visual; these differ. Tests for geometric figure are by visually observed super-position of reference bodies. The variant possibilities of rotating and sliding the superposed test body enables us to separate geometric figure into factors of shape and surface magnitude. Perspectival geometric features of shape and size are founded on visual

operations of sighting through or around bodies of proven superpositional shape and size. Perspectival geometric predicables are different from but dependent upon super-positional ones.

CHAPTER 18. SURFACES AND BODY-BOUNDARIES

1. *Alternative metaphysics for surfaces: Category or construction?* Material surfaces may plausibly be placed within a category related to but separate from that of body. Surfaces may be alternatively placed as body-boundaries, which are constructions from bodies. An argument against surfaces being constructions from bodies is that surfaces may be identified without reference to bodies and may not even be of bodies. Further evidence that surfaces are not body boundaries: Adjacent bodies in contact at apparently different surfaces share a common boundary. Body-surfaces, either as a dependent category or as constructed boundaries, are pertinent to the following examination of space and time.

2. *Criteria for surfaces.* Surfaces and bodies have the same E-test. Surfaces are individuated by smoothing over at a point of contact. On when bodies do and don't have surfaces. Surface distinctness is achieved by separately covering over two points of contact with distinct bodies. This secures a dependence of the category of surfaces upon the category of bodies. Surface identity is proven by tracing from one point of smoothing-over to another without losing contact with matter. Surfaces may be said to "persist" only in a borrowed and doubtful sense. There are reasons to think the surfaces are "two-dimensional" and no reasons to think they are not. Objection and reply: Surface distinction testing is almost always successfully applicable, hence not effective for enforcing interesting distinctions. Objection and reply: The use of covering-over as distinctness test for surfaces fails to explain why some surfaces are apparently necessarily distinct, in that they cannot be parts of any one surface, e.g., the surfaces of distantly separated bodies and the inside and outside surfaces of a shell.

3. *Boundaries* We essay an analysis of boundaries--including body-boundaries (which may be surfaces), lines and points--prospectively as constructions from bounded things. Euclid and Aristotle on points, lines and surfaces. Boundaries are "of" things they bound. A boundary exists when one thing bounds another. The boundary should additionally "separate" the two parts it concurrently bounds in the sense that one could not "get to" the one part "from" the other without being "at" the boundary. The boundary separating a pair of parts must be of an "ontological kind" distinct from those parts and of the whole they form. Two things "intersect" just in case a part of a boundary separating two parts of one is part of a boundary separating two parts of the other. We wish a "mereological" explanation of boundaries in test-theoretic terms of separation and identification of parts and wholes. The idea is that two objects altogether

constitute a third object of which they are parts. The two things may or may not "overlap" a fourth thing which is a part of each and a part of the whole. The two things define a boundary just in case there is no such fourth object. C_1 and C_2 of a categorial kind C "overlap" just in case they are separable parts of a "whole" C_3, where, furthermore, whenever C_1 (or C_2) is separated from C_3, it is also separated from C_2 (or C_1). By our stipulation, cases (i) and (ii) are covered by the condition that the overlap of C_1 and C_2 is itself a C also part of C_3. Call this a "categorial overlap". An overlap is a boundary just in cases where the two objects of the overlap do not have this kind of common part; a boundary is an overlap which is not "categorial". Objection and reply: A body-boundary, by this characterization, can simply be another thing too small to be a body, a molecule, perhaps! If body-boundaries and surfaces are different (as I suppose), it remains that they may be related in this way: body-boundaries are body-pairs whose members lie adjacent along surface parts. The constructed boundaries would themselves be analogically separated with "facing" surface-separations along line-boundaries. This is consistent with our idea that bodies themselves are three-dimensional and their boundaries two-dimensional. Our account makes no objection to the aristotelian view that the boundaries of surfaces are lines and the boundaries of lines are points; indeed, we see no way of producing lines and points except as boundaries. "Instants" are boundary-analogues separating continuous happenings.

CHAPTER 19. *VISIBILIA*

1. *Ontologizing perception.* Doubts about the rule that ontology follows perception. It is not obvious either that audition isn't tactation localized to the ear or that sounds are objects. Tastes are tangible phenomena having olfactory features.

2. *Four questions about objects of sight in their relationships to bodies.* It is reasonable to hold that objects of sight do stand in categories apart from the category of body. A categorial distinction between luminous and material phenomena is confirmed by invisible stuff. Problems arise from the considerations that material things may be seen to have visual features and that visual things may have geometric features. Guiding questions: (i) Are visual and tangible things said to be "seen" in the same sense? (ii) In the same "way"? (iii) Are their cognate properties of color the same? (iv) Are those features of size, shape, and position which both material and luminous things are seen to have the same as the ones bodies may be shown to have tactually; otherwise, how are they related? Our answers are that material and visual things are seen in the same sense but that the sight of matter is mediated by the sight of light; the cognate visual properties ascribed to bodies are different from but related to those ascribed to visual things; finally, visually ascertained geometric features

of both material and visual things are dependent upon hence different from their tactual cognates.

3. *Proto-criteria for visibilia and other luminous categories.* ${}_{L}E$: Looking to see in the direction of a displaced occluder. ${}_{L}\Lambda$: Luminous phenomena are "located in" a direction from a point of perspective. ${}_{L}T$: Tracing the boundaries of a material aperture. Separating light by frequencies affords an alternative procedure for individuating luminous phenomena. ${}_{L}\Delta$: Differential occlusion. ${}_{V}I$: Keeping in sight over a change in perspective. General distinctness testing for *visibilia* is by "parallax".

4. *Predicables of visibilia. Visibilia*, like bodies, have various features. Some color predicables of *visibilia* and corresponding visual properties of bodies are tested for by tactual-cum-kinesthetic operations from perspectives, with the use of apertures. These are primary predicables of *visibilia.*

5. *Visibilia are conceptually dependent upon bodies: An elaboration of answers.* The category of *visibilia* is conceptually dependent upon the category of bodies. Further elaboration upon our answers to the questions raised in #2 about the relations between cognate predicables of bodies and of *visibilia.* Because we determine the color of *visibilia* by observing the light of which they are composed, whereas we determine the color of bodies by observing the light reflected from their surfaces when they are illuminated with white light, these two species of color predicables are distinct. Visual geometry is dominated and controlled by the tactual geometry of apertures. The features in question are distinct nonetheless, since visual geometry involves perspective and tactual geometry does not. Objection and reply: Birds apparently can discriminate visual shapes without any apparatus for making even practical determinations of tactual shape. Views.

CHAPTER 20. PRELIMINARY SPECULATIONS OVER SPACE AND TIME

1. *Mysteries of space and time.*

2. *Space and time are "frameworks of representation" or "forms of intuition" for the individuation and identification of bodies.* Space comprises bodies and other things that stand in spatial relations of removal in directions; time comprises body-changes and other happenings that stand in temporal relations of earlier and later. The "singularity" of space and time is itself a difficulty, one not avoided by speaking always of mutually comparable places and times, rather than of space and time. The first observation that places and intervals are distinguishable only by reference to occupants prompts the leibnizian proposal that space and time are nothing in themselves but are rather to be explained as "relational orders" of occupants. This doesn't yet say what those occupants are and the formula leaves unoccupied space in limbo. Other objections: geometric and dynamical. the "absolutist" alternative is unappealing. We assume that spatial relationships are among bodies and other such things. We start by

stipulating that space-occupants comprise bodies "among other things" and time-occupants include changes in and among bodies. Space and time are frameworks of representation of singular existence; every connected place is a "possibility" for individuation and every duration a "possibility" for identification. Singular existents are identifiable individuals. Singular existents are individuated in places at times and identified over intervals of time. Every such place, occupied or not, is a "possibility" for the existence of an identifiable particular. These "possibilities" are posited within frameworks of representation, which are founded upon procedures or tests of individuation and identification. Objection and reply: "You say we imagine different applications of the same procedure. But what "same procedure" is this?" This order of possibilities is a "generality" of action-kinds, unprovable, but presumed latent, as a fact of life, in the behavior and practical knowledge of such creatures as ourselves. Places and times are called "singular" because they are distinguished from predicables as possibilities of *unique* existence. Leibniz' "principle of sufficient reason" implies that every such possibility of unique existence must be predicatively (descriptively, noumenally) exponible. It appears that we can envisage possibilities Leibniz could not have countenanced. The "ideality" of space and time: Places and durations are conceivable as possibilities of singular existence only in relation to actual objects or happenings. The actual objects in relation to which singular possibilities are conceived are given only "by ostension" to "sensible intuition". The foregoing is an explication of what Kant had in mind when he spoke of space and time as "forms of intuition". Generalizing Space and Time: Our conceptions of space and time are fixed in specifications of the procedures for individuating and identifying bodies; the "singularity" of space and time is traceable to the peculiarities of those procedures. The central question posed by our proposal is simply whether it's true: Are we right in holding that sets of proto-criteria of existence, individuation, separation and identity implicate co-ordinate notions of space and time; specifically, do the named proto-criteria for bodies determine our actual sense of space and time?

3. *From Leibniz to Kant on space and time.* First objection to Leibniz: Predication and contrariety presuppose individuation which must be non-predicatively determined. Second objection to Leibniz: Our ideas of space and time enable us to conceive possibilities which cannot be descriptively schematized. A corollary third objection is that a supposed leibnizian complete description of the world but without temporal determinations might be true of one time and false of another and therefore neither true nor false in itself: Geometric and dynamical objections. Kant required space and time for the sensible presentment of singular phenomena, presupposed for the exercise of the concepts of the understanding.

4. *The need for NOW: The "ideality" of space and time.* Objects are located "there" and "then" by procedures completed in the here-and-now of testing. Various temporal possibilities may be conceived relative to such an identification. The identifications that sustain our sense of time also provide a "measure of change". Time-testing requires that subject have reflective knowledge of what he is doing and practical knowledge of what he has done. Objection: There are no NOW's in Nature. Objection answered: Determinations of where a body was-then are "objective" in that the procedure by which that is now established enables one to envisage other possibilities. Objection: 'This possibility to which you now appeal exists only "in conception"! The right conclusion here is, not that spatial and temporal relations are subjective, but rather that space and time are "ideal" in Kant's sense.
5. *Temporal succession is "external".* There is a view that our sense of time is, most fundamentally, a sense of the flow of our own experience. This view has false consequences for our sense of outer succession. An argument that our sense of time-past cannot most fundamentally relate to our experiences: The sense of partness would have to be in the experience, as in "deja vu"; but, without an external reference to what has happened, "deja vu" would be just another eerie feeling.`
6. *Stratification.* That stratification among the concepts of existence, individuality, distinctness, identity and generality, which is a feature of our analysis, is most evident in our spatial and temporal conceptualization of physical things. A stratification of our spatial conceptions. Places cannot be "identified" across time. Reflection upon these capabilities gives us our sense of space. Our temporal conceptions are based upon $_{B}$I-testing. A condition for a subject acquiring that procedure is that he be capable of knowing what he is doing. Reflection may result in the elaboration of a "temporal order", for which is needed a proof that $_{B}$I is "temporal": If $_{B}$I is successfully applicable to an ordered pair of locations, it is not applicable at all to the reversed pair. Places and durations are "constructions".
7. *Accommodating body-space and -time to various constructions and to other categories.* A problem for our program is to account for why things other than bodies are conceptualized in the same spatial and temporal terms as bodies are. We expect to explain the extension of spatial and temporal conceptions to these other things by attention to criterial dependencies upon procedures for individuating and identifying bodies. Happenings are likely "constructions". *Visibilia* are individuated by reference to where they are seen from in body-space. This approach to the explanation of "singularity", while not strictly provable, seems altogether plausible.
8. *Projects and restrictions.* The central task is to effect a partial construction of spatial and (most especially) the temporal "orders" in terms of the categorial basis for bodies. We shall seek solutions to problems having to do with the

dimensionality of bodies and other space-occupants and with the "linearity" and the a-symmetry of the temporal relation of "before-and-after", in terms of proto-criteria. We shall also look for "soft" explanations of why we should have become subject to these procedures, rather than others. The investigation will be "a-priori" with data drawn from those everyday ways of thinking we all acquire by the time we enter high-school. The restriction to "commonsense" is unavoidable but more consequential. In comparison with Kant: Conceptual stratification; the "Unity" of Space and Time; determination; hypothetical judgement and causation; other differences follow.

CHAPTER 21. PRELIMINARIES FOR GEOMETRY AND HYPOTHETICAL DETERMINATIONS OF SPACE

1. *Spatial notions arise from reflection upon the categorial basis of body.* Review of formalities: We seek to show that procedures in the categorial basis for bodies satisfy conditions that determine an approximation to our previously defined idea of a space-around. We concentrate on a space of places, with little attention to the more abstract and general idea of a space of points. Notions of space arise from reflection upon the categorial basis for bodies. The main task of this chapter is to give adequate explications of ideas that lie at the foundations of elementary geometry, chiefly co-incidence, transformation, and dimension. We shall also, briefly and tentatively, consider some "properties of space". That will bring us, I believe, into the field of physical hypothesis. Spatial determinations of bodies which implicate body-identifications do not routinely carry over to space.

2. *Pre-euclidean geometry: contact and coincidence.* The use of bodies as geometric standards for making various spatial determinations presupposes that we can bring pairs of bodies into contact as separable parts of connected larger wholes. Any pair of contacting bodies defines a boundary of coincidence or "cross-section". A boundary of coincidence is defined by bodies in contact at "spots." A problem for observation. A further problem: Contacting bodies may cover more than what is in contact along the boundary of coincidence. Spot-covering always separates a spot from something else on the same body. Solution: Pairs of bodies define a boundary of coincidence if each is a spot cover for the other. By this analysis, a boundary of coincidence is a boundary of a kind defined by reference to a pair of bodies; it is additionally presupposed that the bodies be identifiable; the actual procedure for showing that the boundary is of this kind is a restriction upon the separation procedure for surfaces. We now wish to introduce procedures for determining the coincidence of "figures" on bodies. Covered spots on bodies may be separated from complementary spots on those bodies by boundary lines. The boundary lines of the two spots respectively covered in a boundary of coincidence of a pair of contacting bodies "coincide". We need a conception of point coincidence to

gain the general case of line-coincidence. Points are fixed as boundaries of whole spot-boundary lines. Two points "coincide" if they are on "intersecting lines". Two lines coincide if each of them is part of a line boundary for reciprocally covered spot-surfaces in a boundary of coincidence and each is point-bounded by coinciding intersections. We call any set of lines and points constructed from spots on a single body a "figure" on the body.

3. *Pre-euclidean geometry: "slides" and "transformations".* Contacting bodies may be slid across one another in various ways. Slides "deemed" to leave no pair of coincident figures coincident are called "translations"; slides "deemed" to leave some coincident subfigures coincident are called "rotations" (a "generalization"). These definitions cannot be test-theoretically expounded. A figure, F, that is deemed to be such that, whenever any subfigure, F_1, can be brought into coincidence with another figure F_2, then any subfigure s_m of F either can be brought into coincidence with a part of F_2 or a part of F_2 can be brought into coincidence with s_m, is called "homeomeric". Coincident homeomeric lines which are deemed to be such that any rotation that leaves all pairs of initially coincident subfigures still coincident are called "straights". Applications of geometries which employ those notions are "provisional" because of the adduced difficulties of proving them out by test. Sight as an Alternative.

4. *Pre-euclidean geometry: dimensionality.* We wish to explain why we think that "bodies" must be "three-dimensional". Explanations of the dimensionality of bodies do not routinely provide answers to questions about the dimensionality of space. Bodies before space. "Dimensions", in what sense? An object is n-dimensional if it has a part that is separated from its rest only by an n-1 dimensional boundary. We now formulate our question in this way: What is it in our conception of bodies, fixed with the categorial basis, that makes us suppose that every body can be separated into connected parts by two-dimensional configurations and has at least one such part that cannot be separated except by such a configuration? A problem: Bodies taken either visually or tactually, seem to be only two-dimensional. Some unavailing explanations. First, principles of physics. Second, tactual discontinuities are three-branched. Third, the kinesthetic sensations induced by operations with bodies constitutes a three-dimensional "group". Fourth, vision is given depth binocularly. Fifth: The "axis" of a rotation is defined by a "cross-product" of vectors; the operation of taking cross-products ceases to be associative when we go above three dimensions; it may seem from this that there can be no group of rotations of objects of more than three dimensions, hence no sense of rotation at all. Sixth, orientation is controlled by the three semi-circular canals in each inner ear. Seventh, depth is perceived as visual fading. First Solution: Bodies may be built up from hence are partitionable into surface-covers which coincide along two-dimensional boundaries. Second solution: The "third dimension"

may arise from coordination of sight and touch. Relationships of right-left, up-down, and near-far among *visibilia* may be interchanged pairwise by alterations in perspective. The interchangeability of up-down, near-far and right-left among two-dimensional *visibilia* shows that the description of these relationships cannot be given except within a perspective space which is in "some sense" three-dimensional. This space is a space of bodies. There are uncertainties over whether it is a "space-around". The sense in which this space is "three-dimensional" is at least "close to" what we have adapted. Solution: Bodies serve as differential occluders for defining a visual relationship of near-far, interchangeable with right-left or up-down by rotations of that same body. We conclude that bodies in themselves provide the equivalent of a three-dimensional space of perspectives and in that sense are themselves visually "three-dimensional". The body itself acts as a differential occluder for purposes of defining relations of near and far. Objection: the visually defined fore and after parts of a body, which were essential to the analysis, are themselves separated by a drawn one-dimensional line. Answer: The whole seen front part of the body (and not its unseen boundary) serves as a two-dimensional occluder for separating the *visibilia* reflected there from all the others reflected at the parts unseen from the first perspective. First Difficulty: The argument goes through only if proximate *visibilia* are reflected by proximate parts of the body. Second Difficulty: The explanation goes through only if bodies are potentially visible. This observation may be a killing objection. "Tactual occlusion" in relation to the connection and separation of contacting bodies gives the right solution. Observation: The sense in which bodies are three-dimensional is also "metric". Why only three? A "preliminary" to Mathematics.

5. *Properties of space.* The traditional ascription to space itself of "properties" of divisibility, persistence, dimensionality, uniformity and isotropy is founded on the rule that space must be "adequate" to its occupants and otherwise have no influence upon them. Contemporary theorists have come to the view that space is under the influence of its occupants. From this view, features of space are laws of physics, and the traditional properties of space coincide with some of the principles of the old physics. Properties of uniformity and isotropy are principles of conservation; dimensionality ties up with basic symmetries in the laws of physics and (surprisingly, perhaps), the Principle of Galilean Invariance destroys the presumption of the unity of space. The "persistence" of space is equivalent to a rule that places at different times may be coordinated in a way that preserves relations of immediate proximity; such a coordination would be established among locations successively referred to an identified body presumed "rigid" and at rest, and could be established if the laws of mechanics were invariant over time, as required by the principle of the conservation of energy. The three-dimensionality of the *places* of bodies seems to require the Postulate of Persistent Space; the three-dimensionality of *space* itself is a further

hypothesis that any two reciprocally accessible three-dimensional space occupants may be connected together into a larger three-dimensional space-occupant through linkages with other such things. The relevance of incongruent counterparts. A question whether space may be at once persistent, three-dimensional, constant and unified. Relativistic doubts about unification. We speculate that these various partially inaccessible spaces could be embedded in a unified four-dimensional persisting and regular space, and that hypothesis be confirmable by observations.

CHAPTER 22. ON THE TEMPORAL ORDERING OF HAPPENINGS
1. *Impressions of passage.* Time's passage is a metaphor. Aging isn't changing.
2. *Space-time and the order of changes.* Temporal facts are established only with the observation of changes or persistences in or among changeable things, like bodies. Changes of position and direction are especially prominent, and have pride of place in physics. Relative directions and positions of bodies are determined, relative to other bodies used as "positional frames of reference", by operations of rotating, extending and translating along reference bodies. These operations provide no representation of succession or other change. The determination of components of position and direction as functions of local time, itself taken as "independent parameter", suggests a pseudo-temporal "frame of reference", with time pictured as one "co-ordinate" among four. Space-time is a pictorial representation of relationships among changes and other "events", not of things. The imagined use of this representation is everywhere controlled by the old aristotelian maxim that changes don't change. An immediate consequence of this is that no literal sense can be given to rotation or translation of space-time origins or of other represented "events". The imagined use of space-time is designed as a representation of "what goes on". It does not represent what it presupposes, the "need for now", in the determination of local time. That fact is registered in the prohibitions implied by the rule that changes do not change. The use of space-time preserves a difference between space and time. Claims for "duality" of space and time: Digression on Taylor and Mayo. There are no "directions" in space-time.
3. *Why time's a "line".* Space-time pictures time as a line. Validating the picture: bodies are identified at interval-bounding "points" of individuation. A pictorial resume of our ordinary preconceptions. This conforms to our rule that changes don't change.
4. *On the question of why nothing's going to happen yesterday.* "You can't get back". Philosophers of science have found the problem in the fact that the classical laws of mechanics are invariant over interchanges of plus and minus time. This way of posing the question of why nothing's going to happen yesterday suggests misunderstandings of physics and confusions in the use of space-time. The temporal order of events cannot define the order of time. Our

problem is to explain why the temporal relation of before-and-after is asymmetric. Temporally asymmetric physical laws, interesting and important as they may be, do not provide an answer to our question because, presupposing a distinction between earlier and later, they may account for temporal order of events but not for the order of local time. Digression on the Physics of Time. As a first step toward a solution to the problem of explaining why the relation of earlier-and-later is asymmetric we shall try to show that $_{\mathrm{B}}$I is "temporal", and that anyone who successfully applied $_{\mathrm{B}}$I to a pair of locations L_1,L_2 would know that $_{\mathrm{B}}$I was not applicable to those locations taken in reverse.

5. *What I have done.* We must explain how one could "now" have a sense of what objectively was or will be so. Objection: Pedagogical guesswork is no substitute for logic. Answer: We are concerned with conditions "logically" implicated in the concept. We "depict" the satisfaction of those conditions by imagining how one "might" acquire the concept. We are interested to fix a transition from "practical" to "reflective" knowledge, gained, as we imagine, by the resolution of a conflict in the understanding. We are concerned with knowledge of time-when. Subject has reflective knowledge of something of his own activities. Our budget of assumptions. "Three" before "four": Coming to know that one has said "three" in counting to four is coming to know that "three" comes before "four". This is reflective "present" knowledge of what one has done.

6. *Objectifying time.* One's sense of what he has done in the course of bringing a body from somewhere transfigures into an outward sense of where that body was. Identifying bodies, by this account, is a matter of separating bodies on condition that one knows where he has been. This "new" way of distinguishing things affords chances for "new" kinds of mistake. The test, which is $_{\mathrm{B}}$I, is following a body from where it was. Progress and the Prospect of Error. Duration: Saying a part is "shorter than" saying a whole. Our expressible knowledge of what has happened is "logical". Why the past wants particulars: A defense. Two counter-examples countered. A connection with truth.

7. *Solution to the problem of why yesterday wasn't going to be today or of why one now knows that what was there then was not then previously somewhere now.* We try to show that anyone who successfully applies $_{\mathrm{B}}$I from L_3 to L_4 must know that he could not have applied $_{\mathrm{B}}$I from L_4 to L_3, and that the asymmetry of "before-and-after" is built into the $_{\mathrm{B}}$I procedure upon which its sense is based. We wish to show that A knew he didn't say "four" before "three". If one could have said "four" before "three" in a regular four-count done in the regular way, then, impossibly, a counted whole would have been a proper part of one of its proper parts. The asymmetry of "before" for counting carries over to $_{\mathrm{B}}$I testing, upon which the earlier- and later relation for locations is founded. We have been investigating the (everyone's) conception of local time and our explanation does not conflict with "goedelian" possibilities.

8. *And then and then.* Our subject has still not come to a sense that one previous location of a present object was "earlier" than another location of that object. A conjunction of successive "is-was"s, each coherent in itself, may confuse the understanding. "Earlier" and "later": Again, new words for new work.

APPENDIX G: OF TIME AND TENSE

1. *A budget of problems.* Facts of time mock the pretensions of abstract thought. On the impossible need to observe what's not so now. For us, this problem becomes one of explaining the possibility of differently tensed assertions of the same one statement. We have argued that the conception of dated fact cannot altogether supersede its expression in tensed assertion. This poses, for us, the task of describing the "conceptual evolution of tense". Differences among the tenses stand to mark distinctions in what can be known at a time, and in this way give rise to classical issues over future indetermination. The alleged "priority of causes" is also "problematic".

2. *A resume' of distinctions.* Assertions are utterances; statements are products of successful assertion; facts are objects of knowledge, formulable in true statements. "Dated facts" are objects at a time. "Temporal statements" are ones whose criteria either are or presuppose tests applicable only over temporally restricted ranges of locations. "Tensed assertions" token-reflexively indicate ranges of testing occasions relative to the occasions of utterance. Facts are said to be "past", "present" or "future" in relation to occasions of assertion and "earlier" or "later" in relation to the temporal statements in which they are formulated.

3. *The conceptual evolution of tense.* On the grammar of time: Tense indications and time specifications may, in natural languages, be distributed across several "parts of speech". Names. Mood indicators. Applicatives. Predicates. Two innovations of paraphrase: Quantifications over times and temporal operators. The "present-perfect" is a "first" expression of our sense of time. Recollections of the simple past are our most primitive expressions of tense proper. Why the past is "prior to" the present. Expression of (present) expectation in regard to what one presently observes to be going on are in what I call the "present intentional". Present intentional assertions presuppose objectified time and a sense of the simple past. Expressions in the simple present mark a contrast with other times. The use of the simple present marks a punctual boundary. The indication of a boundary between what has been and what is going to be presupposes the use of the simple past and of the present intentional. Our imagined subject learns to say that such and such which he has observed is going to be earlier than the such and such he observes: That's the sense of "now". There is a genuine doubt whether anyone could indicate future

occasions of testing. A working case. The future tense indicates that we should use one side of a bilateral identification as a track for the other.
4. "*Future contingents*". Recollections of Aristotle. A schematic condition of success for assertion is that the asserted fact could be known at the time of assertion. Our question is over what restrictions on success for assertions in the future-tense are entailed by our ignorance of the future. Severe restrictions on future-tensed assertion might seem to be required by considerations of an abstract kind having to do with truth and with referential lacking. An argument that, since statements cannot be said to be true before the formulated fact, they do not exist before the fact, Rebutted. An argument that future happenings do not exist and therefore cannot be stated, Rebutted. Genuine restrictions upon successful assertion in the future tense must be argued piecemeal. We have knowledge of what will be from observations of present-tendencies. Prospects of success in tensed assertion improve with an increase in knowledge.
5. *The priority of causes.* We wish to explain our sense that cases of posterior causation must be abnormal, if possible at all. Arguments against posterior causation from the conceivability of the non-occurrence of a later cause depend for their plausibility upon uncertain relationships between causation and knowledge. The asseveration that posterior causation could not be fitted into the network of other temporal concepts we employ is not compelling. Kantians might maintain that causation is a "category" constitutive of our concept of an object and that the order of time is the order of causation. We have argued against this on grounds that "C causes E" is "opaque". The formula that the occurrence of a cause is a "sufficient condition" for the occurrence of an effect is unavailing. Causes and effects are identified as such only in relation to the "understanding". It is not enough to say that earlier causes are better known than the later effects. The old idea that the concept of causation is rooted in our concept of agency is not enough to solve the problem. We shall find our solution from a consideration of the "pattern" of causal explanation. The first thing to say here is that we are concerned only with efficient causation, which is of happenings by happenings. The need for a "nexus". The classical position is strengthened by our observation about the "opacity" of causation. Restatement of the problem. Three cases of posterior causation. In all such cases, the alleged earlier effects have only an incidental place in the nexus. Our problem now becomes one of finding something in the causal nexus that at once allows for such possibilities and accounts for our sense that they must be abnormal, incomplete, and incidental. The identification of happenings requires the identification of the bodies or other such things they happen to. Causal explanation traces connections within a happening. When the trace is between time-separated sub-happenings, we depend upon our ability to track the constituent bodies from earlier to later, and that determines the normal order of

explanation. Prior effects are only incidental to the train of identifications by which the nexus is bound.

CHAPTER 23. CONSTRUCTIONS IN SPACE AND TIME

1. *Constructions in space and in time.* We will canvass a selection of reference-analogues definable within categorial bases for which the identification procedure is "temporal". These phenomena are all of them apparently "in space and time", and our efforts here add to the explanation of the "singularity" of space and time. Phenomena conceptualized in spatial and temporal ways comprise the field of natural science.

2. *Reconnoitering the terrain.* Some kinds are identifiable at or from place to place but only over a definite stretch of time. Views are identifiable from time to indefinite time but only at or from definite places. Kinds fixed to a time and place. Kinds which exist over time but are not talked about spatially at all. Statements. Kinds which are fixed to a time but which we do not conceive of as being in space. Kinds which are variously located in space but not talked about in temporal terms at all. Things fixed in place but not talked about in temporal terms at all. Finally, there are a variety of kinds which are certainly not thought of as "being" in space or time, but which are talked about in spatial and/or temporal terms.

3. *On the sense in which constructions are nothing new.* Constructions are not referents, but "conditions on referents"; these conditions are indicated by the realization of reference-analogues. Referents may "satisfy" reference-analogues. The elaboration of reference-analogues are consequential conceptual advances.

4. *Of times and places.* Location-reference-analogues, for which we already have an "analysis", may or may not be resolvable into complementary time- and place-reference-analogues. A condition sufficient and seemingly necessary for such location-factoring is that the identification procedure in the underlying categorial basis should be "temporal". Times are shown to "exist" by application of temporal I-procedures, "individuated" by *L*-projections of those procedures, "separated" by linked successive applications of that procedure, "identified" by bilateral identifications from synchronous locations and "delimited" in intervals. Places are shown to "exist" by temporally identifying things into locations "at a time", "individuated" by individuating things in such locations, and "separated" by making separations "at the same time".

5. *Aggregations.* Any delimited pair of things constitutes an aggregation. Aggregations are "identified" by identifying two members of a delimited separation. Aggregations may be nondescript or "organized". The constituents of an aggregation may or may not all be parts of another "aggregated object" of the same category. Aggregation-identity is often "sort-specified".

6. *Quantities of stuff.* Every body incorporates a certain definite quantity of stuff. Quantities of stuff were classically conceived of as eternal and unchanging. Quantities of stuff are aggregatable into bodies. Bodies may acquire, lose, and interchange stuff. Quantities of stuff were, and perhaps still are, conceived of as definite "somethings". It may seem that they are a "category" separate from, if dependent upon, that of body. Quantities of stuff are not individuated in the manner of bodies but by a body-delimitation. They are not a category but a construction within the category of bodies.
7. *Events.* "Events" are of many kinds. We argue that "events" are constructions by observing that, if they were referents instead, we could not distinguish on-goings from snaps and, second, by finding suitable "media" for apparent exceptions. Objection: You are crossing between bodies and quantities of stuff. Answer: The sudden coming-into-existence of the body is a change in the stuff, not of the body; the body is not to be equated with its sudden appearance. The observation that events are "with respect to" objects of various categories is further evidence that events are indeed constructions and not a category apart. Indications of criterial analogues. Criteria for "On-goings". Criteria for one-thing terminations. Criteria for one-thing snaps. The easy short way to handle "many-thing" "events" of all varieties is to think of them as "to" or "of" aggregations, *viz* as constructions on constructions. We may extend the analysis to events that happen to other constructions.

CHAPTER 24. BODIES ARE BASIC: A CONCEPTUALISTIC MATERIALISM

1. *Conceptualistic materialism.* We shall defend a version of "materialism". Our version of "materialism" is not "reductive". Aristotle's materialism is a precedent. Our thesis is that criteria for statements about anything presuppose body-identifications. Our basic intensions are restricted to body references.
2. *Some things that the dictum that bodies are basic does not claim.* First, we certainly do not claim that bodies are all there is or, even, that body references are all there are. Second, "haptic" perception is not claimed to be physiologically prior to other perceptual modalities. Third, we do not hold that statements about bodies are somehow "better" than statements about other things. That bodies are basic does not, fourthly, imply that the conceptualization of bodies is required by all forms of human conceptualization. Fifth, I do not maintain that everything in the body basis is independent of anything that may be found in the criterial bases of other categories.
3. *The thesis is a "theory".* The thesis that bodies are basic, if true, is "transcendental" in the sense that its formulation is part of the fact it formulates. If the thesis is true, then it is only "contingently necessary". We would try to refute rival claims that other familiar objects of experience are basic by showing that our actual conceptions of those things are body-dependent. Revisionary

alternatives are more challenging. The thesis if true is a principle of theory. It may confirm the theory.

4. *How to prove it?* The thesis is undemonstrated. We must depend upon "heuristic" arguments and evidence. The first consideration is that we conceptualize in spatial and temporal terms. The notion of an occasion is matter-dependent. We shall review a circle of "considerations". We shall also review "considerations" of the structure of our theory in particular. We assume that the conceptualization of other material phenomena can be shown to be dependent upon the conceptualization of bodies.

5. *Considerations of language and testing.* The second consideration draws attention to the need for materials at hand. The third consideration has to do with "temporality". This recalls, as a fourth consideration, our need, earlier adverted to, for occasions, which are tactually accessible connected places at times. A fifth consideration is that statement-testing seems to require the identification of bodies. Tests successfully applied produce perceptions. Material phenomena, we argue, are the "least equivocal" objects of "direct perception". It also appears that tactually determinable features of things "dominate" or "control" cognate features taken in by other senses. These points together constitute a strong sixth consideration in favor of the basicness of material phenomena. We argue, as a seventh consideration, that only contingent existents, such as bodies, can be basic, because "necessary existents" exist in the fact of conceptualization itself. Necessary existents have "primary" or other "necessary locations". Necessity or contingency of existence is a "fact of thought". Antecedent modes of conceptualization provide locations for necessary existents. A "conceptual scheme" for necessary existents must be less basic than one for contingent existents. Our thesis in regard to necessary existence is not new.

6. *Systematic considerations.* That we make statements of the same forms about different subjects--a focal assumption for any theory of logical inference--suggests that there is an exemplar upon which our conceptualization of other things is modeled. As a ninth consideration in support of the thesis that bodies are basic, we argue, from the observation that only material phenomena are less abstract than their locations, that the likely exemplar is material. The fact that no language can dispose of more than a limited variety of basic syntactic and other form-indicating apparatus underlies the logician's postulate that statements about different subjects may be classified under common "forms". Common form certifies the need for basic models. Our theory projects the etymologies of everyday speech. "Location" is the linchpin of the analogy. Our construction of location-reference analogues within formally similar categorial bases, as a tenth consideration, lends further plausibility to the thesis. We represent categories by specifications of proto-criteria of the same classifiable kinds. It is plausible to assume that criteria for more abstract categories are introduced by adapting

criteria from more basic categories. The comparative readiness with which the different proto-criterial orders of testing, in their various dependency relationships, can be separated for material phenomena is an eleventh consideration in support of the thesis that bodies are basic, for such a "stratification" should be more evident in more basic categories. Non-basic objects are apt to be conceptually incomplete, and their categorial bases apt to be "truncated". As a twelfth consideration, we duly note that the body-basis is a model of completeness. Our last consideration is that demonstrative reference, essential in our conceptual scheme, if not actually restricted to material phenomena, cannot be achieved except by use of material phenomena.

INDEX OF NAMES

TOPICAL INDEX
WITH
GLOSSARY ATTACHMENTS*

*Principal references and glossary attachments are in **bold-face**.

199. R. Wójcicki, *Theory of Logical Calculi.* Basic Theory of Consequence Operations. 1988 ISBN 90-277-2785-6
200. J. Hintikka and M.B. Hintikka, *The Logic of Epistemology and the Epistemology of Logic.* Selected Essays. 1989 ISBN 0-7923-0040-8; Pb 0-7923-0041-6
201. E. Agazzi (ed.), *Probability in the Sciences.* 1988 ISBN 90-277-2808-9
202. M. Meyer (ed.), *From Metaphysics to Rhetoric.* 1989 ISBN 90-277-2814-3
203. R.L. Tieszen, *Mathematical Intuition.* Phenomenology and Mathematical Knowledge. 1989 ISBN 0-7923-0131-5
204. A. Melnick, *Space, Time, and Thought in Kant.* 1989 ISBN 0-7923-0135-8
205. D.W. Smith, *The Circle of Acquaintance.* Perception, Consciousness, and Empathy. 1989 ISBN 0-7923-0252-4
206. M.H. Salmon (ed.), *The Philosophy of Logical Mechanism.* Essays in Honor of Arthur W. Burks. With his Responses, and with a Bibliography of Burk's Work. 1990 ISBN 0-7923-0325-3
207. M. Kusch, *Language as Calculus vs. Language as Universal Medium.* A Study in Husserl, Heidegger, and Gadamer. 1989 ISBN 0-7923-0333-4
208. T.C. Meyering, *Historical Roots of Cognitive Science.* The Rise of a Cognitive Theory of Perception from Antiquity to the Nineteenth Century. 1989 ISBN 0-7923-0349-0
209. P. Kosso, *Observability and Observation in Physical Science.* 1989 ISBN 0-7923-0389-X
210. J. Kmita, *Essays on the Theory of Scientific Cognition.* 1990 ISBN 0-7923-0441-1
211. W. Sieg (ed.), *Acting and Reflecting.* The Interdisciplinary Turn in Philosophy. 1990 ISBN 0-7923-0512-4
212. J. Karpiński, *Causality in Sociological Research.* 1990 ISBN 0-7923-0546-9
213. H.A. Lewis (ed.), *Peter Geach: Philosophical Encounters.* 1991 ISBN 0-7923-0823-9
214. M. Ter Hark, *Beyond the Inner and the Outer.* Wittgenstein's Philosophy of Psychology. 1990 ISBN 0-7923-0850-6
215. M. Gosselin, *Nominalism and Contemporary Nominalism.* Ontological and Epistemological Implications of the Work of W.V.O. Quine and of N. Goodman. 1990 ISBN 0-7923-0904-9
216. J.H. Fetzer, D. Shatz and G. Schlesinger (eds.), *Definitions and Definability.* Philosophical Perspectives. 1991 ISBN 0-7923-1046-2
217. E. Agazzi and A. Cordero (eds.), *Philosophy and the Origin and Evolution of the Universe.* 1991 ISBN 0-7923-1322-4
218. M. Kusch, *Foucault's Strata and Fields.* An Investigation into Archaeological and Genealogical Science Studies. 1991 ISBN 0-7923-1462-X
219. C.J. Posy, *Kant's Philosophy of Mathematics.* Modern Essays. 1992 ISBN 0-7923-1495-6
220. G. Van de Vijver, *New Perspectives on Cybernetics.* Self-Organization, Autonomy and Connectionism. 1992 ISBN 0-7923-1519-7
221. J.C. Nyíri, *Tradition and Individuality.* Essays. 1992 ISBN 0-7923-1566-9
222. R. Howell, *Kant's Transcendental Deduction.* An Analysis of Main Themes in His Critical Philosophy. 1992 ISBN 0-7923-1571-5

223. A. García de la Sienra, *The Logical Foundations of the Marxian Theory of Value*. 1992 ISBN 0-7923-1778-5
224. D.S. Shwayder, *Statement and Referent*. An Inquiry into the Foundations of our Conceptual Order. 1992 ISBN 0-7923-1803-X

Previous volumes are still available.

KLUWER ACADEMIC PUBLISHERS – DORDRECHT / BOSTON / LONDON